Computational Intelligence and Data Sciences

Computational Intelligence and Data Sciences
Paradigms in Biomedical Engineering

Edited by
Ayodeji Olalekan Salau, Shruti Jain, and
Meenakshi Sood

CRC Press
Taylor & Francis Group
Boca Raton London New York

CRC Press is an imprint of the
Taylor & Francis Group, an **informa** business

First edition published 2022
by CRC Press
6000 Broken Sound Parkway NW, Suite 300, Boca Raton, FL 33487-2742

and by CRC Press
2 Park Square, Milton Park, Abingdon, Oxon, OX14 4RN

© 2022 selection and editorial matter, Ayodeji Olalekan Salau, Shruti Jain and Meenakshi Sood; individual chapters, the contributors

CRC Press is an imprint of Taylor & Francis Group, LLC

ISBN: 9781032123134 (hbk)
ISBN: 9781032123172 (pbk)
ISBN: 9781003224068 (ebk)

DOI: 10.1201/9781003224068

Typeset in Times
by codeMantra

Contents

Preface...vii
Acknowledgments..ix
Editors...xi
Contributors xiii

Chapter 1   Performance of Diverse Machine Learning Algorithms for
Heart Disease Prognosis 1

*Dhruv Kaliraman, Gauri Kamath, Suchitra Khoje, and
Prajakta Pardeshi*

Chapter 2   Intelligent Ovarian Detection and Classification in
Ultrasound Images Using Machine Learning Techniques........... 23

V. Kiruthika, S. Sathiya, and M.M. Ramya

Chapter 3   On Effective Use of Feature Engineering for Improving the
Predictive Capability of Machine Learning Models.................... 53

M. R. Pooja

Chapter 4   Artificial Intelligence Emergence in Disruptive Technology........ 63

*J. E. T. Akinsola and M. A. Adeagbo, K. A. Oladapo,
S. A. Akinsehinde, and F. O. Onipede*

Chapter 5   An Optimal Diabetic Features-Based Intelligent System to
Predict Diabetic Retinal Disease... 91

M. Shanmuga Eswari and S. Balamurali

Chapter 6   Cross-Recurrence Quantification Analysis for Distinguishing
Emotions Induced by Indian Classical Music 107

*M. Sushrutha Bharadwaj, V. G. Sangam, Shantala Hegde,
and Anand Prem Rajan*

Chapter 7   Pattern Recognition and Classification of Remotely Sensed
Satellite Imagery 123

Pramit Pandit, K. S. Kiran, and Bishvajit Bakshi

Chapter 8 Viability of Information and Correspondence Innovation
for the Improvement of Communication Abilities in the
Healthcare Industry 141

Pinki Paul and Balgopal Singh

Chapter 9 Application of 5G/6G Smart Systems to Overcome Pandemic
and Disaster Situations 155

*Jayanta Kumar Ray, Sanjib Sil, Rabindranath Bera, and
Quazi Mohmmad Alfred*

Chapter 10 Risk Perception, Risk Management, and Safety Assessments:
A Review of an Explosion in the Fireworks Industry 177

N. Indumathi, R. Ramalakshmi, N. Selvapalam, and V. Ajith

Chapter 11 High-Utility Itemset Mining: Fundamentals, Properties,
Techniques and Research Scope 195

V. Jeevika Tharini and B.L. Shivakumar

Chapter 12 A Corpus Based Quantitative Analysis of Gurmukhi Script 211

Gurjot Singh Mahi and Amandeep Verma

Chapter 13 An Analysis of Protein Interaction and Its Methods,
Metabolite Pathway and Drug Discovery 237

P. Lakshmi and D. Ramyachitra

Chapter 14 Biosensors for Disease Diagnosis 253

*Ramneet Kaur, Dibita Mandal, Juveria Ansari,
Prachi R. Londhe, Vedika Potdar, and Vishakkha Dash*

Index 267

Preface

Computational intelligence is closely related to artificial intelligence where heuristic as well as metaheuristic algorithms are designed to provide better and optimized solutions in a reasonable amount of time. These algorithms have been effectively used in a variety of biomedical, bioinformatics, and biological science application domains in health informatics and computer science. The practice of recent biomedical research most times requires sophisticated technologies to manage patient information, plan diagnostics, prognostics, procedures, interpretations, and investigations. This provides a conceptual foundation as well as practical inspiration for computer science, decision science, information science, cognitive science, and biomedicine, which are all rapidly growing engineering and scientific areas.

Computational intelligence approaches are gaining attraction in the field of health informatics as a way to improve people's health. In this book, we focus on the applications of computational intelligence techniques in the domain of biomedical engineering and computer science. The applications of computational intelligence techniques in the domain of biomedical engineering is the subject of this book. In the healthcare sector, biomedical engineers develop algorithms that use artificial intelligence and corresponding hardware for decision modules for diagnosis and prognosis of diseases such as arrhythmia, cancer, and diabetes and other health-related issues in humans for early and more accurate detection and prevention. The use of intelligent strategies to undertake all of these actions could lead to more efficient outcomes.

Ayodeji Olalekan Salau
Shruti Jain
Meenakshi Sood

Acknowledgments

We want to extend our gratitude to all the chapter authors for their sincere and timely support to make this book a grand success. We are equally thankful to all CRC executive board members for their kind approval granted to us as Editors of this book. We want to extend our sincere thanks to Dr. Gagandeep Singh, and Miss. Aditi Mittal from CRC for their valuable suggestions and encouragement throughout the project.

It is with immense pleasure that the Editors of this book express their thanks to our colleagues for their support, love, and motivation during this project. We are grateful to all the reviewers for their timely review and consent, which helped us improve the quality of the book.

We may have inadvertently left out many others, and we sincerely thank all of them for their support.

Ayodeji Olalekan Salau
Shruti Jain
Meenakshi Sood

Editors

Dr. Ayodeji Olalekan Salau received his B.Eng. in Electrical/Computer Engineering from the Federal University of Technology, Minna, Nigeria. He received his M.Sc. and Ph.D. degrees from the Obafemi Awolowo University, Ile-Ife, Nigeria. His research interests include computer vision, image processing, signal processing, machine learning, power systems engineering, and nuclear engineering. Dr. Salau serves as a reviewer for several reputable international journals. His research has been published in reputable international conferences, books, and major international journals. He is a registered Engineer with the Council for the Regulation of Engineering in Nigeria (COREN), a member of the International Association of Engineers (IAENG), and a recipient of the Quarterly Franklin Membership with ID number CR32878 given by the Editorial Board of London Journals Press in 2020 for top-quality research output. More recently, Dr. Salau's paper was awarded the best paper of the year 2019 in Cogent Engineering. In addition, he is the recipient of the International Research Award on New Science Inventions (N) under the category of "Best Researcher Award" given by ScienceFather in 2020. Currently, Dr. Salau works in the Department of Electrical/Electronics and Computer Engineering at Afe Babalola University.

Dr. Shruti Jain is an Associate Professor in the Department of Electronics and Communication Engineering at Jaypee University of Information Technology, Waknaghat, HP, India, and received her Doctor of Science (D.Sc.) degree in Electronics and Communication Engineering. She has 16 years of teaching experience and has filed five patents, out of which one patent is granted and four are published. She has published more than 15 book chapters and 100 research papers in reputed indexed journals and international conferences. She has also published 11 books. She has completed two government-sponsored projects. She has guided six Ph.D. students and now has two registered students. She has also guided 11 M.Tech. scholars and more than 90 B.Tech. undergrads. Her research interests are in image and signal processing, soft computing, bio-inspired computing, and computer-aided design of FPGA and VLSI circuits. She is a senior member of IEEE, life member and Editor-in-Chief of the Biomedical Engineering Society of India, and a member of the International Association of Engineers. She is a member of the editorial board of many reputed journals. She is also a reviewer of many journals and a member of the technical program committees of different conferences. She was awarded a Nation Builder Award in 2018–2019.

Dr. Meenakshi Sood is currently an Associate Professor in CDC and Department of Electronics and Communication Engineering, National Institute of Technical Teachers' Training & Research (Ministry of Human Resource Development, Govt. of India), Chandigarh, India. She has teaching experience of around 20 years and worked in various institutes of repute. She received her Ph.D. in Biomedical Signal Processing and is a Gold Medalist and has been awarded Academic Award for her performance in Master of Engineering (Hons.) from Panjab University, Chandigarh. She has guided four Ph.D. scholars, around 20 M.Tech. scholars, and more than 100 B.Tech. undergrads. Her research areas of interest are image and signal processing, bio-inspired computing, antenna design, metamaterials, soft computing techniques, and curriculum design and development. She has two government-sponsored projects currently running under her and has published more than 100 research papers in reputed indexed journals and international conferences. She has edited three books and authored study materials for ICDOEL, HP University. She is a senior member of IEEE and a life member of International Technical Societies and BMSEI. She is also an editor in reputed journals and a Member of the Expert Committee for Evaluation of Impact of DST-FIST Scheme. She was selected as a GSE member of Rotary International and visited the USA in Exchange Program.

Contributors

M. A. Adeagbo
Department of Mathematics
and Computer Sciences
First Technical University
Ibadan, Nigeria

S. A. Akinsehinde
Software Development
The Amateur Polymath
Lagos, Nigeria

J.E.T. Akinsola
Department of Mathematics and
Computer Sciences
First Technical University,
Ibadan, Nigeria

Quazi Mohmmad Alfred
ECE Department,
Aliah University,
Kolkata, India

Juveria Ansari
Department of Biotechnology G. N.
Khalsa College of Arts, Science
and Commerce
Mumbai University
Matunga East, Mumbai, India

V. Ajith
Department of Mechanical
Engineering
Kalasalingam Academy of Research
and Education
Virudhunagar, Tamilnadu, India

Bishvajit Bakshi
Department of Agricultural Statistics,
Applied Mathematics and
Computer Science
University of Agricultural Sciences
Bangalore, India

S. Balamurali
Department of Computer Applications
Kalasalingam Academy of Research
and Education
Srivilliputhur, India

Rabindranath Bera
ECE Department,
Sikkim Manipal Institute of
Technology
Sikkim Manipal University,
Majitar, Rangpo, Sikkim, India

M. Sushrutha Bharadwaj
Department of Medical Electronics
Engineering
Dayananda Sagar College of
Engineering
Bangalore, India
School of Biosciences and Technology
Vellore Institute of Technology
Vellore, India

Vishakkha Dash
Department of Biotechnology G. N.
Khalsa College of Arts, Science
and Commerce
Mumbai University
Matunga East Mumbai, India

M. Shanmuga Eswari
Department of Computer Applications
Kalasalingam Academy of Research
 and Education
Srivilliputhur, India

Shantala Hegde
Music Cognition Laboratory and
 Clinical Neuropsychology and
 Cognitive Neuroscience Center,
Department of Clinical Psychology
National Institute of Mental Health
 and Neurosciences
Bangalore, India

N. Indumathi
Department of Computer Applications
Kalasalingam Academy of Research
 and Education
Virudhunagar, Tamilnadu, India

Dhruv Kaliraman
School of Computer Engineering and
 Technology
MIT WPU
Pune, India

Gauri Kamath
School of Electronics and
 Communication Engineering
MIT WPU
Pune, India

Ramneet Kaur
Department of Life Sciences, School
 of Bio Sciences
Regional Institute of Management and
 Technology University
Mandi Gobindgarh, India

Suchitra Khoje
School of Electronics and
 Communication Engineering
MIT WPU
Pune, India

K.S. Kiran
Department of Agricultural Statistics,
 Applied Mathematics and
 Computer Science
University of Agricultural Sciences
Bengaluru, India

V. Kiruthika
Department of Electronics and
 Communication Engineering
Hindustan Institute of Technology and
 Science
Chennai, India

P. Lakshmi
Department of Computer Science
Bharathiar University
Coimbatore, India

Prachi R. Londhe
Department of Biotechnology G. N.
 Khalsa College of Arts, Science
 and Commerce
Mumbai University
Matunga East Mumbai, India

Gurjot Singh Mahi
Department of Computer
 Science
Punjabi University
Patiala, India

Dibita Mandal
Department of Biotechnology G. N.
 Khalsa College of Arts, Science
 and Commerce
Mumbai University
Matunga East Mumbai, India

K. A. Oladapo
Department of Computer
 Science
Babcock University
Ilishan-Remo, Nigeria

F. O. Onipede
Department of Mathematics and
    Computer Sciences
First Technical University
Ibadan, Nigeria

Pramit Pandit
Department of Agricultural
    Statistics
Bidhan Chandra Krishi
    Viswavidyalaya
Mohanpur, India

Prajakta Pardeshi
School of Electronics and
    Communication Engineering
MIT WPU
Pune, India

Pinki Paul
Faculty of Management Studies
Wisdom, Banasthali Vidyapith
Vanasthali, India

Vedika Potdar
Department of Biotechnology G. N.
    Khalsa College of Arts, Science
    and Commerce
Mumbai University
Matunga East Mumbai, India

M. R. Pooja
Department of Computer Science
    and Engineering
Vidyavardhaka College of
    Engineering
Mysuru, India

Anand Prem Rajan
School of Biosciences and Technology
Vellore Institute of Technology
Vellore, India

R. Ramalakshmi
Department of Computer Science and
    Engineering
Kalasalingam Academy of Research
    and Education
Virudhunagar, Tamilnadu, India

M.M. Ramya
Centre for Automation and Robotics
Hindustan Institute of Technology and
    Science
Chennai, India

D. Ramyachitra
Department of Computer Science
Bharathiar University
Coimbatore, India

Jayanta Kumar Ray
ECE Department,
Sikkim Manipal Institute of
    Technology,
Sikkim Manipal University,
Majitar, Rangpo, Sikkim, India

V.G. Sangam
Department of Medical Electronics
    Engineering
Dayananda Sagar College of
    Engineering
Bangalore, India

S. Sathiya
Department of Obstetrics and
    Gynaecology
Chettinad Hospital and Research
    Institute
Chennai, India

N. Selvapalam
Department of Chemistry
Kalasalingam Academy of Research
    and Education
Virudhunagar, Tamilnadu, India

B.L. Shivakumar
Sri Ramakrishna College of Arts and
    Science
Coimbatore, India

Sanjib Sil
A.K. Choudhury School of
    Information Technology,
University of Calcutta,
Kolkata, India

Balgopal Singh
Faculty of Management Studies
Wisdom, Banasthali Vidyapith
Vanasthali, India

V. Jeevika Tharini
Sri Ramakrishna College of Arts and
    Science
Coimbatore, India

Amandeep Verma
Department of Computer Science
Punjabi University
Patiala, India

1 Performance of Diverse Machine Learning Algorithms for Heart Disease Prognosis

Dhruv Kaliraman, Gauri Kamath,
Suchitra Khoje, and Prajakta Pardeshi
MIT WPU

CONTENTS

1.1 Introduction 1
1.2 Literature Review 3
1.3 Materials and Methods 5
    1.3.1 Data.. 5
    1.3.2 Outlier Detection 6
    1.3.3 Data Preprocessing 6
    1.3.4 Dimensionality Reduction 7
    1.3.5 Ensemble Methods of Machine Learning 8
1.4 Proposed Approach for the Classification Model...................................... 8
    1.4.1 Logistic Regression.. 8
    1.4.2 Random Forest.. 8
    1.4.3 Gradient Boosting.. 9
    1.4.4 Extra-Trees Classifier.. 10
    1.4.5 AdaBoost 11
    1.4.6 MLP 13
    1.4.7 Decision Tree Classifier.. 14
1.5 Results.. 15
1.6 Conclusions... 19
References... 21

1.1 INTRODUCTION

Heart failure is the prime cause of death. It is one of the most chronic illnesses, and it can lead to disabilities and pose financial problems to patients. As per World Health Organization records, 17.5 million individuals die every year from cardiovascular disease [1]. The prognosis of heart disease is challenging for doctors as some of the symptoms experienced can be related to other illnesses or may be indicators of aging [2]. When the arteries of the heart lose the ability to transport blood that is rich in oxygen, heart disease is likely to occur. A common cause is plaque buildup in the lining of larger coronary arteries. It may partially or entirely block the blood flow in the heart's large arteries. This condition may occur as a result of an illness or accident that changes the way the heart arteries function [3]. Electrocardiogram (ECG), Holter screening, echocardiogram, stress examination, cardiac catheterization, cardiac computerized tomography (CT) scan, and cardiac magnetic resonance imaging are some of the medical tests that doctors and experts run to detect cardiovascular disease [4].

Diagnosis is a difficult, and critical process must be completed correctly and quickly. The availability of high-quality treatments at reasonable prices is a major concern for healthcare organizations such as hospitals and emergency centers [5]. However, if the coronary disease is diagnosed early enough, it can be successfully treated by a combination of dietary modifications, medical treatments, and surgical procedures [3]. The complications of heart disease can be decreased, and the heart's rhythm can be increased with the proper therapy [6].

Factors: After a lot of research, experts have classified the risk factors that can cause heart disease into two categories: risk factors that can be controlled and managed, and risk factors that will remain unaffected even after the treatment. Risk factors that don't have a scope of improvement include family background, ethnicity, and age. High levels of blood pressure, cholesterol, frequent alcohol intake, and physical inactivity are all risk factors that can be controlled to a certain extent. Hypertension is a condition that can harm the blood arteries, making it a highly likely risk factor for heart disease. Blood arteries may be damaged by high blood pressure. Tobacco consumption of any type increases the risk of CVD. Chemicals used to prepare tobacco products too have detrimental effects on the blood vessels. When high levels of cholesterol are detected in the body, heart disease is most likely to occur. Obesity or being overweight raises the risk of heart failure as well [7]. The precise timing of disease diagnosis determines the extent to which the disease can be controlled. The proposed research aims to diagnose these heart conditions early to prevent catastrophic effects [8].

Health researchers have produced a vast collection of medical evidence that can be analyzed, and useful information can be extracted from it. Data mining techniques are methods for retrieving useful information from vast amounts of data [9]. Large networks of data in a medical database are discrete [10]. As a consequence, making decisions based on discrete data becomes a daunting challenge. Machine learning (ML), a subfield of data mining, excels at handling massive, well-formatted, normalized datasets. ML is a tool that can be used to diagnose, track, and forecast different diseases in the medical field [11]. The goal is to make the process easier and to deliver successful care to patients while avoiding serious repercussions [12]. The role of ML in detecting hidden discrete patterns and analyzing the data is critical. Following data processing and dimensionality reduction, ML methods aid in the early detection and speedy diagnosis of heart

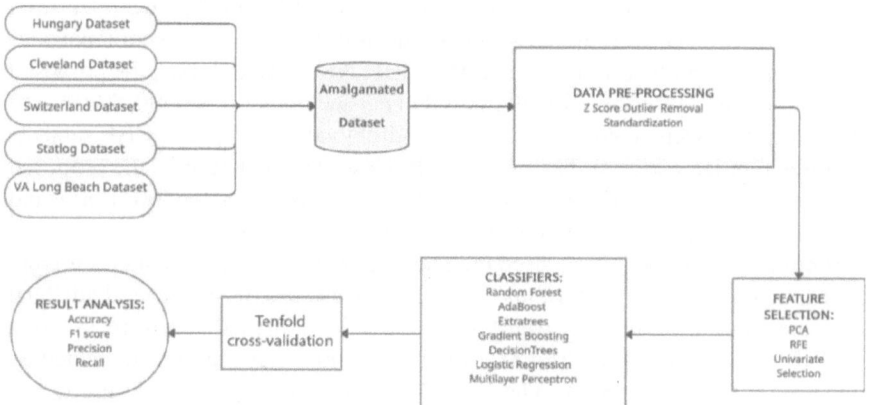

FIGURE 1.1   Prediction model flowchart.

disease. This chapter aims at testing the efficacy and the potential of numerous ML and deep learning [13] techniques for predicting cardiac disease at an early level (Figure 1.1).

1.2   LITERATURE REVIEW

Bayu and Sun [14] suggested a new method to build a double-tier ensemble. Random forest, gradient boosting, and extreme gradient boosting were the three ensemble learners that were merged with the help of a stacked architecture. To determine which feature set was the most important for each dataset, a particle swarm optimization-based attribute selection was performed. They also carried out a double-layered statistical test to buttress their postulations and to show that they were not based on suppositions. They also implemented tenfold cross-validation to improve their results.

Emmanuel et al. [15] aimed at implementing dimensionality reduction and a feature extraction technique by searching attributes that can cause cardiovascular disease. Phenomenal results were obtained when chi-square analysis and principal component analysis (PCA) were applied together to random forest and the accuracy that was obtained was 98.7% using Cleveland, 99.0% using Hungarian, and 99.4% using Cleveland–Hungarian datasets. According to the outcomes obtained from different models, the amalgamation of chi-square and PCA produced stronger results. The models were evaluated based on the accuracy, recall, precision, f1 ranking, Matthews correlation coefficient, and finally Cohen's kappa coefficient.

Ludi et al. [16] focused on congestive heart failure detection that suggests an ensemble methodology and employs heart rate variability data as well as deep neural networks. The databases employed in this study were the BIDMC Congestive Heart Failure Database (BIDMC-CHF), Congestive Heart Failure RR Interval Database (CHF-RR), MIT-BIH Normal Sinus Rhythm (NSR) Database, Fantasia Database (FD), and Normal Sinus Rhythm RR Interval Database

(NSR-RR). After extracting the expert features of RR intervals, a deep learning feature extraction network based on a long short-term memory convolutional neural network was built. Taking the BIDMC-CHF, NSR, and FD data, the proposed approach obtained 99.85%, 99.41%, and 99.17% accuracy on $N=500$, 1,000, and 2,000 duration RRIs, respectively, with blindfold validation (three CHF subjects and three regular subjects). Taking the NSR-RR and CHF-RR data, the proposed approach obtained 83.84%, 87.54%, and 85.71% precision on $N=500$, 1,000, and 2,000 duration RRIs, respectively, with blindfold validation.

Senthilkumar et al. [17] focused on a new approach of recognizing prominent features to enhance the performance of heart disease prognosis. Several attribute combinations and numerous well-known classification methods were utilized. The hybrid random forest with linear model was implemented, which yielded an enhanced accuracy of 88.7%. The UCI Cleveland dataset was used, and it was obtained from a UCI ML repository. HRFLM used an ANN with backpropagation as the input, as well as 13 clinical features. To shortlist the features that can most likely cause cardiovascular disease, statistical analysis was carried out. R studio rattle was used to carry out the classification. The dataset's classification rule was used to generate the results. The linear model strategy helped achieve better results than the random forest classifier and decision tree classifier. To boost the performance of the model, the RF and LM methods were combined, and the HRFLM approach was proposed.

Muhammad et al. [18] proposed a cardiovascular disease prognosis model for a CDSS (clinical decision support system) that involves DBSCAN (density-based spatial clustering of applications with noise) to detect and delete outliers, SMOTE-ENN to balance the training samples distribution, and XGBoost to forecast cardiovascular disease. Accuracy (acc), precision (pre), recall/sensitivity/-true-positive rate (rec/sec/R), f-measure (f), MCC, false-positive rate (FPR), false-negative rate (FNR), and true-negative rate (TNR) were all measured using tenfold cross-validation. With accuracies of 95.90% for the Statlog dataset and 98.40% for the Cleveland dataset, the suggested framework surpassed other simulations and prior research results.

In the implementation carried out in their paper, Pronab et al. [19] aimed at introducing various methods to accurately predict cardiovascular illness. A total of five heart disease datasets were used. To compute important functions, the relief and least absolute shrinkage and selection operator (LASSO) techniques are utilized. The main strategy applied in this paper was that the classifiers are combined using bagging and boosting techniques and new hybrid models are generated: decision tree bagging method (DTBM), random forest bagging method (RFBM), K-nearest neighbors bagging method (KNNBM), AdaBoost boosting method (ABBM), and gradient boosting boosting method (GBBM). The model also focused on reducing overfitting using relief and long computation times using LASSO. The data were split, and 80% of them were used for training and 20% for testing. The following efficiency metrics were used to assess the model: accuracy, f1 score, sensitivity, error rate, precision, negative predictive value, and false-positive rate. RFBM was found to be the most useful, with a 99.05%

accuracy rate. In addition, the most appropriate characteristics of a patient with heart disease have been recommended in this diagnostic method.

1.3 MATERIALS AND METHODS

1.3.1 DATA

Five common heart disease datasets: Cleveland, Hungarian, Switzerland, Long Beach VA, and Statlog (Heart) (Table 1.1) datasets, which were previously available separately, but had never been compiled, were merged, and the resultant was utilized in this project to perform heart disease prognosis. This amalgamated dataset has 11 common attributes, and it is the world's biggest heart disease dataset ever used for analysis. There are 1,190 instances in this dataset, along with 11 features.

The motive of collecting and combining these databases is to support and provide assistance to potential research into the prognosis of heart [20] disease and the ML algorithms and data mining techniques related to it. This initiative also aims at improving healthcare services and early diagnosis.

The possible factors that can lead to cardiac illness/cardiovascular disease are age, gender, chest pain, resting blood pressure, cholesterol, fasting blood sugar, resting ECG, maximum heart rate, angina felt after exercise, old peak, ST slope, and target. These attributes are all included in the dataset. This dataset can be found in and downloaded from the Kaggle website (https://www.kaggle.com/sid321axn/heart-statlog-cleveland-hungary-final).

The dataset is divided into training and test sets, with 90% (952 samples) being used for testing and 10% (238 samples) being used for classification.

This amalgamated dataset has not been used by many researchers [21]. Most of the existing research and analysis was done on individual datasets rather than on an amalgamated one which makes the model too specific and not universal. Furthermore, individual datasets have very few records which are not sufficient to make an efficient ML model. Hence, the idea to use an amalgamated dataset is to make a generalized model that trains over a larger dataset.

The following findings were made after data visualization and exploratory data processing. Chest pain, maximum heart rate, and ST slope all have a direct

TABLE 1.1
Dimensions of the Dataset

Dataset	[303×14]
Cleveland	[294×14]
Switzerland	[123×14]
Long Beach VA	[200×14]
Statlog (Heart)	[270×13]
Heart disease dataset (comprehensive)	[1,190×12]

correlation with the target. The data are uniform in nature. There are a total of 1,190 samples in the dataset, with heart disease present in 629 samples and heart disease absent in 561 samples. According to the data, the estimated age of onset of heart disease is 54, with a minimum age of 28 and a maximum age of 77. Heart disorder is more common between the ages of 55 and 80 than between the ages of 30 and 55. The presence of an irregular STT wave in the electrocardiographic findings indicates a greater risk of cardiovascular disease. If a sample is in the category of exercise-induced angina, then that sample has a higher risk of heart disease. According to the analysis conducted, males are more likely than females to have a heart attack and males in this dataset are older than females [22]. These observations aided us to better comprehend the results.

1.3.2 OUTLIER DETECTION

While modeling, it's important to clean the data sample because the dataset can often produce extreme values that are beyond the predicted range. This can leave deleterious effects on the performance of the ML model, which might influence its efficiency as well [23].

The Z score method was used to detect and exclude outliers in the dataset. In the dataset, 28 outlier samples were discovered and were removed for better prediction (Figures 1.2 and 1.3).

1.3.3 DATA PREPROCESSING

Rescaling, a technique in which all the attributes are brought in the same dimension, is very effective, when the data have features that have differing dimensions. Exceptional results are obtained when the features are brought in the range of zero to one. This mechanism is called normalization. When Gaussian instances have different means as well as standard deviations, they are transformed using standardization. These instances are converted to standard Gaussian distributions wherein the mean is 0 and the standard deviation is 1 [24].

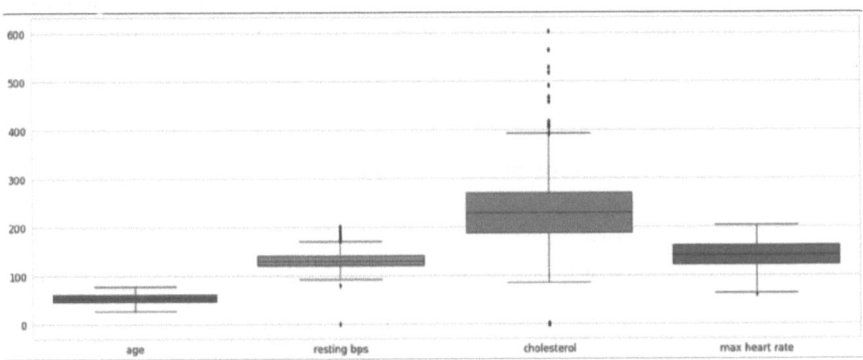

FIGURE 1.2 Outliers.

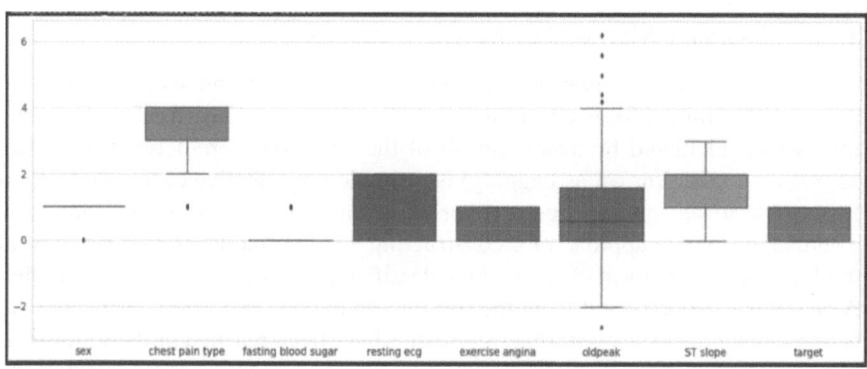

FIGURE 1.3 Outliers.

The need for standardization in most cases is experienced when the features in the data have varying sizes. StandardScaler is more efficient in classification than regression. Standardization is useful for data that have negative values and for the features that have the characteristics of a normal distribution. Since the dataset has features that adhere to a normal distribution, the best results were achieved using StandardScaler.

StandardScaler, MinMaxScaler, and Normalizer methods were tested and implemented. From the results obtained, this research arrives at the conclusion that the StandardScaler scaling technique provides the highest accuracy for this particular dataset. This holds for all of the ML algorithms we've tested.

1.3.4 DIMENSIONALITY REDUCTION

Model output may be harmed by elements that are irrelevant or only partly significant [25]. Feature discovery is a mechanism in which the features are dynamically picked in the data that have the largest effect on the forecast variable or performance that we, as scholars, are interested in. One of the most common problems faced by linear techniques such as linear regression and logistic regression is the presence of irrelevant features in their results. Overfitting is minimized, accuracy is increased, and training time is reduced as features are chosen. Using the univariate method for attribute analysis, the chi-square (2) regression test for non-negative features was implemented. The recursive feature elimination (RFE) method was implemented as well. RFE functions by extracting attributes recursively and creating a construct on the ones that remain [26]. The primary objective of RFE is to determine the instances that focus the most on forecasting the target instance using model accuracy. Furthermore, PCA was also tested, wherein the vectors with a high degree of covariance are taken out and the projection of the vector is employed for curtailing the attribute dimensions. The best results are obtained from PCA after applying all three dimensionality reduction strategies.

1.3.5   ENSEMBLE METHODS OF MACHINE LEARNING

The process in which numerous instances are taken from the testing dataset and the model is trained for each instance is called bootstrap. The final performance forecast is calculated by averaging all of the sub-model predictions. Bootstrap aggregation (also known as bagging) is a simple but effective ensemble method. It's used to lower the variance of algorithms with a lot of variances. Boosting is a general ensemble approach for constructing a strong model from a set of weak models, and the errors made in the first classifier are rectified in the next classifier. A classifier is constructed from the training outcomes, and in this way, boosting is implemented. In the event that full models have been reached or the training set is not perfectly approximated, models are introduced. This chapter implements bagging algorithms such as random forest and extra-trees, as well as boosting algorithms such as stochastic gradient and AdaBoost, for this dataset.

1.4   PROPOSED APPROACH FOR THE CLASSIFICATION MODEL

1.4.1   LOGISTIC REGRESSION

Logistic regression is considered effective while solving binary classification problems. The input values in the form of numbers are represented using a Gaussian distribution function.

The logistic function [27], which is at the heart of the system, is named after it in logistic regression.

The logistic function is represented as:

e: the base of the natural logarithm; x: the value to be transformed.

After applying standardization and PCA, tenfold cross-validation was applied to improve the prediction accuracy and the maximum average accuracy achieved was 83.3% (Figure 1.4; Table 1.2).

1.4.2   RANDOM FOREST

On huge datasets, the random forest algorithm can provide the same result, even though a large number of instance values are absent. The random forest has a two-step methodology: First, it generates a random forest, and then, it makes a guess based on the particular random forest.

Since random forest employs numerous decision trees that act as base learning models, row sampling and feature sampling is carried out on a random basis for each model. This process is known as bootstrap.

The algorithm only recognizes a minor subgroup of the model's attributes at each split in the tree, rather than all of them. This process ensures that the variance is eliminated by averaging it out. The outliers are dealt with in this algorithm

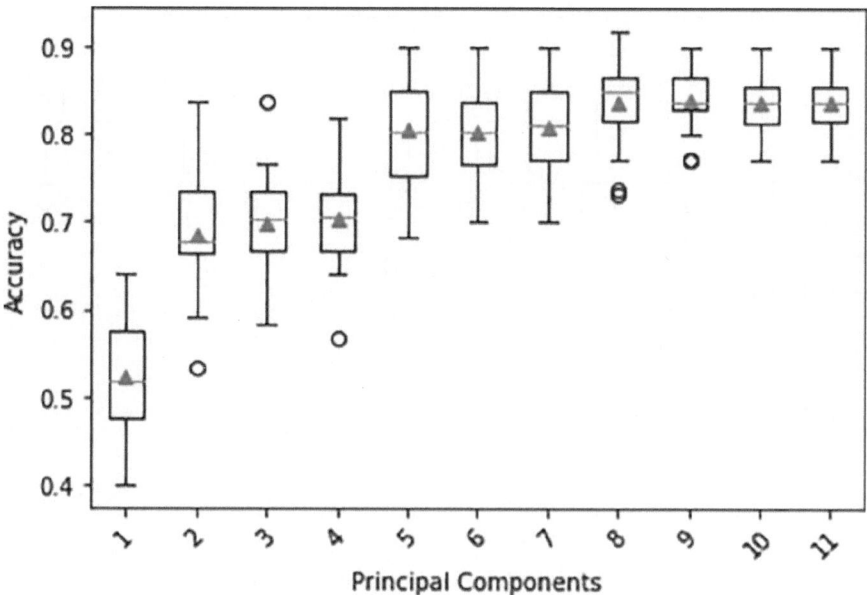

FIGURE 1.4 Box plot: logistic regression.

TABLE 1.2
Confusion Matrix: Logistic Regression

	Positive	Negative
Positive	543 (TP)	103 (FP)
Negative	102 (FN)	504 (TN)

---

by binning them. To achieve low bias and stable variance, the average is taken of the variance. After applying standardization and PCA, tenfold cross-validation was applied to improve the prediction accuracy and the maximum average accuracy achieved was 94.4% (Figure 1.5; Table 1.3).

1.4.3 GRADIENT BOOSTING

In order to solve regression and classification problems, gradient boosting is another very effective ML technique where a prediction model is generated from a group of weak prognosis models, ordinarily decision trees. The learning technique in gradient boosting machines fits new models in a sequential manner to provide a more accurate estimation of the response variable. The basic idea behind this algorithm is to create new base-learners that are maximally correlated with the loss function's negative gradient, which is associated with the entire ensemble [28].

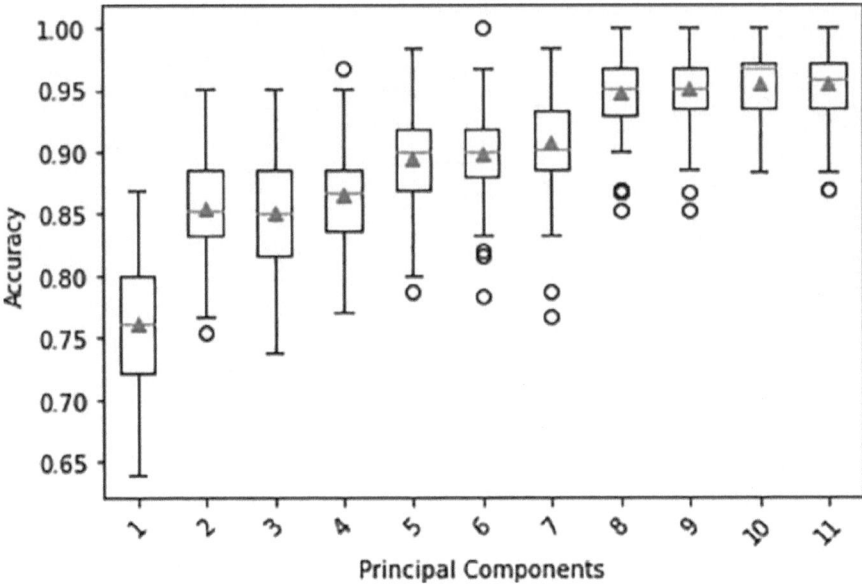

FIGURE 1.5   Box plot: random forest.

TABLE 1.3
Confusion Matrix: Random Forest

	Positive	Negative
Positive	507 (TP)	49 (FP)
Negative	32 (FN)	574 (TN)

An ensemble is simply a group of predictors that work together to generate a final prediction. Ensembles are used since several different predictors attempting to predict the same target variable would do so better than any single predictor alone. After applying standardization and PCA, tenfold cross-validation was applied to improve the prediction accuracy and the maximum average accuracy achieved was 90.4% (Figure 1.6; Table 1.4).

1.4.4   Extra-Trees Classifier

Extra-trees classifier is an ensemble learning strategy that produces a classification outcome by combining the outputs of several uncorrelated decision trees merged in a "forest."

The extra-trees classifier closely resembles the random forest classifier.

It can also produce results equivalent to or even superior than those of the random forest algorithm since it uses a straightforward algorithm to create the

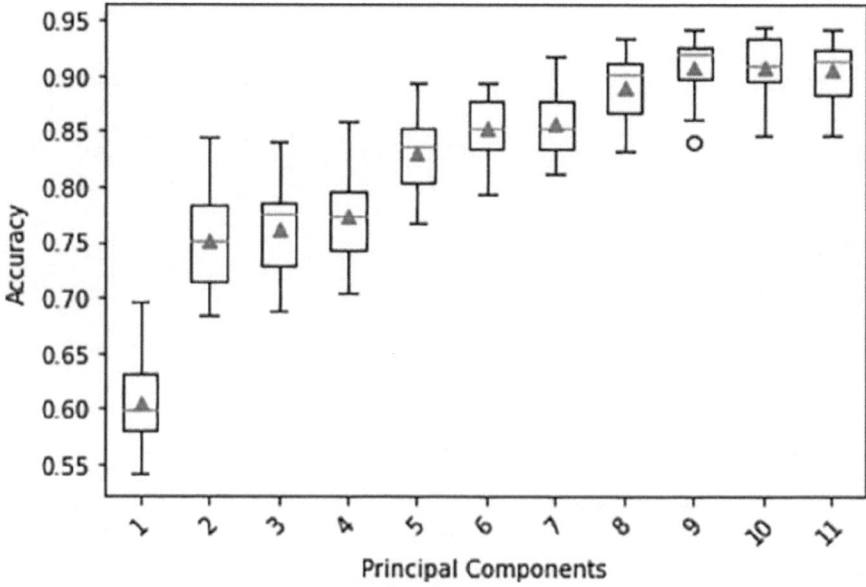

FIGURE 1.6 Box plot: gradient boosting.

TABLE 1.4
Confusion Matrix: Gradient Boosting

	Positive	Negative
Positive	491 (TP)	65 (FP)
Negative	58 (FN)	548 (TN)

decision trees used as part of the ensemble. It works by using the training dataset to produce a significant quantity of unpruned decision trees.

After applying standardization and PCA, tenfold cross-validation was applied to improve the prediction accuracy and the maximum average accuracy achieved was 94% (Figure 1.7; Table 1.5).

1.4.5 AdaBoost

AdaBoost is an algorithm that can be used in combination with different types of ML algorithms to enhance the performance of the model altogether. A "weak" learner is an algorithm that performs subpar—its performance is only beyond chance, but just by a small margin. A strong ensemble classifier is generated by combining the instances of the algorithm using various techniques such as bagging and boosting. Other learning algorithms' outputs ("weak learners") are compiled into a weighted summation that reflects the

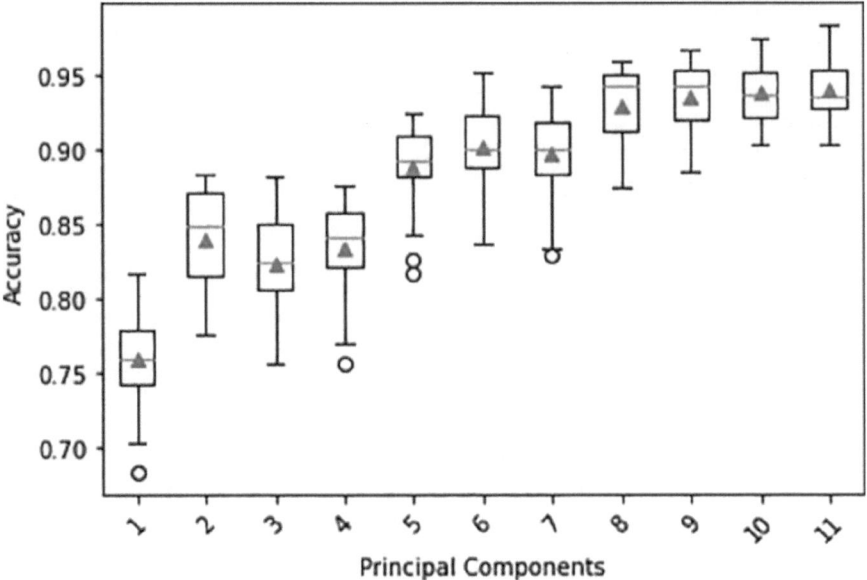

FIGURE 1.7   Box plot: extra-trees.

TABLE 1.5
Confusion Matrix: Extra-Trees

	Positive	Negative
Positive	510 (TP)	46 (FP)
Negative	33 (FN)	573 (TN)

---

boosted classifier's ultimate result. Exceptional results are obtained when AdaBoost is combined with decision trees as the weak learner algorithm. AdaBoost aims at enhancing the predictive ability of the model, reducing computation time and dimensionality, and eliminating unrelated features. So, it selects only the features that make a significant contribution to increasing the accuracy as well as the efficiency of the models. This algorithm follows a methodology in which a model is first created from the training data and then a second model is generated that rectifies the errors of the previous model. Before the training range is perfectly estimated or before the maximum number of models is met, models are added. The weights are reassigned to each case, with higher weights allocated to instances that were incorrectly labeled. This is called adaptive boosting. After applying standardization and PCA, tenfold cross-validation was applied to improve the prediction accuracy and the maximum average accuracy achieved was 84.03%, as shown in Figure 1.8 (Table 1.6).

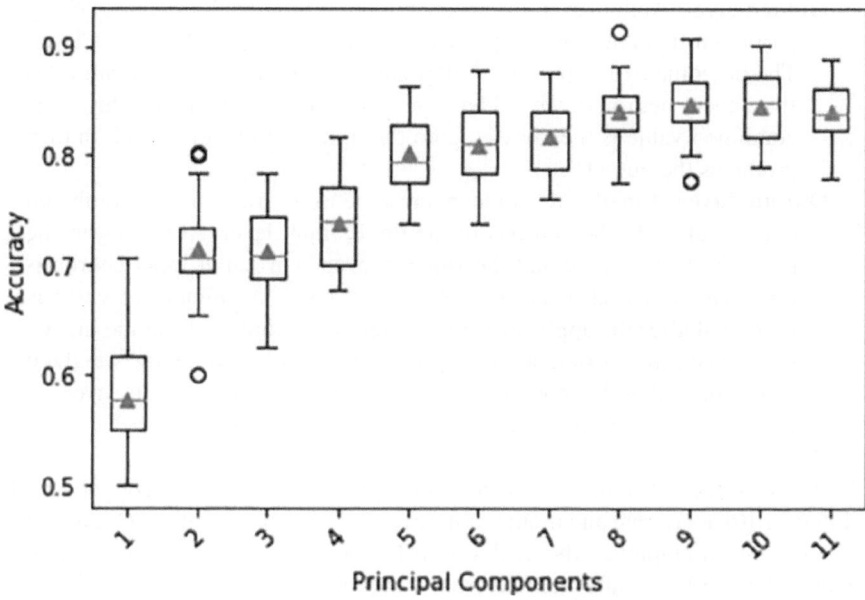

FIGURE 1.8 Box plot: AdaBoost.

TABLE 1.6
Confusion Matrix: AdaBoost

	Positive	Negative
Positive	452 (TP)	104 (FP)
Negative	101 (FN)	505 (TN)

1.4.6 MLP

A multilayer perceptron (MLP) is a perceptron that combines several perceptrons in various layers to tackle complex problems. The MLP is made of at least three layers: input layer, hidden layer, and output layer. One of the most effective and widely used techniques for training models is MLP. The layers are made up of neurons, and each neuron has its activation mechanism. Each layer's output becomes the next layer's entry.

Input layer: The purpose of an input layer in the MLP is to take a one-dimensional input vector and apply data standardization to convert the values between −1 and 1. The standardized numbers with a fixed input known as the bias of a number are provided to the hidden layer. The product of the weight and the bias value is sent to the neuron.

Hidden layer: At this layer, the product of the weight and the value of every particular input neuron is represented at each neuron of the hidden layer. The outcome that is produced after summing up the weighted numbers from every neuron in the hidden layer is called the combined value. This combined value is then passed onto the transfer function, which in turn produces the output.

Output layer: Finally, the hidden layer neurons transfer the combined output values to the neurons in the final output layer. At this layer, the product of the weight and the value of every particular input neuron is represented at each neuron of the output layer. A cumulative value is produced after the application of the weighted numbers from the hidden layer. The transfer function then produces the output when the weighted sum is passed as the input. Ultimately, the neurons in the output layer are provided with the consolidated outputs from the hidden layer.

To calculate the optimum initialization mode (Figure 1.9), batch size and epochs (Figure 1.10), learn rate and momentum (Figure 1.11), and optimizer (Figure 1.12), the perceptron employed the grid search CV method and the highest accuracy obtained was 82.35% after tenfold cross-validation (Figures 1.13–1.16).

1.4.7 DECISION TREE CLASSIFIER

Decision tree classifiers are used for visual and descriptive decision-making. It employs a decision-tree-like model, as the name suggests. The algorithm focuses on determining different methods to divide a dataset based on various conditions. It is a supervised technique that is non-parametric and can be used for classification as well as regression. By moving down the tree from the root to any leaf node, decision trees categorize the instances or, in simple words, classify them into multiple or binary classes. The classification for the instances is given by the leaf node. Every node in a particular decision tree represents a potential cause for a feature, and each edge descending from that node represents one of the potential case's solutions. For every subtree that is rooted at new nodes, the recursive process is carried out again and again. After applying

```
Best: 0.823548 using {'init_mode': 'uniform'}
0.823548 (0.043940) with: {'init_mode': 'uniform'}
0.812364 (0.038826) with: {'init_mode': 'lecun_uniform'}
0.819246 (0.041104) with: {'init_mode': 'normal'}
0.401127 (0.130083) with: {'init_mode': 'zero'}
0.819260 (0.052184) with: {'init_mode': 'glorot_normal'}
0.805504 (0.048953) with: {'init_mode': 'glorot_uniform'}
0.808915 (0.045052) with: {'init_mode': 'he_normal'}
0.802888 (0.048808) with: {'init_mode': 'he_uniform'}
```

FIGURE 1.9   GridSearchCV for init mode.

```
Best: 0.821810 using {'batch_size': 80, 'epochs': 200}
0.744474 (0.048196) with: {'batch_size': 5, 'epochs': 10}
0.777093 (0.078563) with: {'batch_size': 5, 'epochs': 50}
0.808893 (0.056127) with: {'batch_size': 5, 'epochs': 100}
0.802034 (0.089229) with: {'batch_size': 5, 'epochs': 150}
0.820122 (0.041121) with: {'batch_size': 5, 'epochs': 200}
0.765053 (0.072143) with: {'batch_size': 10, 'epochs': 10}
0.808901 (0.058321) with: {'batch_size': 10, 'epochs': 50}
0.799447 (0.056233) with: {'batch_size': 10, 'epochs': 100}
0.814080 (0.036962) with: {'batch_size': 10, 'epochs': 150}
0.801216 (0.059364) with: {'batch_size': 10, 'epochs': 200}
0.709984 (0.107581) with: {'batch_size': 20, 'epochs': 10}
0.795248 (0.067332) with: {'batch_size': 20, 'epochs': 50}
0.789935 (0.074053) with: {'batch_size': 20, 'epochs': 100}
0.820130 (0.052083) with: {'batch_size': 20, 'epochs': 150}
0.802896 (0.048331) with: {'batch_size': 20, 'epochs': 200}
0.728949 (0.083561) with: {'batch_size': 40, 'epochs': 10}
0.802019 (0.043100) with: {'batch_size': 40, 'epochs': 50}
0.814979 (0.050858) with: {'batch_size': 40, 'epochs': 100}
0.815768 (0.051649) with: {'batch_size': 40, 'epochs': 150}
0.820144 (0.043484) with: {'batch_size': 40, 'epochs': 200}
0.708304 (0.068878) with: {'batch_size': 60, 'epochs': 10}
0.800324 (0.044838) with: {'batch_size': 60, 'epochs': 50}
0.810625 (0.040061) with: {'batch_size': 60, 'epochs': 100}
0.815805 (0.047664) with: {'batch_size': 60, 'epochs': 150}
0.810610 (0.045494) with: {'batch_size': 60, 'epochs': 200}
0.704804 (0.061426) with: {'batch_size': 80, 'epochs': 10}
0.759925 (0.062088) with: {'batch_size': 80, 'epochs': 50}
0.812393 (0.055666) with: {'batch_size': 80, 'epochs': 100}
0.817507 (0.051108) with: {'batch_size': 80, 'epochs': 150}
0.821810 (0.050858) with: {'batch_size': 80, 'epochs': 200}
0.699013 (0.114658) with: {'batch_size': 100, 'epochs': 10}
0.754045 (0.078334) with: {'batch_size': 100, 'epochs': 50}
0.757339 (0.094216) with: {'batch_size': 100, 'epochs': 100}
0.815841 (0.045442) with: {'batch_size': 100, 'epochs': 150}
0.816681 (0.045830) with: {'batch_size': 100, 'epochs': 200}
```

FIGURE 1.10 GridSearchCV for batch size and epochs.

standardization and PCA, tenfold cross-validation was applied to improve the prediction accuracy and the maximum average accuracy achieved was 89.5% (Figure 1.17; Table 1.7).

1.5 RESULTS

The implementation of the heart disease prediction system is derived from seven ML algorithms: logistic regression, gradient boosting, random forest classifier, extra-trees classifier, decision tree classifier, adaptive boosting, and MLP.

The following results were obtained from the amalgamated dataset (Tables 1.8 and 1.9).

```
Best: 0.820116 using {'learn_rate': 0.1, 'momentum': 0.8}
0.812371 (0.021299) with: {'learn_rate': 0.001, 'momentum': 0.0}
0.800325 (0.020653) with: {'learn_rate': 0.001, 'momentum': 0.2}
0.814078 (0.031245) with: {'learn_rate': 0.001, 'momentum': 0.4}
0.808071 (0.019309) with: {'learn_rate': 0.001, 'momentum': 0.6}
0.816671 (0.024487) with: {'learn_rate': 0.001, 'momentum': 0.8}
0.804623 (0.020141) with: {'learn_rate': 0.001, 'momentum': 0.9}
0.802881 (0.054008) with: {'learn_rate': 0.01, 'momentum': 0.0}
0.799464 (0.020443) with: {'learn_rate': 0.01, 'momentum': 0.2}
0.812360 (0.035589) with: {'learn_rate': 0.01, 'momentum': 0.4}
0.790005 (0.025264) with: {'learn_rate': 0.01, 'momentum': 0.6}
0.816662 (0.031800) with: {'learn_rate': 0.01, 'momentum': 0.8}
0.776222 (0.044586) with: {'learn_rate': 0.01, 'momentum': 0.9}
0.763335 (0.045483) with: {'learn_rate': 0.1, 'momentum': 0.0}
0.780548 (0.002342) with: {'learn_rate': 0.1, 'momentum': 0.2}
0.788285 (0.011936) with: {'learn_rate': 0.1, 'momentum': 0.4}
0.802030 (0.036348) with: {'learn_rate': 0.1, 'momentum': 0.6}
0.820116 (0.020219) with: {'learn_rate': 0.1, 'momentum': 0.8}
0.795144 (0.030060) with: {'learn_rate': 0.1, 'momentum': 0.9}
0.771975 (0.031771) with: {'learn_rate': 0.2, 'momentum': 0.0}
0.802912 (0.018086) with: {'learn_rate': 0.2, 'momentum': 0.2}
0.792558 (0.037223) with: {'learn_rate': 0.2, 'momentum': 0.4}
0.814946 (0.026542) with: {'learn_rate': 0.2, 'momentum': 0.6}
0.795180 (0.013741) with: {'learn_rate': 0.2, 'momentum': 0.8}
0.815794 (0.036350) with: {'learn_rate': 0.2, 'momentum': 0.9}
0.770217 (0.043632) with: {'learn_rate': 0.3, 'momentum': 0.0}
0.812366 (0.024439) with: {'learn_rate': 0.3, 'momentum': 0.2}
0.812368 (0.024609) with: {'learn_rate': 0.3, 'momentum': 0.4}
0.790855 (0.022412) with: {'learn_rate': 0.3, 'momentum': 0.6}
0.790871 (0.022881) with: {'learn_rate': 0.3, 'momentum': 0.8}
0.814937 (0.032391) with: {'learn_rate': 0.3, 'momentum': 0.9}
```

FIGURE 1.11   GridSearchCV for batch learn rate and momentum.

```
Best: 0.813240 using {'optimizer': 'Adam'}
0.631020 (0.125583) with: {'optimizer': 'SGD'}
0.693744 (0.160149) with: {'optimizer': 'RMSprop'}
0.627468 (0.113306) with: {'optimizer': 'Adagrad'}
0.506631 (0.164596) with: {'optimizer': 'Adadelta'}
0.813240 (0.039538) with: {'optimizer': 'Adam'}
0.792595 (0.056014) with: {'optimizer': 'Adamax'}
0.751989 (0.112936) with: {'optimizer': 'Nadam'}
```

FIGURE 1.12   GridSearchCV for optimizer.

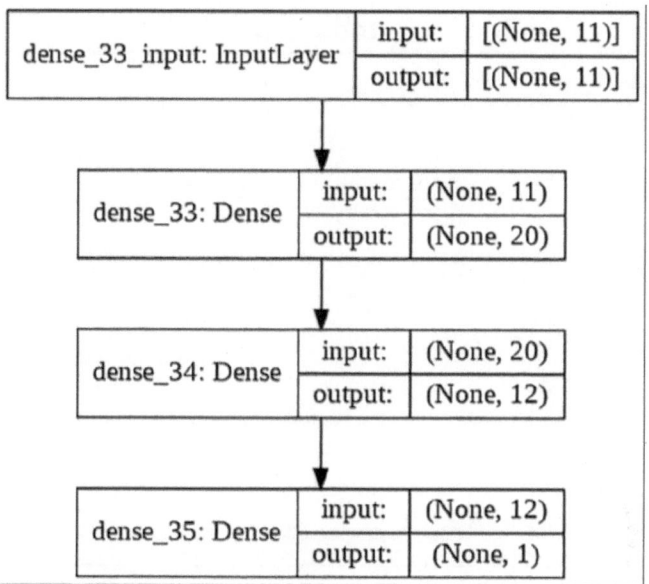

FIGURE 1.13   Number of neurons in layers.

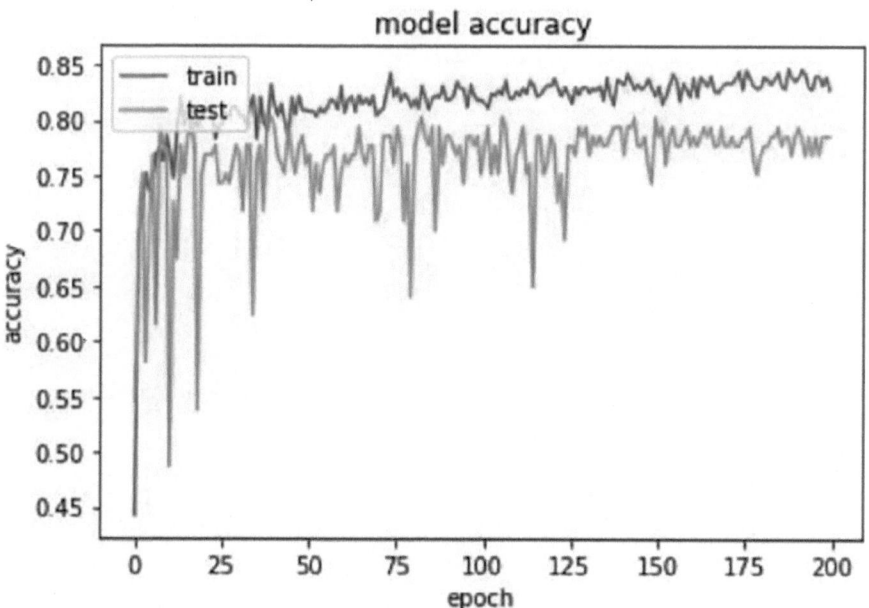

FIGURE 1.14   Epochs vs accuracy.

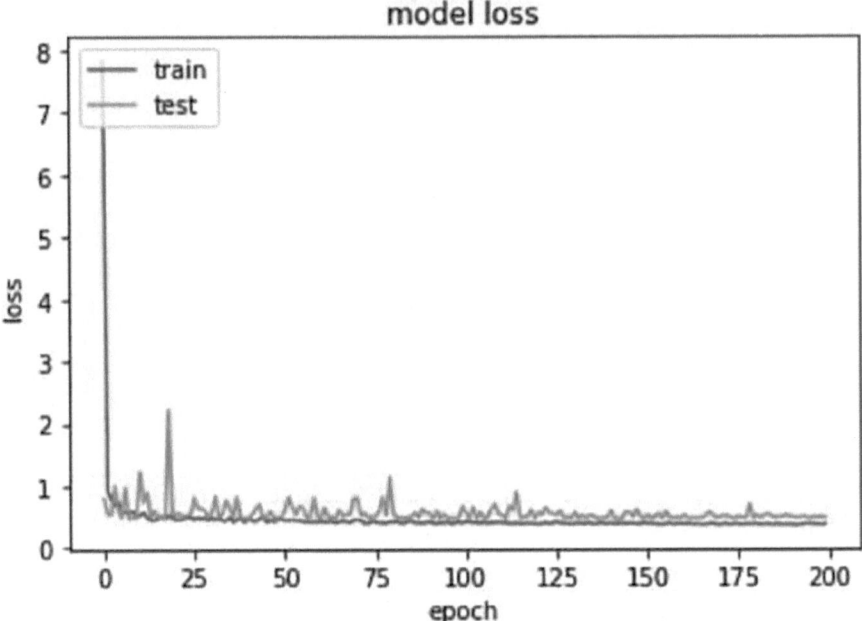

FIGURE 1.15   Epochs vs model loss.

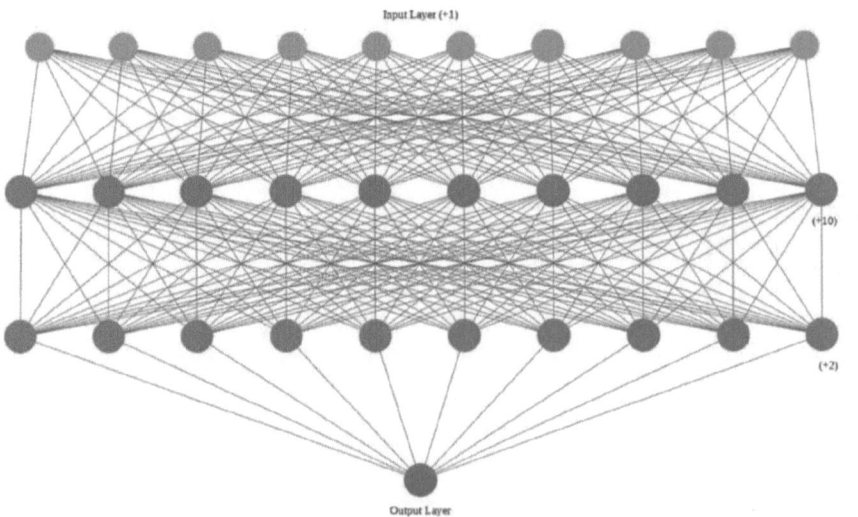

FIGURE 1.16   Multilayer perceptron architecture.

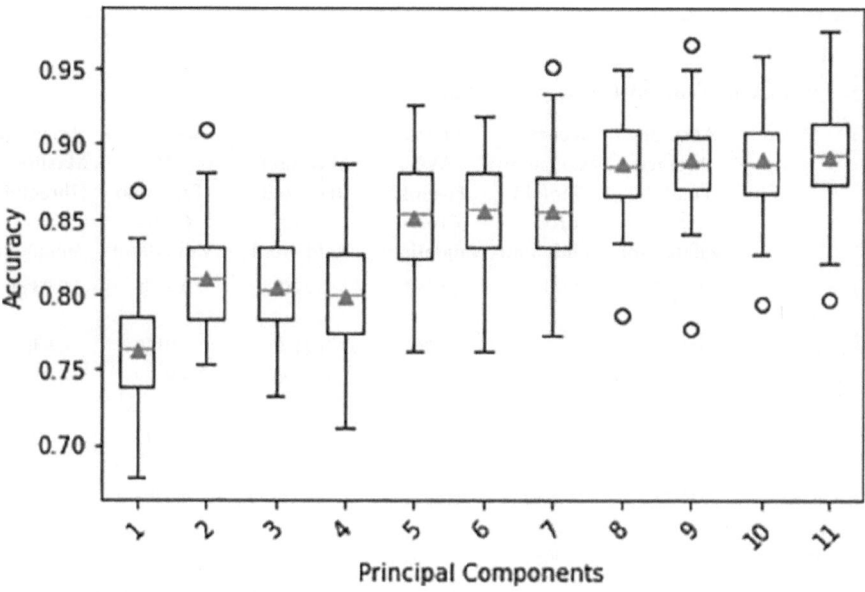

FIGURE 1.17 Box plot: decision trees.

TABLE 1.7
Confusion Matrix: Decision Trees

	Positive	Negative
Positive	501 (TP)	55 (FP)
Negative	63 (FN)	543 (TN)

1.6 CONCLUSIONS

This chapter focuses on the implementation of diverse ML algorithms to observe the performance of the cardiovascular disease prognosis system. The amalgamated dataset with a total of 1,190 instances was obtained from Statlog, Hungarian, Switzerland, Long Beach VA, and Cleveland datasets. Normalization was performed using StandardScaler, which provided the highest accuracy out of the other standardization techniques: MinMaxScaler and Normalizer. Feature selection was experimented using different techniques: PCA, RFE, and the univariate method. PCA helped extract the highest accuracy in comparison with the other techniques. The outcomes demonstrated that the random forest classifier had a better performance than all the other ML techniques. The chapter also employs the tenfold cross-validation method to enhance the prediction accuracy. The metrics used are recall, accuracy, precision, and f1 score to test the ML models. This system will help diagnose heart disease at a preliminary stage. Furthermore, it

TABLE 1.8
K-Fold Cross-Validation Comparison

	Accuracy Average (Tenfold Cross-Validation)	Accuracy Maximum (Tenfold Cross-Validation)	Accuracy AVG (Fivefold Cross-Validation)	Accuracy Maximum (Fivefold Cross-Validation)	Accuracy AVG (Threefold Cross-Validation)	Accuracy Maximum (Threefold Cross-Validation)
Random forest	0.942	1.00	0.9191	0.9482	0.9059	0.9354
Extra-trees	0.9363	0.9827	0.9207	0.9482	0.9112	0.9405
Gradient boosting	0.9021	0.9655	0.8948	0.9396	0.8852	0.9198
Decision trees	0.8852	0.9743	0.8771	0.9224	0.8613	0.8943
Logistic regression	0.8322	0.9482	0.8308	0.8626	0.8293	0.8578
AdaBoost	0.8403	0.9051	0.8361	0.8927	0.8339	0.8708

TABLE 1.9
Performance Metrics

	Accuracy Average (PCA)	Accuracy Maximum (PCA)	Precision Average (PCA)	Recall Average (PCA)	f1 Score Average (PCA)
Random forest	0.9420(11)	1.00(10)	0.9297(10)	0.9538(10)	0.9413(11)
Multilayer perceptron	0.8235				
Extra-trees	0.9363(10)	0.9827(8)	0.9265(11)	0.9596(10)	0.9392(11)
Gradient boosting	0.9021(10)	0.9655(11)	0.9029(10)	0.9174(11)	0.9093(10)
Decision trees	0.8852(10)	0.9743(9)	0.8948(10)	0.8901(11)	0.8942(10)
Logistic regression	0.8322(9)	0.9482(10)	0.8418(9)	0.8373(9)	0.8388(9)
AdaBoost	0.8403(9)	0.9051(8)	0.8507(10)	0.8522(9)	0.8475(9)

will help in controlling and managing the illness and will have a significant effect on the survival rate. It will reduce the expenses and introduce efficient models, which in turn will increase the availability of assistance.ML, while used as a decision enhancement tool, must not be utilized without human interference and human recommendations. It does not substitute doctors and nurses, but it can make their efforts more efficacious and structured. Similar prediction systems can be built to diagnose other chronic diseases such as cancer and diabetes at an early stage.

REFERENCES

1. https://www.who.int/news-room/fact-sheets/detail/cardiovascular-diseases-(cvds).
2. https://www.nia.nih.gov/health/heart-health-and-aging.
3. https://www.nhlbi.nih.gov/health-topics/coronary-heart-disease#:~:text=Coronary%20 heart%20disease%20is%20often, large%20arteries%20of%20the%20heart.
4. https://www.mayoclinic.org/diseases-conditions/heart-disease/diagnosis-treatment/ drc-20353124.
5. Committee on Diagnostic Error in Health Care; Board on Health Care Services; Institute of Medicine; The National Academies of Sciences, Engineering, and Medicine; Balogh, E.P., Miller, B.T., Ball, J.R., editors. *Improving Diagnosis in Health Care.* Washington, DC: National Academies Press (US); 2015 Dec 29. 2, The Diagnostic Process. Available from: https://www.ncbi.nlm.nih.gov/books/NBK338596/
6. https://www.nhlbi.nih.gov/health-topics/coronary-heart-disease.
7. Hajar, R. Risk factors for coronary artery disease: historical perspectives. *Heart Views.* 2017 Jul–Sep; 18(3), 109–114. Doi: 10.4103/HEARTVIEWS.HEARTVIEWS_106_17. PMID: 29184622; PMCID: PMC5686931.
8. Yan, Y., Zhang, J.W., Zang, G.Y., Pu, J. The primary use of artificial intelligence in cardiovascular diseases: what kind of potential role does artificial intelligence play in future medicine? *J Geriatr Cardiol.* 2019 Aug; 16(8), 585–591. Doi: 10.11909/j. issn.1671–5411.2019.08.010. PMID: 31555325; PMCID: PMC6748906.
9. Pandey, S. Data mining techniques for medical data: a review, 2016. Doi: 10.1109/SCOPES.2016.7955586.
10. Bush, R.A., Kuelbs, C., Ryu, J., Jiang, W., Chiang, G. Structured data entry in the electronic medical record: perspectives of pediatric specialty physicians and surgeons. *J Med Syst.* 2017 May; 41(5), 75. Doi: 10.1007/s10916-017-0716-5. Epub 2017 Mar 21. PMID:28324321; PMCID: PMC5510605.
11. Salau, A.O., Jain, S. Adaptive diagnostic machine learning technique for classification of cell decisions for AKT protein. *Infor Med Unlocked.* 2020; 23(1), 1–9. Doi: 10.1016/j.imu.2021.100511.
12. Juneja, S., Juneja, A., Anand, R. Healthcare 4.0 digitizing healthcare using big data for performance improvisation. *J Comput Theor Nanosci*, 2020; 17(9–10), 4408–4410.
13. Sindhwani, N., Verma, S., Bajaj, T., Anand, R. Comparative analysis of intelligent driving and safety assistance systems using YOLO and SSD model of deep learning. *Int J Inform Syst Model Design (IJISMD).* 2021; 12(1), 131–146.
14. Tama, B.A., Im, S. Lee, S. Improving an intelligent detection system for coronary heart disease using a two-tier classifier ensemble. *BioMed Res Int.* 2021; 2020, Article ID 9816142, 10 p, Doi: 10.1155/2020/9816142.
15. Gárate-Escamila, A.K., El Hassani, A.H., An- drès, E. Classification models for heart disease prediction using feature selection and PCA. *Inform Med Unlocked.* 2020; 19, 100330, ISSN 2352-9148, Doi: 10.1016/j.imu.2020.100330.
16. Wang, L., Zhou, W., Chang, Q., Chen, J., Zhou, X. Deep ensemble detection of congestive heart failure using short-term RR intervals. *IEEE Access.* 2019; 7, 69559–69574, Doi: 10.1109/ACCESS.2019.2912226.
17. Mohan, S., Thirumalai, C., Srivastava, G. Effective heart disease prediction using hybrid machine learning techniques. *IEEE Access.* 2019; 7, 81542–81554, Doi: 10.1109/AC-CESS.2019.2923707.

18. Fitriyani, N.L., Syafrudin, M., Al-fian, G., Rhee, J. HDPM: an effective heart disease prediction model for a clinical decision support system. *IEEE Access.* 2020; 8, 133034–133050, Doi: 10.1109/AC- CESS.2020.3010511.

19. Ghosh, P., Azam, S., Jonkman, M., et al., Efficient prediction of cardiovascular disease using machine learning algorithms with relief and LASSO feature selection techniques. *IEEE Access.* 2021; 9, 19304–19326, Doi: 10.1109/ACCESS.2021.3053759.

20. Javeed, A., Zhou, S., Yongjian, L., Qasim, I., Noor, A., Nour, R. An intelligent learning system based on random search algorithm and optimized random forest model for improved heart disease detection. *IEEE Access.* 2019; 7, 180235–180243, Doi: 10.1109/ACCESS.2019.2952107.

21. Sonawane, J.S., Patil, D.R. Prediction of heart disease using multilayer perceptron neural network. *International Conference on Information Communication and Embedded Systems (ICICES2014),* 2014, 1–6, Doi: 10.1109/ICICES.2014.7033860.

22. https://www.cdc.gov/heartdisease/facts.htm.

23. https://www.healthissuesindia.com/2019/09/28/a-fifty-percent-rise-in-heart-disease.

24. https://www.world-heart-federation.org/resources/cardiovascular-diseases-cvds-global-facts-figures/.

25. Pan, Y., Fu, M., Cheng, B., Tao, X., Guo, J. Enhanced deep learning assisted convolutional neural network for heart disease prediction on the internet of medical things platform. *IEEE Access.* 2020; 8, 189503–189512, Doi: 10.1109/ACCESS.2020.3026214.

26. Mathur, P., Srivastava, S., Xu, X., Mehta, J.L. Artificial Intelligence, machine learning, and cardiovascular disease. *Clin Med Insights Cardiol.* 2020; 14, 1179546820927404. Published 2020 Sep 9. Doi: 10.1177/1179546820927404.

27. PMC5408160/ Benjamin, E.J., Blaha, M.J., Chiuve, S.E., et al. Heart disease and stroke statistics-2017 update: a report from the American heart association [published correction appears in Circulation. 2017 Mar 7; 135(10), e646] [published correction appears in Circulation. 2017 Sep 5;136(10), e196]. *Circulation.* 2017; 135(10), e146–e603. Doi: 10.1161/CIR.0000000000000485.

28. Powar, A., Shilvant, S., Pawar, V., Parab, V., Shetgaonkar, P., Aswale, S. Data mining & Artificial Intelligence techniques for prediction of heart disorders: a survey. *2019 International Conference on Vision Towards Emerging Trends in Communication and Networking (ViTECoN)*, 2019, 1–7, Doi: 10.1109/ViTECoN.2019.8899547.

2 Intelligent Ovarian Detection and Classification in Ultrasound Images Using Machine Learning Techniques

V. Kiruthika
Hindustan Institute of Technology and Science

S. Sathiya
Chettinad Hospital and Research Institute

M.M. Ramya
Hindustan Institute of Technology and Science

CONTENTS

2.1 Introduction 24
2.2 Materials and Methods 27
    2.2.1 Datasets...27
    2.2.2 Methodology 28
        2.2.2.1 Preprocessing 28
        2.2.2.2 Feature Extraction...31
        2.2.2.3 Machine Learning-Based Ovarian Detection................33
        2.2.2.4 Intelligent System for Ovarian Classification (ISOC) 34
        2.2.2.5 Performance Metrics...36
2.3 Results...37
    2.3.1 Preprocessing...37
    2.3.2 Feature Extraction 38
        2.3.2.1 Intensity Features...38
        2.3.2.2 Texture Features...38

DOI: 10.1201/9781003224068-2

    2.3.3   Machine Learning-Based Ovarian Detection (MLOD) 40
    2.3.4   Intelligent System for Ovarian Classification 42
            2.3.4.1   Classification Using ANN.. 42
            2.3.4.2   Classification Using LDA 42
            2.3.4.3   Classification Using SVM.. 44
2.4 Discussion 44
2.5 Conclusions 48
Acknowledgements 49
References 49

2.1   INTRODUCTION

Female infertility is a major and thoughtful concern in today's world and contributes to about 37% worldwide and 12.5% in India. Among the recognizable causes, it is reported that ovulatory disorders are a main cause of female infertility, contributing to 25% in majority of the infertile women (WHO Technical Report Series, 1992). Ovaries have a fluid filled sac called ovarian follicle in which the mature egg is present. In a normal ovary, the mature egg is released from the follicle during the ovulation process. Ovulatory disorders are due to the failure or irregularity in the ovulation process that occurs in the ovary, resulting in ovarian cysts or polycysts. An ovary with an ovarian cyst is called cystic ovary, and an ovary with polycysts is called polycystic ovary.

Diagnostic imaging is of paramount importance in confirming the presence of the disease or abnormality along with the clinical judgements. In the present scenario, various female pelvic imaging modalities are available for diagnosing ovulatory disorders. Among them, diagnostic ultrasound is mostly preferred as it is completely free from radiation and administration of contrast.

In diagnostic ultrasound, transvaginal ultrasonography is the first and primary choice for diagnosis of ovulatory disorders (Hamm, 1994). The transvaginal scan gives precise information about the ovarian reserve, status of the growth of the follicle, size of the follicle, response of follicular growth due to hormonal stimulations, expected time for the rupture of the follicles, size, shape, location and number of follicles/cysts. Hormones such as follicle-stimulating hormone (FSH), luteinizing hormone (LH) (Luderer, 2014), prolactin and thyroid-stimulating hormone (TSH) (Goswami et al., 2009) are vital in infertility treatment. Anaemia (Singh et al., 2006) and type II diabetes (Szaboova and Devendra, 2015) are also a common cause of infertility. Appropriate blood tests will give a clear diagnostic value of the required hormone parameters that would help in diagnosis and treatment.

Computer-aided diagnosis (CAD) ensures an imperative role in intelligent detection and classification. It helps the physician enhance the diagnostic opinion and proceed with appropriate treatment. Ultrasound technology offers a real-time assessment of organs in the pelvis (Benacerraf et al., 2015). In case of ovarian follicle/cyst detection, transvaginal ultrasound is the first and primary choice for diagnosis. Magnetic resonance imaging (MRI) or computed tomography (CT) is chosen only if ultrasound findings are not clear or if there is a requirement of differential diagnosis of lesions and tumour staging (Jain and Salau, 2019). CAD systems involve the following phases: (i) preprocessing of input, (ii) extraction of features and (iii) classification.

Preprocessing is defined as a procedure of accomplishing some computations on the image before an important image processing step. Colour space transformation helps specify and identify information in an image in a more intuitive way (Ford and Roberts, 1998). Ultrasound images are generally susceptible to speckle noise (Oleg and Allen, 2006). The existence of speckle noise in images disturbs the fine specifics and boundaries of the image. Speckle reduction using wavelet transform was performed, which ensured that the ultrasound images are completely denoised, thereby preserving the image information (Kumar and Indranil, 2014).

Clustering techniques have a wide range of applications in many areas of engineering (Pham and Afify, 2007). K-means clustering and fuzzy c-means clustering are the two common clustering techniques that are preferred for medical image segmentation. K-means clustering is applied in medical images for segmentation and feature extraction (Hae et al., 2016). However, feature extraction based on intensity alone may not yield effective segmentation.

Statistical texture features were proposed and are useful in aerial and medical images (Haralick et al., 1973). Texture analysis is useful in medical images to segment the anatomical structures, detect lesions and distinguish between a healthy and affected tissue (Castellano et al., 2004). Texture-based segmentation helps group regions based on the statistical parameters which vary for different regions in an image.

Studies reveal that demographic and diagnostic information was also taken into consideration along with the image interpretation for diagnosis and decision-making. In diagnosis of any kind of medical abnormality, in addition to diagnostic imaging the specific blood tests and demographic information pertaining to a patient are of profound importance. They help the medical experts understand and get a broad idea about the underlying problem in detail. So, medical experts always perform their diagnosis and treatment based on image interpretations along with diagnostic information obtained from the blood tests and demographic information collected from the patient.

In medical image diagnostics, the use of machine learning (ML) algorithm has a great role in enhancing the accuracy of detection and classification of abnormalities. An artificial neural network (ANN) model having seven input features such as height, weight, age, gender, serum creatinine, texture features and mean value of texture features was developed to classify the level of chronic kidney disease. Here the extracted texture features include homogeneity, energy, contrast and correlation (Madhanlal et al., 2017). Demographic features such as gender, age and body mass index were included along with the extracted tongue colour and texture features for diagnosing the presence or absence of diabetes by making use of principal component analysis (PCA) and support vector machine (SVM)

(Jianfeng et al., 2017). These studies reveal that the combination of image characteristics along with demographic and diagnostic data greatly contributed to better classification.

Classification problems attempted using two or more ML algorithms in order to find the best performing classifier for medical diagnosis are extensively studied. Liver lesions acquired using MRI were classified using ANN, SVM, logistic regression and K-nearest neighbour (KNN) classifier. ANN outperformed other classifiers (Zhenjiang et al., 2017). Lung cancer images were classified as normal and abnormal. Here a combined input of fractal and texture features was given to six different classifiers such as decision trees, discriminant classifier, Naïve Bayes, KNN, ANN and SVM. SVM outperformed all the other classifiers (Punithavathy et al., 2019). The above studies show that a particular classifier performs well and yields the highest classification in diagnosis of a particular medical abnormality. So, it is always better to develop and try two or more than two ML algorithms for a study because a suitable algorithm that performs efficient classification with high accuracy can be identified.

In general, ovulatory disorders can be efficiently diagnosed by medical procedures using a combination of transvaginal ultrasonography, appropriate blood tests and relevant demographic information from the patient. Reliability is not ensured in manual analysis of ovarian ultrasound images. It is laborious, and the interpretation purely depends on the proficiency of the physician by whom the scan is performed. However, only multi-speciality medical centres offer such expertise. Expert opinion can be availed by the individuals in rural areas only if they commute to urban regions. Further incorrect opinion and false interpretation are possible due to inter-observer difference and the likelihood of wrong diagnosis. To overcome these challenges, an automatic system for ovarian detection needs to be developed in order to support the radiologist in decision-making as well as in making clear visual interpretation of the ovarian ultrasound images.

Classification of ovarian images as cystic, polycystic and normal forms a vital part in infertility treatment. During image interpretation, an ambiguity arises between the classification of a normal ovarian follicle and an ovarian cyst due to common features. Ovarian cyst, if not detected at right time, can lead to ovarian cancer. Similarly, polycystic ovaries can also lead to severe infertility in women. Misclassification may occur if diagnosis solely relies on image characteristics obtained using ultrasound scans. Classification of ovaries is indispensable for the medical experts for subsequent treatment and decision-making. Also, diagnosis of ovulatory disorders is carried out using demographic and diagnostic data along with image interpretations.

The follicular/cystic regions in an ultrasound image at times combine with the background, and it creates a problem for the physician to appropriately identify the follicle/cyst. Misinterpretation of the shape, location and size of the follicle/cyst is possible due to camouflaging that leads to wrong diagnosis. In case of polycystic ovaries, the polycysts are very small and are clumped together. This is challenging for the reason that two or more cysts may be grouped as one. Interfollicular/intercystic regions are also difficult to be recognized in such cases. In ovarian classification, same appearance and sharing of common features between normal follicles and cysts creates ambiguity in classification. Both may have similar size and shape at times. This makes the investigation of ovarian follicles/cysts a thought-provoking and demanding task.

Ovarian detection and classification methods were developed by some researchers. Edge-based segmentation followed by SVM classification was performed for detection of polycystic ovarian syndrome (PCOS) (Sheikdavood and Palanivel, 2016). Ovarian follicle segmentation was performed using k-means clustering (Kiruthika and Ramya, 2014). Classification of polycystic ovaries alone was performed using competitive neural network (CNN), which resulted in an accuracy of 80.84% (Dewi et al., 2018). An overall average accuracy of 94% was obtained in texture and intensity-based ovarian classification (TIOC) (Kiruthika et al., 2018). The ovarian images were classified as polycystic, cystic and normal using SVM. A technique called active contours without edges was used for segmenting the follicle. Geometric features were given to the SVM classifier (Hiremath and Tegnoor, 2012). But the demographic and diagnostic data were not used here as input for classification. There is also a lack of generic framework for automated classification of ovaries. Hence, there is a serious requirement for an intelligent system to detect and classify the ovarian follicle/cyst by combining the image features and the chosen demographic and diagnostic features. So, in this study a methodology to implement an intelligent system for ovarian classification (ISOC) is focused.

2.2 MATERIALS AND METHODS

2.2.1 DATASETS

A database of 115 datasets comprising 41 normal datasets, 41 datasets from PCOS patients and 33 datasets from cystic patients was created. Dataset distribution is represented in Table 2.1.

Three images, one from each cluster, were used for the presentation purpose in this research.

Figure 2.1 shows the sample dataset containing normal (Image a), polycystic (Image b) and cystic (Image c) ovarian follicles to validate the methodology.

TABLE 2.1
Dataset Distribution

Ovary Type	No. of Datasets	
	Training	Testing
Normal ovary	30	11
Polycystic ovary	29	12
Cystic ovary	23	10

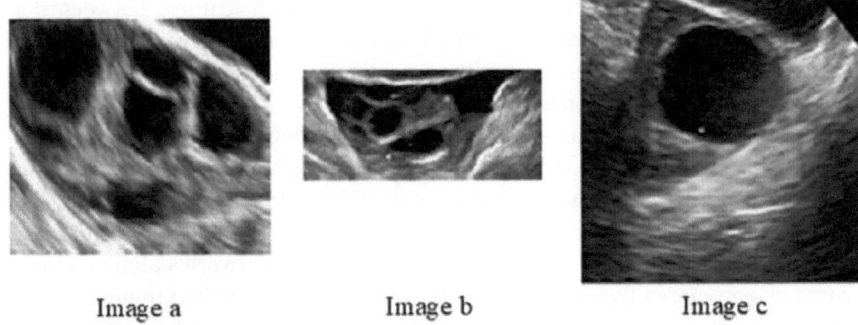

Image a   Image b   Image c

FIGURE 2.1 Sample image dataset.

The dimensions of images a, b and c were $107 \times 119 \times 3$, $54 \times 113 \times 3$ and $132 \times 129 \times 3$, respectively. The methodology was developed in MATLAB R2018a.

Demographic and diagnostic datasets were also collected along with the ovarian follicle/cystic ultrasound images. The sample demographic and diagnostic datasets in Table 2.2 correspond to sample image datasets in Figure 2.1.

Careful investigation was done in selecting the demographic and diagnostic features. The influence of each parameter in classification was thoroughly studied (Luderer, 2014; Goswami et al., 2009; Szaboova and Devendra, 2015; Singh et al., 2006). Finally, based on the study, eight features were selected. The selected features were day of the menstrual cycle, size of the follicle, the number of follicles, FSH, LH, prolactin, TSH and random blood sugar (RBS).

2.2.2 METHODOLOGY

The methodology used in this study was to facilitate the automatic detection of follicles/cysts and classify the ovarian image as normal, polycystic or cystic using ultrasound image, demographic and diagnostic data, as depicted in Figure 2.2.

Preprocessing techniques such as colour space transformation and wavelet transform were performed to reduce computational intricacies and remove the speckle noise, respectively. Intensity and texture features were taken from the despeckled image and fed to the ML-based ovarian classifier using ANN for intelligent detection. An intelligent methodology was developed using a combination of image, selected demographic and diagnostic features for effective ovarian classification. Three different classifiers, namely, ANN, LDA and SVM, were developed, and their performance metrics such as specificity, sensitivity, precision, accuracy, F-measure, Matthews correlation coefficient (MCC) and receiver operating characteristic (ROC) curve were calculated for determining a suitable classifier.

2.2.2.1 Preprocessing
Preprocessing methods such as colour space transformation and wavelet transform were applied on input images for extracting the required information, enhancing images and diminishing the noise.

TABLE 2.2

Sample Demographic and Diagnostic Dataset

Data	Ovary Type		
	Normal	Polycystic	Cystic
Age	30	24	23
Sex	Female	Female	Female
Education	<12th	UG	UG
Occupation	Housewife	Professional	Professional
Income	Nil	>20,000	>20,000
Weight	58 kg	80 kg	43 kg
Height	158 cm	156 cm	140 cm
Body mass index	23	32.87	22.12
Marital status	Married	Married	Married
No. of children	0	0	0
Type of family	Nuclear	Joint	Joint
Day of menstrual cycle	10	12	20
Number of follicles/cysts	5	8	1
Size of the follicles/cysts	2.1 cm	1.03 cm	3.2 cm
FSH	3.21	3.42	2.64
LH	4.36	5.12	5.69
Prolactin	10.26	10.23	16.24
TSH	1.14	3.14	3.13
Random blood sugar	76	88	101
Blood group	O+ve	O+ve	O+ve
Haemoglobin	10.2	8.9	9.6

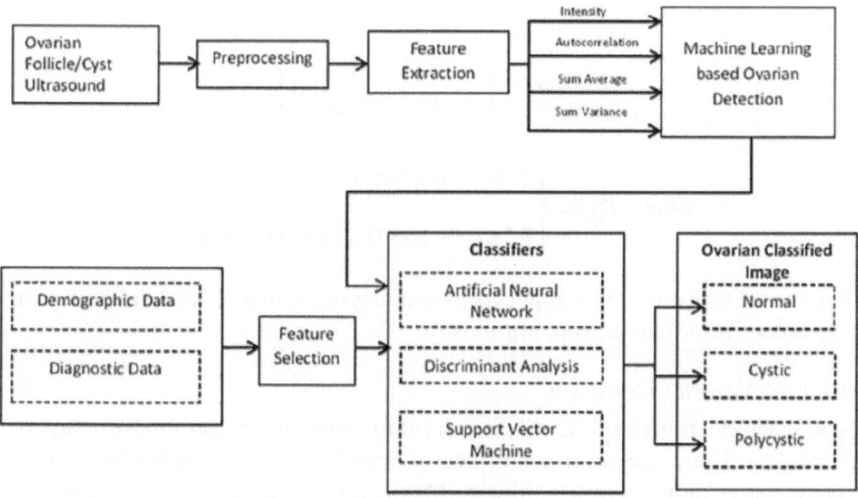

FIGURE 2.2 Flow diagram of the ISOC methodology.

2.2.2.1.1 Colour Space Transformation

Colour space is a definite association of colours. There are different types of colour spaces, which include RGB, cyan–magenta–yellow–key (CMYK), hue–saturation–value (HSV), hue–saturation–luminance (HSL) and Commission Internationale de l'Eclairage (CIE)-based colour spaces. Colour space conversion is performed depending on the requirement. It helps specify and identify information in an image in a more intuitive way.

The CIE colour spaces were defined in 1976 by the International Commission on Illumination. This system classifies colours according to human visual system. CIELuv and CIELab are the two CIE-based colour spaces which are device independent.

The $L*a*b*$ colour space is otherwise named as CIELAB or CIE $L*a*b*$. It is a derivative from CIE XYZ tristimulus values. It comprises a luminosity layer '$L*$', chromaticity layer '$a*$' and chromaticity layer '$b*$'. The chromaticity layer '$a*$' shows the region of colour fall in red–green axis, and the chromaticity layer '$b*$' shows the region of colour fall in the blue–yellow axis. The '$a*$' and '$b*$' layers contain all the information pertaining to the colours. The value of $L*$ ranges from 0 signifying the darkest black and to 100 signifying the brightest white. True neutral grey values at $a*=0$ and $b*=0$ are represented by the colour channels (Ford and Roberts, 1998).

The mathematical expressions of $L*$, $a*$ and $b*$ are given as follows:

$$L^* = \begin{cases} 116(Y/Y_n)^{1/3} - 16 \text{ if}(y/y_n) > 0.008856 \\ 903.3(Y/Y_n) \text{ if}(Y/Y_n) \leq 0.008856 \end{cases} \tag{2.1}$$

$$a^* = 500 * \left(f\left(\frac{X}{X_n}\right) - f\left(\frac{Y}{Y_n}\right) \right) \tag{2.2}$$

$$b^* = 200 * \left(f\left(\frac{Y}{Y_n}\right) - f\left(\frac{Z}{Z_n}\right) \right) \tag{2.3}$$

$$\text{where } f(t) = \begin{cases} t^{1/3} \text{if } t > 0.008856 \\ 7.787^* t + 16/16 \text{ if } t \leq 0.008856 \end{cases} \tag{2.4}$$

Colour space transformation supports in quantifying visual differences and separating colour information from luminosity.

2.2.2.1.2 Wavelet Denoising

Discrete wavelet transform (DWT) is helpful in analysing or decomposing signals and images. There are several families of wavelet, such as Daubechies, Haar, biorthogonal, Coiflets, symlets, Morlet, Mexican hat, and other real and complex wavelets. Daubechies wavelet is the most commonly used wavelet for signal and image processing applications.

2.2.2.1.2.1 Daubechies Wavelet   DWT using Daubechies wavelet comes under the orthogonal family of wavelets. It was framed in 1988 by Ingrid Daubechies. Daubechies wavelet is used in despeckling of medical ultrasound images (Ashish et al., 2010; Khare and Tiwary, 2007).

A Daubechies wavelet is defined as (de Vries, 2006):

$$\psi(x) = \sqrt{2} \sum_{k=0}^{2N-1} \left(-1^k\right) h_{2N-1-k} \varphi(2x-k)$$

(2.5)

where $h_0 \ldots h_{2N-1}$ are the constant filter coefficients satisfying the conditions

$$\sum_{k=0}^{N-1} h_{2k} = \frac{1}{\sqrt{2}} = \sum_{k=0}^{N-1} h_{2k+1}$$

(2.6)

and the scaling function $\varphi(x)$ is given by

$$\phi(x) = \sqrt{2} \sum_{k=0}^{2N-1} h_k \phi(2x-k)$$

(2.7)

Every wavelet contains numerous vanishing moments identical to the number of coefficients. The ability of wavelet to characterize the polynomial behaviour or the data available in a signal is limited by the vanishing moments (de Vries, 2006). The decay of a function towards infinity is determined by the vanishing moments (Chun-Lin, 2010). Daubechies wavelets have different orders represented by N ranging from db1 to dbN. Commonly used Daubechies wavelets are db1–db10. The Daubechies wavelet db1 has one vanishing moment definitely encoding constant signal components or polynomials of one coefficient. Similarly, db2 has two vanishing moments signifying the encoding of constant and linear signal components in a polynomial. This denotes that the vanishing moments determine the number of coefficients in a polynomial to be encoded. It facilitates highlighting various characteristics and preserving the sharpness of the image. The Daubechies wavelets possess an added feature that the finite linear combinations of the Daubechies wavelets yield local pointwise depictions of low-degree polynomials (Kessler et al., 2008). Daubechies wavelets are localized in temporal domain and approximately localized in the frequency domain (Alistair et al., 1995). Wavelet transform which is a suitable technique that neither consumes more computation time nor degrades the quality of the image was used in ovarian image denoising.

2.2.2.2   Feature Extraction

Image features are the distinct attributes in an image that help resolve the computational problems by providing evident information pertaining to the image content. In automatic detection, features help differentiate between the desired and the undesired regions. Image features are characterized by pixel intensity, colour, morphology and texture. The ovarian follicle images are greyscale images, and hence, colour as an image feature will not be useful for detection. Similarly, these images do not have precise boundary, shape or size. So morphological image features do not contribute to detection. Intensity and texture feature will help in detection of follicles/cysts.

2.2.2.2.1　Intensity Features

Intensity serves as a significant feature if the desired and the background regions share distinct pixel intensities. In ovarian images, the pixel intensities within the follicular regions were the same or near to one another and different or far from the pixel intensities in the non-follicular regions. So, each pixel intensity will exactly fall into only one cluster. So partitional clustering which is an intensity-based segmentation divides the data objects into non-overlapping clusters. This will help in the classification of follicular and non-follicular regions. Moreover, there is a requirement for the developed algorithm to consume less time and memory space so that computational complexity is avoided and the results will be obtained faster. Partitional clustering satisfies this condition also. K-means clustering is an extensively used partitional clustering algorithm, which belongs to unsupervised classification. In this study, hierarchical clustering was not suitable because it does not help group different regions distinctly and consumes more time for computation. K-means clustering was used in medical images for segmentation and feature extraction (Ramanjot et al., 2011; Samundeeswari et al., 2016; Hae et al., 2016).

In this study, a given ovarian ultrasound image having 'n' objects will be classified into 'k' clusters. The intensity was used as a feature to separate the ovarian image into three clusters specifically, such as germinal centres, interfollicular regions and mantle zone using k-means clustering. Euclidean distance was applied as the distance measure and was calculated between each cluster centre and each data point, and the data points were assigned to the cluster which has the minimum Euclidean distance. K-means algorithm was chosen as it was efficient to segment the given image into specified number of clusters. It also indicated the cluster which contained the desired information. It enabled stressing the predominant features of the given image. Intensity when combined with texture enhances effective feature extraction. So, texture-based segmentation was combined along with intensity-based segmentation.

2.2.2.2.2　Texture Features

Texture-based segmentation splits an image as a collection of disconnected regions depending on the texture properties so that every region in the image is similar to specific texture characteristics.

In an image, local features of each pixel are computed by statistical methods to examine the spatial distribution of the grey values. The relationship between the grey levels of the image is described using either first- or second-order statistical texture analysis (TA) (Haralick, 1973).

In ovarian detection, there was a necessity to analyse the correlation between the neighbourhood pixel values in order to discriminate between the non-follicular, follicular and interfollicular regions. First-order statistical TA does not facilitate this and was not suitable for detection and classification. So, second-order statistical methods were preferred for enhancing the efficiency.

Texture-based segmentation approach using Haralick texture features is a common and powerful technique in medical image analysis. Haralick texture features were preferred as they consider the spatial relationship between neighbourhood pixels, which was relevant here for discrimination of various regions in an ovarian ultrasound image. The grey level co-occurrence matrix (GLCM) is a specific scheme for texture feature extraction, which is advantageous in image segmentation and classification.

In this study, statistical texture methods helped discriminate various regions and facilitate facile analysis. In ovarian ultrasound images, after despeckling, feature extraction techniques based on statistical analysis of texture were applied to extract features that segmented the image into follicular and non-follicular or cystic and non-cystic regions. Besides feature extraction, computation of statistical values also aided in separating the different features of the image and enabled the classification of follicular region from its background. Texture features also helped emphasize the information pertaining to the tonal variations and spatial distribution of the segmented regions. Texture features facilitated indicating the regions of significant change. The given dataset was analysed using fourteen Haralick texture features. In ovarian image analysis, appropriate texture features were chosen depending on texture feature's ability to evade the overlying of non-follicular and follicular regions. Combined intensity and texture features were used for differentiating between the non-follicular or non-cystic regions and follicular/cystic regions.

2.2.2.3   Machine Learning-Based Ovarian Detection

The advent of ML algorithms in ovarian detection would certainly improve the accuracy and robustness. Hence, ML-based ovarian detection (MLOD) using ANN was developed for automatic detection.

2.2.2.3.1   Dataset Preparation

Intensity and the chosen principal texture features constituted dataset for this study. The input to the classifier comprises of four combined features, which include three texture features and one intensity feature. On the whole, 520 samples that comprise both follicular and non-follicular regions were taken for the study. A ratio of 50:50 was used here. Training and testing datasets were taken in a ratio of 70:30.

2.2.2.3.2   ANN-Based Ovarian Detection

The detection of abnormalities, tumours and cancers is the significant application of ANN in medical image processing. Studies reveal that classification of medical abnormalities using ANN with texture features as input has proved to yield effective and promising results. Combination of texture features and ANN has been used in classification of lung cancer (Almas and Bariu, 2012), classification of liver lesions (Zhenjiang et al., 2017) and classification of mammogram images (Biswas et al., 2016).

In the present study, there is a need for the proposed model to be easy for implementation and at the same time give a prospect for instantaneous analysis of the obtained results since the intended application of the system is in rural areas. Backpropagation network was applied in this study since it has the advantage of simple and easy implementation in addition to its ability to yield accurate results.

The performance of ANN solely depends on the selection of network parameters. A detailed examination was carried out in selecting the network parameters, which comprise the number of inputs, learning rate, momentum and hidden and output neurons. Effectual functioning of the network can be ensured only if optimal network parameters are chosen. The lowest normalized mean squared error (MSE) was computed to choose the relevant network parameters.

Intensity and the chosen texture features were used for developing the classifier which enabled an intelligent automatic ovarian detection. Untrained data were used for testing, which helped predict the classifier's performance, and the accuracy was calculated for the network's output. The presence or absence of follicle/cyst was determined as the classifier output.

2.2.2.4   Intelligent System for Ovarian Classification (ISOC)

ML algorithms such as ANN, discriminant classifier and SVM were developed in the ISOC methodology to identify a suitable algorithm for efficient classification of ovaries as cystic, normal and polycystic. So along with the image-based features, if the demographic and diagnostic features were included, it would facilitate a better classification. So, it was proposed to include the demographic and diagnostic features also for classification. Careful investigation was done in selecting the demographic and diagnostic features. The influence of each parameter on classification was thoroughly studied. Finally, based on the study, eight features were selected. The selected features were day of the menstrual cycle, size of the follicle, the number of follicles, FSH, prolactin, TSH, LH and RBS.

2.2.2.4.1   Dataset Preparation

A database comprising of 115 datasets were created. In the collected datasets, ten datasets were found to be incomplete with the required information such as day of the cycle and the values of FSH and LH. Hence, these were considered as outliers and the total number of datasets considered for ISOC was 105. In order to increase the training efficiency, multiple samples of input were taken. So, a total of 630 samples were used, out of which 442 were utilized in training and the remaining 188 were utilized in testing. Demographic and diagnostic data were used as features to improve classification accuracy. The input to train the classifiers includes the output of the MLOD classifier along with the selected demographic and diagnostic variables. The classifier classified the given input as normal, cystic or polycystic. The effective functionality of the classifiers was calculated through tenfold cross-validation (CV) technique.

2.2.2.4.2   Classification Using ANN

An ANN-based ovarian classifier was developed. Studies were carried out to fix suitable network parameters during the automatic ovarian detection. The same network parameters worked well for ANN-based ovarian classifier also.

Feedforward backpropagation network using gradient descent with momentum training function was applied. Hyperbolic tangent sigmoid function was used as transfer function. A momentum value of 0.2 and a learning rate value of 0.02 were fixed as network parameters.

2.2.2.4.3 Classification Using Discriminant Analysis (DA)

DA is a method to find discriminant functions that are useful in building a predictive model by making use of the labelled training data. Linear discriminant analysis (LDA) is applied to predict a linear grouping of features for categorizing two or more classes of groups or categories. DA is broadly used in medical diagnostics for decision-making (Wernecke, 1994).

During classification, it is necessary to remove the redundant predictors without affecting the model's predictive power in order to make the model simpler and robust. Regularization helps compute a minor group of predictors due to which an efficient model can be developed. The redundant predictors are found and eliminated using the two regularization parameters gamma (γ) and delta (δ) (Guo et al., 2007).

Gamma is the amount of regularization applied during the estimation of the covariance matrix of the predictors. It is a scalar which ranges from 0 to 1. It offers better control over the covariance matrix structure as compared to the discriminant type. Delta is a linear coefficient threshold and is a non-negative scalar. More predictors can be eliminated if the delta is set to a higher value.

In medical images, different variables related to a patient's health condition are learnt for the purpose of predicting which variables best predict and classify the type of abnormality. In this study, LDA is useful to predict the linear combination of the image, diagnostic and demographic features for classification of ovary as normal, polycystic and cystic. The discriminant type and model parameters for this study were determined by predicting the lowest MSE and the highest accuracy.

2.2.2.4.4 Classification Using SVM

SVM is a linear supervised classifier. It tries to categorize the new data by predicting an optimal hyperplane with the largest margin. SVM is applied in various medical applications such as cancer classification, speech recognition, bioinformatics and medical data mining. Kernel trick method is used by SVM for data transformation, and depending upon this transformation, an optimal boundary is found between the probable outputs. In SVM, distance measure between the support vectors and the new data is defined by kernel. The selection of the kernel is evident in predicting the performance of SVM (Huang et al., 2018). Similarly, solvers are required for solving quadratic programming (QP) optimization problems. The two standard QP solvers are sequential minimal optimization (SMO) and iterative single data algorithm (ISDA). SVM is useful in various tumour classifications of breast (Chien and Chuin, 2012), brain (Hari et al., 2014) and liver (Virmani et al., 2013).

In this study, classifying the ovary as normal, cystic and polycystic was inevitable. So, a three-class classification using SVM was developed. SVM was preferred because the number of input features used here was more and SVM is capable enough of establishing a correlation between those input features of a particular labelled class which may be either normal, cystic or polycystic and classifying the test data belonging to that class perfectly. This was because SVM was able to find a hyperplane with largest margin that will separate all these three classes and achieve a higher performance in classifying the given ovarian ultrasound image. Misclassification of ovaries was also completely avoided as it did not converge in local minima. Moreover, it was highly fault tolerant and showed a significant improvement in ovarian classification accuracy.

Here, a linear kernel was used. This was because the linear kernel when used in SVM learnt and tested the given ovarian dataset much faster as compared to the non-linear SVMs. It operated well with a large number of features used in this study. Similarly, pertaining to choice of solvers, SMO and ISDA were used and tested. In ovarian classification problem, there was a necessity to reduce the chance of misclassification of ovaries by minimizing the error function. But the solver 'ISDA' uses the gradient ascent technique and so it was not possible to reduce the misclassification of ovaries. Moreover, solver ISDA was not suitable for this study because the ovarian classification problem is not a binary classification as well. So, SMO solver was used here. The other model parameters were carefully chosen by estimating the lowest generalized error and the highest accuracy.

2.2.2.5 Performance Metrics

False acceptance rate (FAR) and false rejection rate (FRR) were used for computing the effectiveness of the implemented algorithms. The fraction of incorrect acceptances is given by FAR, and the fraction of incorrect rejections is given by FRR.

Moreover, the classifier's efficiency was also assessed using performance metrics such as specificity, sensitivity, precision, accuracy, F-measure and MCC. These indices were computed using the false-positive (FP), true-positive (TP), false-negative (FN) and true-negative (TN) values from the images.

Sensitivity: Quantitative prediction of positive outcomes. It is expressed as

$$\text{Sensitivity} = \frac{TP}{TP + FN} \tag{2.8}$$

Specificity: Quantitative prediction of negative outcomes. It is expressed as

$$\text{Specificity} = \frac{TN}{TN + FP} \tag{2.9}$$

Precision: Fraction of true-positive outcomes against all probable outcomes. It is expressed as

$$\text{Precision} = \frac{TP}{TP + FP} \tag{2.10}$$

Accuracy: Fraction of true outcomes indicated by the true positives and true negatives. It is expressed as

$$\text{Accuracy} = \frac{\text{TP+TN}}{\text{TP+FP+TN+FN}} \qquad (2.11)$$

F-measure: Harmonic average of recall and precision.

$$F-\text{measure} = \frac{2 \times \text{Precision} \times \text{Recall}}{\text{Precision+Recall}} \qquad (2.12)$$

MCC: It helps verify the performance of the classifier and ranges between −1 and +1. Better performance is depicted with larger values of MCC and is expressed below:

$$\text{MCC} = \frac{(\text{TP} \times \text{TN})-(\text{FN} \times \text{FP})}{\sqrt{(\text{TP+FN})(\text{TP+FP})(\text{TN+FN})(\text{TN+FP})}} \qquad (2.13)$$

High efficiency of classification algorithm is demonstrated if the values of performance measures are closer to 100%. Likewise, precise efficiency of the classifier is depicted if MCC value is nearer to 1 as this reveals that the ground truth corroborates with the obtained value. On the contrary, if MCC value is nearer to zero, it indicates misclassification.

2.3   RESULTS

2.3.1   PREPROCESSING

The ovarian RGB image was transformed to $L*a*b*$ colour space because computation became complex in RGB colour space as it was difficult to spot the luminosity information. This process reduced the computational time by 17%. $L*$(luminosity layer) was chosen for successive processing as it had better visualization and a good relationship with perceived lightness. Single-level two-dimensional DWT with Daubechies wavelet (db2) was used on luminosity layer for the purpose of speckle noise removal. A better classification between the follicle/cyst and the background/noise was obtained as DWT facilitated the normalization of the image in the wavelet space. Here, single-level decomposition yielded a better denoised image. Higher-level decomposition resulted in the loss of image information. Daubechies wavelet was chosen as it had the capacity of solving self-similarity problems, thereby helping to differentiate between the noise and the follicular regions. Balanced frequency responses are present in the Daubechies wavelet, which gave smoothness and preserved the edges in the ovarian images. A brighter interpretation of the follicles/cysts was provided by the approximation coefficient. It indicated smooth colour variations, where minute differences were clearly represented as edges among the smooth variations. Figure 2.3 represents the wavelet transformed images a, b and c.

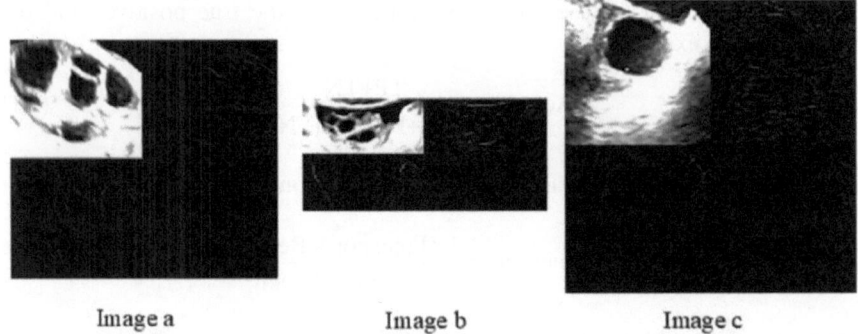

Image a   Image b   Image c

FIGURE 2.3 Wavelet transformed images.

Thus, DWT helped in noise removal, which was revealed from the peak signal-to-noise ratio (PSNR) values that are greater than 30 db.

2.3.2 FEATURE EXTRACTION

2.3.2.1 Intensity Features

K-means clustering, which is an intensity-based segmentation, was applied to the approximation band of the wavelet transformed image.

Cluster numbers varying between 2 and 10 were tried. A better segmentation was obtained with three clusters as compared to selecting the other cluster numbers. The results with other cluster numbers yielded either an under-segmented or an over-segmented image. This algorithm helped group the samples having the same intensity by which the non-follicular, interfollicular and follicular regions were clustered distinctly. Figure 2.4 shows the ovarian image in clusters 1–3.

A better segmentation of the ovarian image was obtained in the cluster with the germinal centre. The intensity-based clustering did not segment the interfollicular/intercystic regions properly since the classification rule was solely based on intensity. So, texture-based segmentation was performed, and those features that facilitated an efficient detection were combined with the intensity to give an enhanced detection of follicles/cysts.

2.3.2.2 Texture Features

Texture feature extraction was performed on the approximation band of the wavelet transformed image. Fourteen Haralick parameters were extracted and analysed.

Three Haralick features, namely autocorrelation (f1), sum average (f2) and sum variance (f3), were chosen since they clearly distinguished between the follicular/cystic area and its background, thus contributing to better accuracy. This is clearly revealed from Figure 2.5. These features enabled the elimination of undesired areas of the image, thereby highlighting the prominent attributes and interfollicular regions of the image. The non-selected features had an overlapping between the desired and the undesired regions.

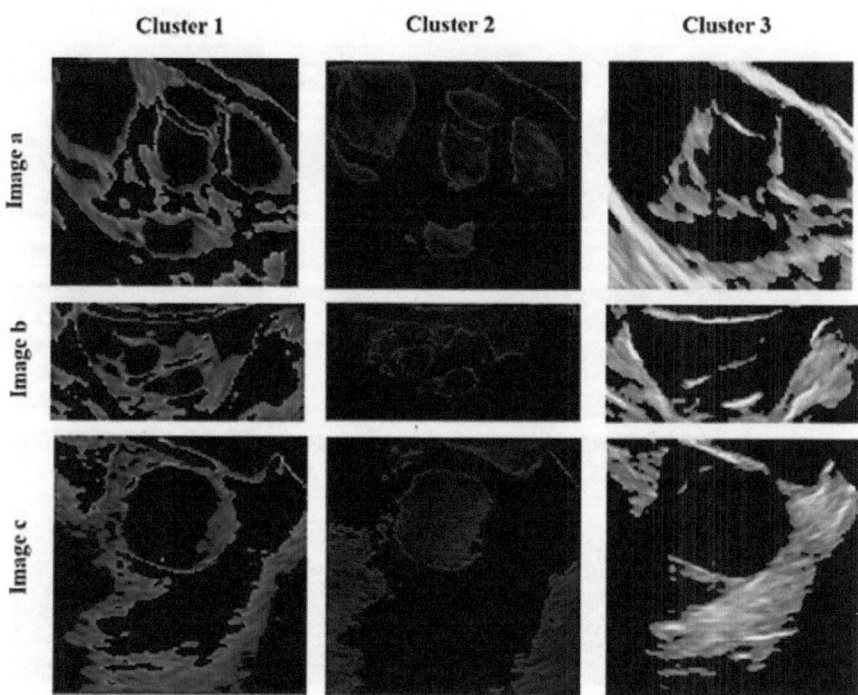

FIGURE 2.4   Ovarian image in three clusters.

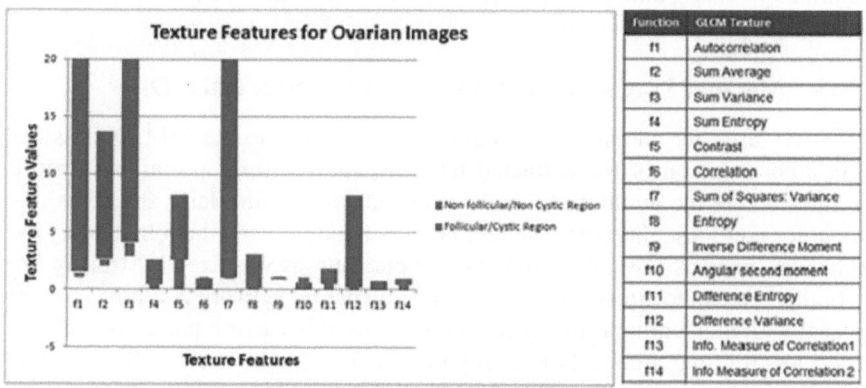

FIGURE 2.5   Selection of Haralick texture features.

The results of texture feature extraction are represented in Figure 2.6. To enhance the effectiveness in ovarian detection, texture and intensity-based ovarian detection (TIOD) which uses effectual combination of intensity and the three extracted texture features was performed. Accurate detection of follicular, non-follicular and interfollicular regions was successfully accomplished since the

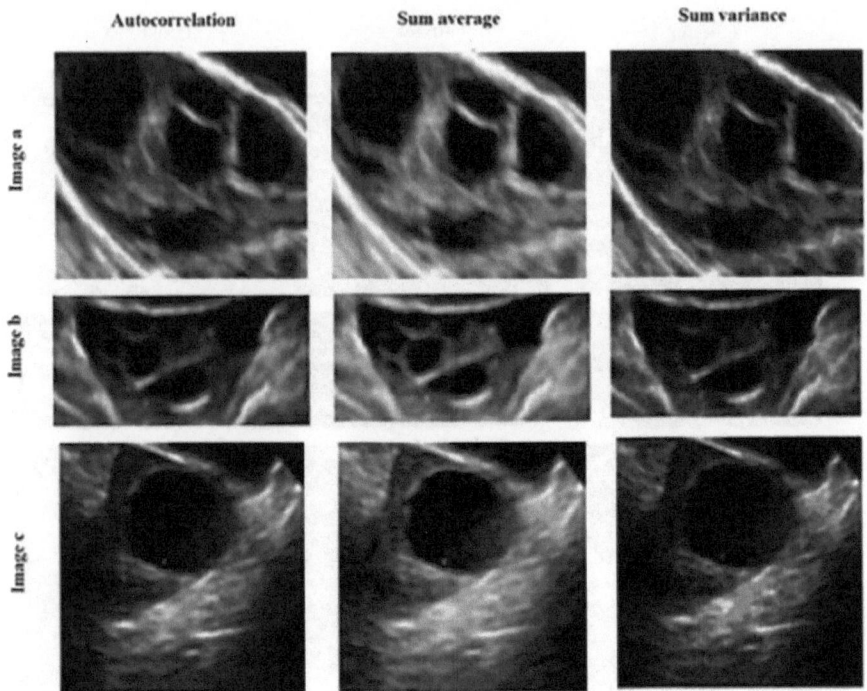

FIGURE 2.6   Extraction of texture features.

combined intensity and texture features were used. It also helped eliminate the undesired regions in the image.

2.3.3 Machine Learning-Based Ovarian Detection (MLOD)

ANN was used for automatic ovarian detection. Inputs to the MLOD classifier were a combination of the extracted intensity and the three prominent texture parameters. So, the number of inputs was fixed as four. Similarly, the network's output for non-follicular/non-cystic region was fixed as 0 and the follicular/cystic region was fixed as 1. So, the number of output neurons was fixed as two.

In this system, feedforward backpropagation neural network is used. Gradient descent with momentum training function was used because it had faster convergence, less training time and the ability to slide through the local minima. In this network with respect to the hidden layers, hyperbolic tangent sigmoid neurons were used and, with respect to the output layer, linear neurons used. In this system, there was a need to foresee the possibility of classification as follicular/cystic or non-follicular/non-cystic region. So, the usage of sigmoid neuron served as the right choice. Similarly, linear neurons were used as the model operates within nominal parameters.

Different momentum and learning rate values varying between 0.1 and 0.9 and between 0.01 and 0.9, respectively, were tried for testing the efficiency of the classifier. The final value for training was fixed based on the condition of lowest MSE. The lowest MSE was obtained for a momentum value of 0.2 and a learning rate value of 0.02, which is clear from Figure 2.7.

The number of hidden neurons were varied from 1 to 8, and the number of neurons for which the lowermost MSE was achieved was finally chosen. Hidden neurons with three numbers yielded the lowest value of MSE. Overfitting occurred with the increase in the number of hidden neurons, which also resulted in an increase in MSE. So, a classifier with four neurons as input, three neurons as hidden and two neurons as output was developed taking gradient descent with momentum as training function with momentum and learning rate values of 0.2 and 0.02, respectively.

Significant outcome was achieved in detection and classification due to the grouping of linear and sigmoidal activation functions. Classification of the desired region from the background was obtained due to the successive updation of weights, resulting in error minimization. The performance of the MLOD training algorithm is shown in Table 2.3.

A testing accuracy of 96% was obtained when the classifier was tested for all the test images with combined features. Figure 2.8 shows the segmented image obtained as the output of the network.

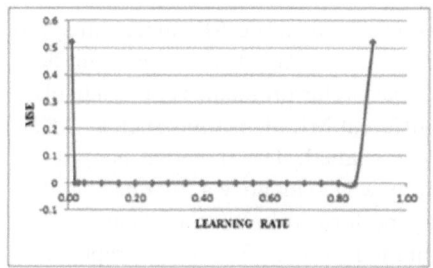 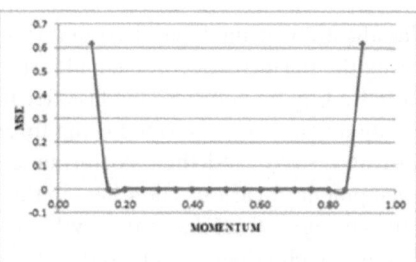

FIGURE 2.7   MSE by learning rate and momentum in MLOD.

TABLE 2.3
Performance of the MLOD Training Algorithm

Parameters	Value
Number of epochs	1000
Training time	0.01 seconds
Training performance	0.0186
Validation performance	0.026519
Gradient	0.0192
Learning rate	0.02

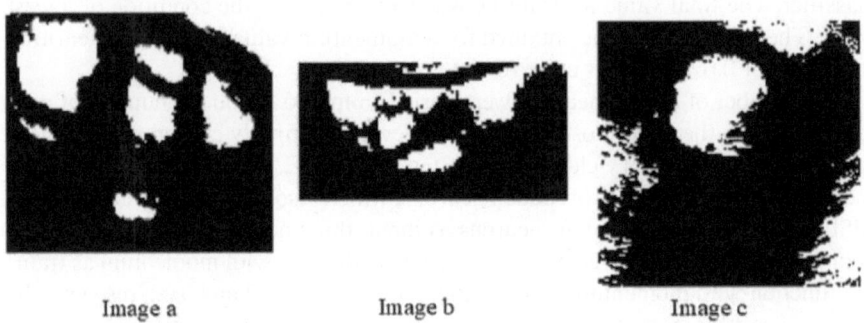

Image a　　　　　　　　Image b　　　　　　　　Image c

FIGURE 2.8 Classified image using ANN.

The ANN helped in automatically classifying the non-follicular/non-cystic regions and follicular/cystic regions. This output served as one of the feature inputs to the classifier that was developed for ovarian classification. The efficiency of the detection was improved using the MLOD algorithm.

2.3.4 INTELLIGENT SYSTEM FOR OVARIAN CLASSIFICATION

Classification of medical abnormalities using two or more ML algorithms and testing the performance of each algorithm for its accuracy had a significant benefit in finding an appropriate classifier for a particular problem. So, ovarian classification was also performed using three different ML algorithms such as ANN, LDA and SVM, and their performance was studied. The intention behind the development of various classifiers in this study was to identify a suitable classifier with the highest accuracy that will exactly perform ovarian classification. The input to the classifiers comprised eight combined features from demographic and diagnostic dataset and one input from the MLOD classifier.

2.3.4.1 Classification Using ANN

An ANN-based ovarian classifier was developed. Studies were carried out to fix suitable network parameters for ovarian classification. A classifier having nine neurons as input, three hidden neurons and three neurons as output was designed with a momentum value of 0.2, learning rate value of 0.02 and gradient descent with momentum as a training function. A prominent result was achieved in classifying the ovaries due to the grouping of linear and sigmoidal activation functions. Figure 2.9 shows the network training performance.

In ISOC, the probability of ovarian classification has to be predicted as an output that lies between 0 and 1 and so sigmoid neurons helped in classifying the ovary as normal, cystic and polycystic.

2.3.4.2 Classification Using LDA

Discriminant classifier was developed for ovarian classification. A systematic examination was carried out in selecting the model parameters. With respect to

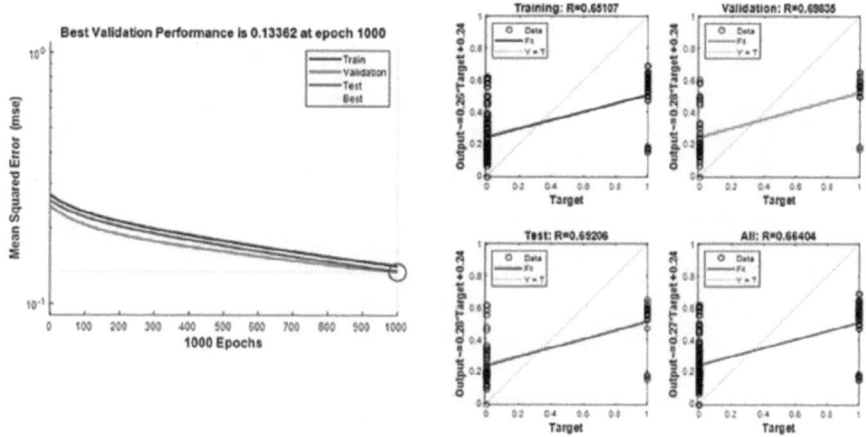

FIGURE 2.9 Performance of ANN classifier for ISOC.

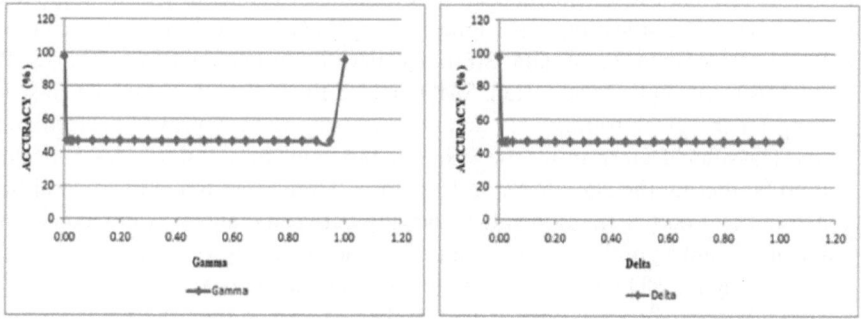

FIGURE 2.10 Accuracy using gamma and delta in LDA.

discriminant type, trials were made using the linear, pseudolinear, diaglinear, quadratic, diagquadratic and pseudoquadratic discriminant types.

In LDA-based ovarian classification problem, singular covariance matrix was used and the discriminant type 'linear' did not work well. So 'pseudolinear' discriminant type was chosen and it yielded the highest accuracy.

In ISOC methodology, there was a necessity to remove unwanted predictors which would lead to misclassification of ovary and reduce the classification accuracy. Different values of delta and gamma varying from 0 to 1 were used to analyse the performance of the classifier. The value which yielded the highest accuracy and lowest generalized error was fixed to be the ultimate value for training. A delta value of 0 and a gamma value of 0 yielded the lowest generalized error and highest accuracy, which is understood from Figure 2.10. These values also helped in regularization, thereby yielding an effective classification system.

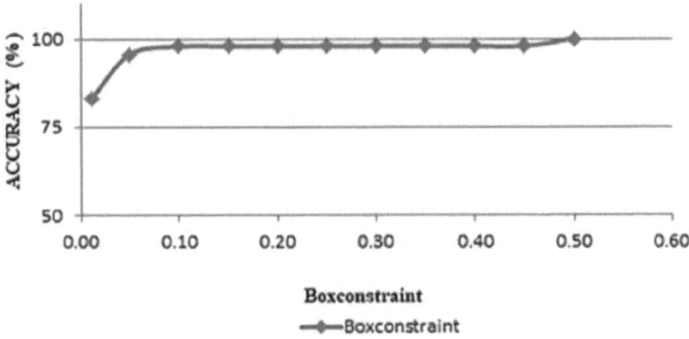

Boxconstraint
—◆—Boxconstraint

FIGURE 2.11   Accuracy using boxconstraint in SVM.

2.3.4.3   Classification Using SVM

The selection of model parameters such as kernel function, solver, box constraint, the number of iterations and outlier fraction was suitably made. With respect to kernel functions, linear, polynomial and radial basis functions were applied to the problem. Classification was highest using linear kernel function because it helped in faster training of data. Similarly, pertaining to choice of solvers, SMO and ISDA were applied and tested. SMO yielded the highest accuracy because its scaling for the given training set was better, resulting in improved performance of the classifier. Determining the value of box constraint contributes considerably to predicting the classifier's performance. Trials were made for several box constraint and outlier fraction values varying between 0 and 1. A box constraint value of 0.5 and an outlier fraction value of 0 yielded the lowest error and highest accuracy. Figure 2.11 shows the accuracy obtained using box constraint.

The chosen box constraint value neither increases the training error nor increases the training time for the developed ISOC system. Similarly, the highest classification accuracy was obtained at 1000 iterations. The developed model was used for classifying the test data.

2.4   DISCUSSION

The performance of the classifiers was compared by computing a confusion matrix. The confusion matrix reveals the extent to which each classifier gives the accurate classification and helps find the best performing classifier. Table 2.4 represents the results of ANN, LDA and SVM classifiers.

From the confusion matrix, with respect to ANN, 24 samples were appropriately classified as cystic, all the 72 samples were appropriately classified as polycystic, and 83 samples were appropriately classified as normal. But five normal samples were incorrectly classified as cystic and four cystic samples were incorrectly classified as normal. The classification accuracy of the ANN classifier was 95%.

With respect to LDA, it was inferred that 28 samples were appropriately classified as cystic and 72 samples were appropriately classified as polycystic.

TABLE 2.4
Classification Results of Intelligent Classifiers

		ANN		LDA		SVM	
Category		T	F	T	F	T	F
Normal ovary	T	83(TP)	5(FN)	84(TP)	4(FN)	88(TP)	0(FN)
	F	4(FP)	96(TN)	0(FP)	100(TN)	0(FP)	116(TN)
Polycystic ovary	T	72(TP)	0(FN)	72(TP)	0(FN)	72(TP)	0(FN)
	F	0(FP)	116(TN)	0(FP)	116(TN)	0(FP)	116(TN)
Cystic ovary	T	24(TP)	4(FN)	28(TP)	0(FN)	28(TP)	0(FN)
	F	5(FP)	155(TN)	4(FP)	156(TN)	0(FP)	160(TN)

Likewise, 84 samples were appropriately classified as normal. Similarly, there was no misclassification for cystic or polycystic sample. But four normal samples were incorrectly classified as cystic, contributing to 2.1% of the entire data. So, 97.9% estimates were right and 2.1% estimates were wrong, representing a better classification with an accuracy of 97.9%.

With respect to SVM, it was inferred that 28 samples were properly classified as cystic, 88 samples were classified as normal, and 72 samples were classified as polycystic. No sample was incorrectly classified, thus yielding an accuracy of 100%.

Adding to this, FAR and FRR values for normal, polycystic and cystic samples were computed for all the three ovarian classification methodologies such as ANN, LDA and SVM. The effectiveness and accuracy of the best performing classifier were predicted by comparing all the three methods. The results of comparison are shown in Figure 2.12.

From the above graphs, it is observed that, in the ANN classifier, the FAR and FRR values for polycystic samples are zero, showing that no polycystic sample was wrongly classified as cystic or normal sample. The FAR value (2.6%) for the cystic samples is high because some of the non-cystic samples were classified as cystic, and the FRR value (2.12%) shows that few cystic samples were classified as normal. With respect to normal samples, FAR value (2.12%) shows that few cystic samples were classified as normal and FRR value (2.6%) shows that some normal samples were classified as cystic.

In LDA, the FAR value (2.12%) for the cystic samples is high because some of the non-cystic samples were classified as cystic, which results in ambiguity, whereas the FRR value for cystic samples is good, showing that all the cystic samples were classified accurately. Also, the FAR value for normal samples is zero, indicating that no cystic or polycystic sample was classified as normal, whereas the FRR value (2.12%) is high, depicting that some normal samples were classified as cystic. On the other hand, the FAR and FRR values for polycystic samples are zero, showing no misclassification. The chance of cystic sample being classified as normal was evaded using this classifier.

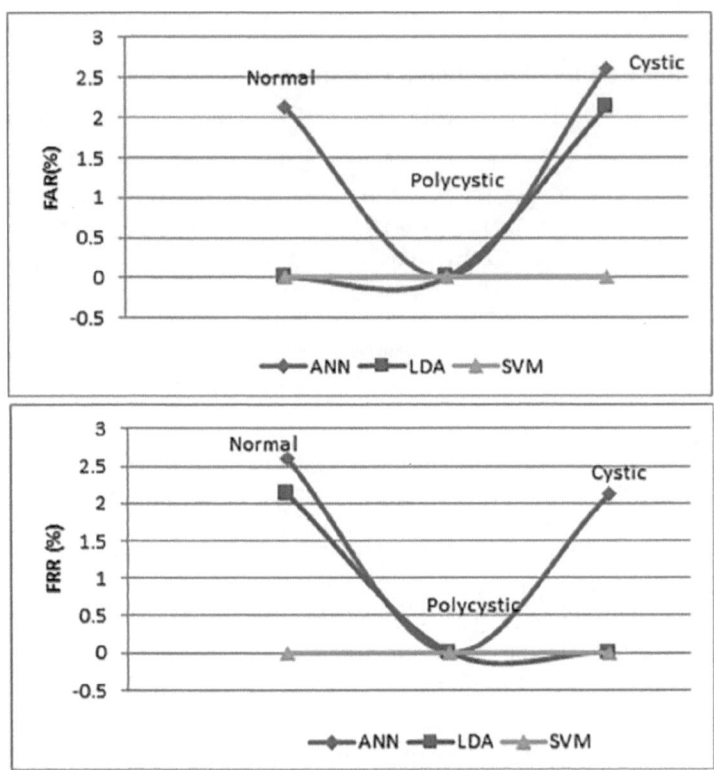

FIGURE 2.12 Comparison of intelligent ovarian classification methodologies.

In SVM classifier, a fine consensus is depicted among the FAR values and the FRR values for all the samples. As an outcome, a general decrease in the FAR and FRR fractions was observed as compared to other ML-based classification methodologies. The percentage values are zero, showing 100% classification accuracy. This reveals that SVM would be a more suitable classification algorithm for the ISOC methodology.

Although the accuracy of the individual intelligent classifiers seems to be high, the overall classification accuracy of the system depends on the detection accuracy obtained from MLOD methodology as well. The misclassification in MLOD will certainly have an impact on the overall performance of the system. Hence, the study necessitates the computation of the overall performance indices of the different classifiers, which includes the performance indices of the MLOD also.

A comparison was made between the various classifiers using the various performance indices to find the overall efficiency of the methodology. This helped identify a suitable classifier. Table 2.5 represents the results of various performance indices.

From Table 2.5, it is inferred that all the performance metrics of SVM classifier are higher when compared to the LDA and ANN classifiers. ANNs can suffer

TABLE 2.5

Performance Comparison of ANN, LDA and SVM Classifiers

Performance Index	ANN			LDA			SVM		
	Normal	Polycystic	Cystic	Normal	Polycystic	Cystic	Normal	Polycystic	Cystic
Sensitivity (%)	91	97	82	92	97	95	96	97	95
Specificity (%)	89	96	95	93	96	98	93	96	98
Accuracy (%)	91	97	92	93	97	97	95	97	98
Precision (%)	89	96	81	94	96	98	94	96	98
F-measure (%)	90	97	81	93	97	96	95	97	96
MCC	0.80	0.93	0.76	0.85	0.93	0.94	0.89	0.93	0.94

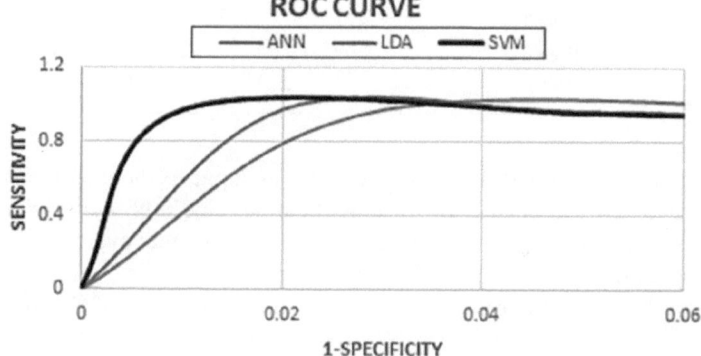

FIGURE 2.13 ROC curve for intelligent ovarian classification methodologies.

from multiple local minima. In this study, common features are shared by the follicular/cystic area and cystic/non-cystic area and ANN gets struck in exactly finding the variations between both regions. So, there is a lack of 2% accuracy. If SVM is used in detection, it converges on a global minimum, allows a better tolerance, adapts to the variations between both regions and may yield improved accuracy.

A comparison was made between all these methods using the ROC curve so as to prove the competence of the best performing classifier. The comparison results are shown in Figure 2.13.

From Figure 2.13, it is revealed that SVM would be a more suitable classification algorithm for the ISOC methodology.

The SVM-based ovarian classifier had classified the ovaries as normal, cystic and polycystic accurately. The combination of image-based features and the

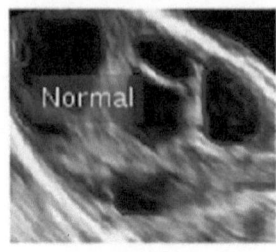
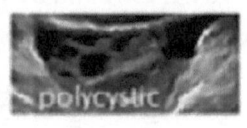
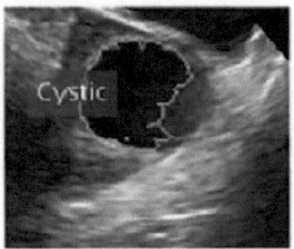

Image a　　　　　　　　　Image b　　　　　　　　　Image c

FIGURE 2.14　Results of the ISOC methodology.

TABLE 2.6
Time Consumption of Various Classifiers

Sample Images	SVM	LDA	ANN
Image a	2.51	3.125	8.12
Image b	2.85	3.56	9.26
Image c	3.12	3.90	10.14

selected demographic and diagnostic features as input to the classifier had successfully avoided the chance of misclassification. The ambiguity of a normal ovary being classified as cystic or vice versa was completely overcome using this classifier even though some features were common in both.

Figure 2.14 shows the results of the ISOC methodology. ISOC had classified the ovaries as normal, polycystic and cystic ovaries using the SVM classifier, which gave an overall maximum classification accuracy of 98%.

Time is an important computational trade-off in the models that were developed. Computational time differed depending upon the size of the test image that is given to the classifier. The ratio of time taken between SVM and ANN is 1:3.25, and the ratio of time taken between SVM and LDA is 1:1.25. As an overall inference, the least amount of time was consumed by the SVM classifier, followed by LDA and ANN. A table on time consumed by various classifiers in seconds is given in Table 2.6.

The results were validated by the medical expert of Department of Obstetrics & Gynaecology, Chettinad Hospital and Research Institute, Chennai.

2.5　CONCLUSIONS

An intelligent methodology was developed using ML technique for effective ovarian detection and classification. Preprocessing techniques such as colour transformation and DWT reduced the computational complexity and removal of speckle noise, respectively. The hybrid system called texture and intensity-based ovarian detection that uses a combination of intensity and textural features made a remarkable difference between the desired and undesired regions, thus enabling a facile analysis of the ovarian image.

The advent of ML algorithm for intelligent ovarian detection facilitated improving the accuracy and robustness in ovarian detection. The development of an intelligent ovarian classifier was tried using three different classifiers such as ANN, LDA and SVM for identifying an appropriate classifier that achieves the highest accuracy. Learning model parameters for the classifiers were chosen properly as they played an important role in predicting the model performance. A combination of image, demographic and diagnostic features served as input to classifiers. SVM outperformed the other classifiers, giving a maximum of 98% accuracy. This intelligent system will help the medical expert reduce the possibility of misinterpretation and certainly support in decision-making process.

ACKNOWLEDGEMENTS

The authors are thankful to the Department of Obstetrics and Gynaecology, Chettinad Hospital and Research Institute, Chennai, for giving the required datasets, providing suggestions and validating the results. The authors also record their special thanks to the management of Hindustan Institute of Technology and Science for the help and motivation given during the entire research work.

REFERENCES

Alistair, C.H.R. and Paul, C.A. (1995). Daubechies wavelets and mathematica. *Computers in Physics*, 9(6): 635–648.

Ashish, K., Manish, K., Yongyeon J., Hongkook, K. and Moongu, J. (2010). Despeckling of medical ultrasound images using Daubechies complex wavelet transform. *Signal Processing*, 90(2): 428–439.

Almas, P. and Bariu, K. S. (2012). Detection and classification of lung cancer using artificial neural network. *International Journal on Advanced Computer Engineering and Communication Technology*, 1(1): 62–67.

Benacerraf, B.R., Abuhamad, A.Z., Bromley, B., Goldstein, S.R., Groszmann, Y., Shipp, T.D. and Timor, T.I.E. (2015). Consider ultrasound first for imaging of the female pelvis. *American Journal of Obstetrics & Gynecology*, 212(4): 450–455.

Biswas, R., Nath, A. and Roy, S., Mammogram classification using gray-level co-occurrence matrix for diagnosis of breast cancer. *Proceedings of 2016 International Conference on Micro-Electronics and Telecommunication Engineering*, Ghaziabad, pp. 161–166, 2016. ISBN: 9781509034116.

Castellano, G., Bonilha, L., Li, L.M. and Cendes, F. (2004). Texture analysis of medical images. *Clinical Radiology*, 59: 1061–1069.

Chien, S.L. and Chuin, M.W. (2012). Support vector machine for breast MR image classification. *Computers and Mathematics with Applications*, 64(5): 1153–1162.

Chun-Lin, L. (2010). A Tutorial of the wavelet transform. http://disp.ee.ntu.edu.tw/tutorial/WaveletTutorial.pdf.

deVries, A. (2006). Wavelets. http://math-it.org/Publikationen/Wavelets.pdf.

Dewi, R.M. and Wisesty, U.N. (2018). Classification of polycystic ovary based on ultrasound images using competitive neural network. *Journal of Physics Conference Series*, 971-012005: 1–8.

Ford, A. and Roberts, A. (1998). Color space conversions. available at URL http://www.wmin.ac.uk/ITRG/docs/coloreq/coloreq.

Goswami, B., Patel, S., Chatterjee, M., Koner, B.C. and Saxena, A. (2009). Correlation of prolactin and thyroid hormone concentration with menstrual patterns in infertile women. *Journal of Reproduction and Infertility*, 10(3): 207–212.

Guo, Y., Hastie, T. and Tibshirani, R. (2007). Regularized linear discriminant analysis and its application in microarrays. *Biostatistics*, 8(1): 86–100.

Hae, J.L., Doo, H.S. and Kwang, B.K. (2016). Effective computer-assisted automatic cervical vertebrae extraction with rehabilitative ultrasound imaging by using K-means clustering. *International Journal of Electrical and Computer Engineering*, 6(6): 2810–2817.

Hamm, B. (1994). Computerized tomography and MR tomography in diagnosis of ovarian tumors. *Der Radiologe*, 34(7):362–369.

Haralick, R.M., Shanmugam, K. and Dinstein, I. (1973). Textural features for Image Classification. *IEEE Transactions on Systems Man and Cybernetics*, 3(6): 610–621.

Hari, B.N., Salankar, S.S. and Bora, V.R. MRI brain cancer classification using support vector machine. *Proceedings of 2014 IEEE Students' Conference on Electrical, Electronics and Computer Science*, Bhopal, pp. 1–6, 2014. ISBN: 9781479925261.

Hiremath, P.S. and Tegnoor, J.R. (2012). Automated ovarian classification in digital ultrasound images using SVM. *International Journal of Engineering Research and Technology*, 1 (6): 1–17.

Huang, Q., Zhang, F. and Li, X. (2018). Machine learning in ultrasound computer-aided diagnostic systems: a survey. *BioMed Research International*, 2018: 1–10.

Jain, S. and Salau, A.O. (2019). Detection of Glaucoma using two dimensional tensor empirical wavelet transform. *SN Applied Sciences*, 1(11): 1417. Doi: 10.1007/s42452-019-1467-3.

Jianfeng, Z., Jiatuo, X., Xiaojuan, H., Qingguang, C., Liping, T., Jingbin, H. and Cui, J. (2017). Diagnostic method of diabetes based on support vector machine and tongue images. *BioMed Research International*, 2017: 1–9.

Kessler, B.M., Payne, G.L. and Polyzou, W.N. (2008). Wavelet notes. https://arxiv.org/pdf/nucl-th/0305025.pdf.

Khare, A. and Tiwary, U.S. (2007). Daubechies complex wavelet transform based technique for denoising of medical images. *International Journal of Image and Graphics*, 7 (4): 663–687.

KiruthikaV. and Ramya, M.M. (2014). Automatic segmentation of ovarian follicle using K-means clustering. *Proceedings of 5th International Conference on Signal and Image Processing*, Jan 8–10; Bangalore, India, pp. 137–141.

Kiruthika, V., Sathiya, S. and Ramya, M.M. (2018). Automatic texture and intensity based ovarian classification. *Journal of Medical Engineering and Technology*, 42(8): 604–616.

Kumar, P.S.J. and Indranil, S. (2014). Speckle reduction of ultrasound image using wavelet transform. *International Journal of Computer Engineering and Applications*, 8(3): 122–129.

Luderer, U. (2014). Ovarian toxicity from reactive oxygen species. *Vitamins & Hormones*, 94: 99–127.

Madhanlal, U., Kalpana, R. and Soundararajan, P. (2017). Assessment of chronic kidney disease using skin texture as a key parameter: for South Indian population. *Healthcare Technology Letters*, 4(6): 223–227.

Oleg, V.M. and Allen, T. (2006). Despeckling of medical ultrasound images. *IEEE Transactions on Ultrasonics, Ferroelectrics and frequency Control*, 53(1): 64–78.

Pham, D.T. and Afify, A.A. (2007). Clustering techniques and their applications in engineering. *Proceedings of the Institution of Mechanical Engineers, Part C: Journal of Mechanical Engineering Science*, 221(11): 1445–1459.

Punithavathy, K., Sumathi, P. and Ramya, M.M. (2019). Performance evaluation of machine learning techniques in lung cancer classification from PET/CT images. *FME Transactions*, 47(3): 418–423.

Ramanjot, K., Lakhwinder, K. and Savita, G. (2011). Enhanced k-mean clustering algorithm for liver image segmentation to extract cyst region. *IJCA Special Issue on Novel Aspects of Digital Imaging Applications*, 1: 59–66.

Samundeeswari, E.S., Saranya, P.K. and Manavalan, R. Segmentation of breast ultrasound image using regularized k-means (ReKM) clustering. *Proceedings of International Conference on Wireless Communications, Signal Processing and Networking*, Chennai, pp. 1379–1383, 2016. ISBN: 9781467393386.

Sheikdavood, K. and Palanivel, R.S. (2016). Analysis of ovarian diseases using ultrasound images. *Journal of Advances in Chemistry*, 12(10): 4449–4454.

Singh, P., Singh, S., Singh, R. and Raghuvanshi, R. (2006). Anaemia as a cause of Infertility: focus on management of Anaemia as first line management of infertility. *The Internet Journal of Gynecology and Obstetrics*, 8 (1): 1–3.

Szaboova, R. and Devendra, S. (2015). Infertility in a young woman with Type 2 diabetes. *London Journal of Primary Care*, 7(3): 55–57.

Virmani, J., Kumar, V., Kalra, N. and Khandelwal, N. (2013). SVM based characterization of liver ultrasound images using wavelet packet texture descriptors. *Journal of Digital Imaging*, 26(3): 530–543.

Wernecke, K.D. (1994). On the application of discriminant analysis in medical diagnostics. In: Bock H.H., Lenski W. and Richter M.M. (eds) *Information Systems and Data Analysis. Studies in Classification, Data Analysis and Knowledge Organization*, pp. 267–279, Springer, Berlin, Heidelberg. ISBN: 9783642468087.

WHO Technical Report Series. (1992). Recent advances in medically assisted conception. *American Journal of Law & Medicine*, 820: 1–111.

Zhenjiang, L., Yu, M., Wei, H., Hongsheng, L., Jian, Z., Wanhu, Li. and Baosheng, L. (2017). Texture-based classification of different single liver lesion based on SPAIR T2W MRI images. *BMC Medical Imaging*, 17(42): 1–9.

3 On Effective Use of Feature Engineering for Improving the Predictive Capability of Machine Learning Models

M. R. Pooja
Vidyavardhaka College of Engineering

CONTENTS

3.1   Introduction 53
3.2   Background.. 54
3.3   Data Description and Preparation 55
3.4   Domain Knowledge and Feature Engineering 55
3.5   Balanced Data Creation Using OCSVM 56
      3.5.1   One-Class SVM.. 56
      3.5.2   Data Preparation for OCSVM 58
3.6   Results and Discussion 59
3.7   Conclusions... 60
Declarations 61
References... 61

3.1   INTRODUCTION

With a prognostic model, the primary focus is on the search for a combination of features that are as robust as possible in predicting the disease outcome. This can greatly benefit both the clinicians and the patients, thereby reducing the overall costs associated with patient care [1–3]. Predictive modeling of diseases is done by deploying tools that are seen to assist the decision-making process [4–6]. A few of them aim at predicting clinical outcome, and the rest focus on classifying patients who are at risk of development of a particular disease condition. Incorporating domain knowledge into the prediction models involves identifying

DOI: 10.1201/9781003224068-3

clinical features that are most relevant to the process of disease classification using prior knowledge about the disease and its contributing factors [7,8]. These features further are likely to enhance the performance of the classifiers deployed. Feature engineering is centric to the use of domain knowledge within clinical prediction models and commonly includes the following tasks:

(i) Combination of multiple features
(ii) Creation of new features
(iii) Extraction of features from the original features.

All the three tasks were targeted in our approach for integrating domain knowledge into the model developed. While the first two tasks involve manual intervention in the inclusion of features derived from those existing on the recommendations of the clinical work as per the literature available, the last task involves the application of a filter-based feature selection machine learning approach that is data-driven and is tailored to suit the nature of data available [9].

3.2　BACKGROUND

Spirometry is regarded as one of the most commonly performed investigation methods to gauge the pulmonary function in patients with chest diseases, such as asthma [10]. However, the methodological deployment of the equipment and performance strategies require high degree of attention toward quality control and these have been well standardized and subjected to constant revisions from time to time. Despite this, most of the studies related to prediction equations for spirometry have lost their utility as they are several decades old and were carried out with equipment and standardized procedures that have changed a lot since then. Added to this, the lung health of the population has seen a dramatic change over years, leading to failure of reliability [11,12]. The health of the lung and its functionality to a large extent is affected by age, height, weight, ethnicity, exposure to environmental factors, socio-economic status along with few other factors remaining unidentified. There remain no "typical" or "normal" values as such that can be applied to a common population. As such, in order to have a specific common scale across varying populations, comparison is made through expected values for patients of specific age, gender and physical characteristics. These expected values are broadly called predicted values and are generated using prediction equations built using regression analysis on data collected from a healthy population.

　　A wide range of such equations for lung functions are developed, which show a considerable difference in the predicted values. Of the several such predictive equations including ATS/ERS, GLI and ARTP, ARTP reference equations for lung function has been used for generating the predicted values for FEV1 and FVC as it is seen to generalize well on several populations and a deeper analysis has shown that it performs well for most spirometry parameters. A model that incorporates different sources of disease indicators is most often suggested as a better approach yielding predictive model for individual risk assessment that is of greater value to both patients and clinicians [1,13–15].

The validation of the feature-driven prediction model is done by deploying MSFET for the extraction of significant features that signify asthma severity indicators. MSFET (discussed in Section 3.3) involves the selection of severity indicators from the available set of features using feature scoring techniques. The process is followed by a comparative analysis of the various ML classifiers. Support vector machine, logistic regression and naive Bayes classifiers were used to validate the performance of the model, and the best performing classifier was adopted for the approach used. The efficiency of the model was evaluated on the complete as well as optimized feature sets obtained via MSFET, and the inferences drawn from an empirical analysis are presented at the end of the section.

3.3   DATA DESCRIPTION AND PREPARATION

We deployed the data documented during a study on the operation of the lungs and its diseases by the University of Innsbruck in the district of Brixlegg in Austria. The dataset included a variety of attributes perceived as the covariates with respect to the lung disease. The data contained the responses recorded for 1549 children. The missing values for the attributes were recorded as −1. The variables with respect to which the disease was expressed include degree of pollution in the environment at the place of residence characterized through three categorical levels—extremely polluted, moderately polluted and highly polluted with ozone—along with other attributes including parental characteristics, such as details with respect to paternal/maternal smoking, besides parental level of education and existence of comorbidities including cold, cough and existing allergies. The gender attribute was encoded as 0 for males and 1 for females. A variable on the presence/absence of bronchial tube disease, one of the most important parameters that help in ascertaining asthma as a disease rather than a symptomatic presentation, was also included in the data collected. PEF, the maximum speed of expiration (airflow at exhalation), as measured with a peak flow meter, was also included. Further, clinical findings in the form of spirometer readings obtained by performing spirometry were recorded for the predominant and most common pulmonary function parameters [16].

3.4   DOMAIN KNOWLEDGE AND FEATURE ENGINEERING

Pulmonary function tests involving results revealing lung functionality play a major role in the process of making a prognosis of the disease and the assessment of treatment effects. However, it could lead to mismanagement of the related disease and the patients affected, in the case of encountering differences in the way the lung function is expressed and interpreted [11]. Of the several respiratory lung equations that thrive to predict the expected levels of PFT parameters for a given height, weight and gender, we try to adopt the ARTP reference equations by introducing few more lung function parameters for effectively predicting asthma predisposition.

Initially, few attributes including identification number, month of birth and examination, day of month and examination were eliminated as we found that they do not significantly contribute to the prediction process. However, the "age" attribute was deduced from the year of examination and the year of birth, included in the raw data. Subjects containing missing values for any of the attributes were eliminated. Additional features indicative of pulmonary function parameters fevp and fvcp, representing the predicted values for FEV1 and FVC, were added to the dataset. The fevp and fvcp were computed and deduced using the existing attributes, "height" and "age" for males and females independently using the formulae published by the Association for Respiratory Technology and Physiology. A regression model from a cohort study where "height" is in meters and "age" is in years expressed as follows for male and female genders separately was used for the figures in the formulae:

(i) Male:

$$fev1 \ = \ 4.30 \ * \ height \ -0.029 \ * \ age \ -2.49 \qquad (3.1)$$

$$fvc \ = \ 5.76 \ * \ height \ -0.026 \ * \ age \ -4.34 \qquad (3.2)$$

(ii) Female:

$$fev1 \ = \ 3.95 \ * \ height \ -0.025 \ * \ age \ -2.60 \qquad (3.3)$$

$$fvc \ = \ 4.43 \ * \ height \ -0.026 \ * \ age \ -2.89 \qquad (3.4)$$

Using fevp and fvcp, we further evaluated the ratio between both, characterizing Tiffeneau-Pinelli index, a common measure used by clinicians for the monitoring of the lung functionality, which was added to the existing set of features. The ratio between the actual fvc recorded by test and the predicted fvc computed as shown above was also included. Thus, the resulting dataset now included a host of features incorporating clinical findings, symptomatic and medical history. Confounding variables such as "age" were eliminated as they could negatively impact the prediction performance by demonstrating false associations between independent and dependent variables. The feature "BMI" was added to the set of existing attributes as it was found to be one of the indicators for the diseases in the children of the age group under study, as per the literature [6,17].

3.5   BALANCED DATA CREATION USING OCSVM

3.5.1   One-Class SVM

One-class SVM offers a different approach to classification when compared to standard algorithms by modeling the distribution of one class only and is one of the most preferred solutions in the case of data exhibiting class imbalance. The one-class SVM is used for training only on one class, which in our case represents the non-asthmatics outnumbering the asthmatics, thereby adopting a strategy to eliminate samples that would be regarded as outliers. The OCSVM learning algorithm attempts to input the data into a high-dimensional feature space while iteratively searching the margin that maximizes the hyperplane which separates at its best the training data from the origin. The OCSVM may be visualized as a regular two-class SVM where all the training data lie in the first class, with the origin taken as the only member of the second class.

Involving a quadratic programming problem, the computational complexity of the learning phase is exhaustive, but once the decision function is decided, it can be used to predict the class label of unseen data effortlessly.

The OCSVM optimization as a solution problem is equivalent to solving the dual quadratic programming problem

$$\min \left(\tfrac{1}{2}\right) \Sigma \; \alpha_i \; \alpha_j \; K\left(x_i, x_j\right) \tag{3.5}$$

with the constraint

$$0 \le \alpha_i \le n \,/\, v_n \tag{3.6}$$

and

$$\sum \alpha_i \; = 1 \tag{3.7}$$

where α is a Lagrange multiplier (or "weight" on instance "I" such that the vectors associated with non-zero weights are called "support vectors" and determine the optimal hyperplane), v is a parameter that controls the trade-off between maximizing the distance of the hyperplane from the origin and the number of data points contained by the hyperplane, "n" is the number of points in the training dataset, and $K\,(x_i, x_j)$ is the kernel function. For non-linear decision boundaries, using the kernel function to project input vectors into a feature space, given a feature map:

$$\Phi: \; X \to \mathrm{RN} \tag{3.8}$$

where Φ maps training vectors from input space to a high-dimensional feature space, we can define the kernel function as:

$$K\left(x, y\right) = \; < \Phi\;(x), \; \Phi(y) \; > \tag{3.9}$$

With this, the feature vectors need not have to be computed explicitly, and in fact, this capability greatly improves computational efficiency to directly compute kernel values.

The adoption of OCSVM as an alternative method to sampling data has resulted in the selection of an ideal representation of the data rather than regular sampling approach that is used for undersampling, as there are all chances of leaving out significant instances that might be contributing to the prediction capability as the choice of samples happens randomly. With sampling, there is no underlying computational restriction when the samples are drawn from the original population, and as such, it might not be a preferred approach to representing the overall population effectively.

3.5.2 DATA PREPARATION FOR OCSVM

The initial data containing 1549 samples consisted of 187 subjects with asthma and the rest 1362 samples represented as non-asthmatics, illustrating a clear problem of class imbalance. Of the 187 subjects with asthma, only 163 samples were selected by eliminating the rest, which contained missing values for most of the attributes. On similar lines, samples with missing values were eliminated in the other group representing non-asthmatics, yielding a complete non-asthmatic dataset. About 12% of the inliers estimating to 163 subjects were drawn from the total non-asthmatics using OCSVM. Overall, 326 instances covering 163 subjects with asthma and 163 non-asthmatics constituted the input dataset.

The input data represent a class-balanced dataset, as the number of asthmatics balances the other group representing the non-asthmatics. One-class SVM is one of the preferred approaches to eliminating outliers and produces highly dense regions of data when all the instances belong to the same class. Further, multistage feature extraction technique was used to extract the predominantly contributing risk factors. The features extracted thereby are further used in the reduced feature set to achieve the task of predicting the asthma disease outcome. A comparative analysis of the prediction results is performed by adopting the traditional classifiers stated above.

Figure 3.1 illustrates the process involved in the creation of balanced dataset. The data are preprocessed by eliminating instances containing missing values. A subset of the data constituting the asthmatics is first drawn, followed by the extraction of non-asthmatics. OCSVM algorithm is then applied on the latter group to draw the most concentrated inliers that approximate the sample size as that of the other group (asthmatics). The two subsets are then merged to form

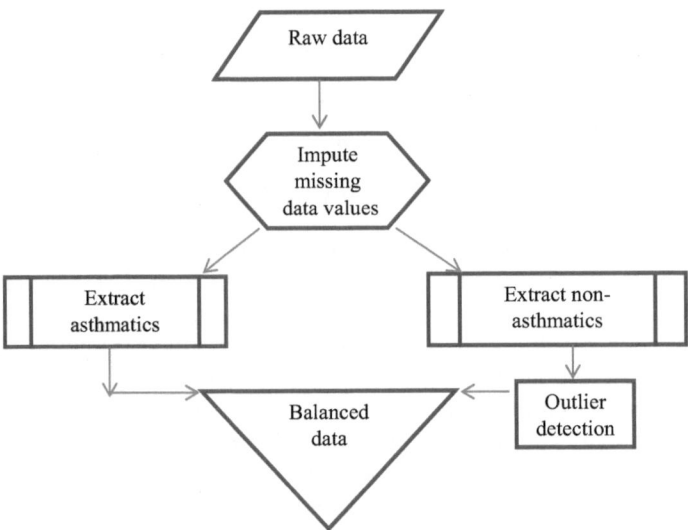

FIGURE 3.1   Balanced data creation.

TABLE 3.1
Ranking Scores

Feature	Information Gain	Gini
PEF	0.131	0.085
Lung_dis	0.130	0.087
FEF50	0.125	0.083
Fsm	0.124	0.069
Cold	0.105	0.054
BMI	0.099	0.067
FVC	0.088	0.059
Tpi	0.055	0.037
Cough	0.054	0.036
Zone	0.033	0.021
aller	0.028	0.018
actp	0.014	0.010

the balanced dataset. The predictors adversely affecting the classification process identified via their negative scores by Relief algorithm were eliminated, and the reduced feature set was subsequently used for the classification task. Table 3.1 shows the feature scores for individual features computed using feature ranking techniques, information gain and Gini decrease, which are ranked by their scores.

3.6   RESULTS AND DISCUSSION

Table 3.2 illustrates the performance evaluation of some of the standard classifiers used on similar data, including SVM, Naïve Bayes and logistic regression. The evaluation was carried out on two different feature sets (optimized vs. raw feature set) with respect to the performance metrics, namely AUC, CA, F1-measure, precision and recall, by using tenfold cross-validation strategy. It was inferred that the optimized combined feature set yielded comparatively better results.

Of all the classifiers, Naïve Bayes classifier performed comparatively well with the optimized feature set. The model was able to predict the disease outcome with a precision of 94.7% and recall of 77.5%, accounting for an F1-measure of 85.1%. The AUC and CA were evaluated to be 92.6% and 86.5%, respectively. Further, when the raw dataset excluding the features generated by feature engineering, but including all the spirometer readings along with the symptomatic and history data was included, the model performed low with the Naïve Bayes classifier performing comparatively better with a reduced F1-measure of 83.3%, accounting for a precision and recall of 93.8% and 74.8%, respectively. Although negligible, the AUC and CA also showed a decline, compared to the results with the optimized feature set.

The sensitivity, which is most desirable from any of the clinical decisions, is quite low with the raw feature set (75%) as compared to the optimized feature set

TABLE 3.2
Performance Evaluation of Optimized vs. Complete Feature Set

	Method	AUC	CA	F1	Precision	Recall
Optimized combined feature set	SVM	0.938	0.850	0.837	0.913	0.773
	Naive Bayes	0.926	0.865	0.851	0.947	0.775
	Logistic regression	0.920	0.840	0.830	0.888	0.779
Complete feature set	SVM	0.933	0.847	0.834	0.906	0.773
	Naive Bayes	0.924	0.850	0.833	0.938	0.748
	Logistic regression	0.919	0.837	0.827	0.882	0.779

(78%). It is evident from this fact that it is of utmost importance to use the optimized features set that can always improve the prediction outcomes and enhance the performance efficiency of the models. With the optimized feature set containing the additional features generated by feature engineering, the classification accuracy was considerably increased [18,19]. Feature engineering further places a significant impact on scenarios involving large data where there are exceedingly a greater number of features and where some of the useful, previously unknown features can be regenerated by using existing features, backed by the knowledge of the relevant domain.

3.7   CONCLUSIONS

A prognostic model driven by feature engineering that combines both data-driven and knowledge-driven feature selection approaches to yield higher accuracies in the prediction model by adopting multistage feature extraction technique that selects the best feature subset possible to capture the latent relationships was developed. The severity the indicators of the disease have been identified from the feature set targeting objective. The adoption of one-class SVM as a strategy to combat class imbalance problem laid a good foundation for the classification models deployed. It is evident from the experimental results that using a combination of medical history data and significant clinical findings results in a better prognostic model. Informative spirometer parameters specific to the disease diagnosis play a vital role in the accurate prediction of the disease, as the same parameters are used to judge a variety of related respiratory diseases and a very small margin between them could be the basis for differentiation. Neither the clinical findings from spirometry nor the medical history alone can result in an optimal performance of the model. Clinical decisions concerning the obstructive lung diseases such as COPD, have a high chance of leading to results that can be misinterpreted with wrong inferences drawn that may have long-term implications, including the targeted therapy that can be mistakenly beset. Hence, we provide data-centric approaches harnessing machine learning techniques to facilitate the disease prediction process that can augment the inferences through clinical findings.

DECLARATIONS

(i) Funding—not applicable.
(ii) Conflicts of interest/competing interest—not applicable.

REFERENCES

1. Sorkness CA, Schatz M, Li JT, Nathan RA, Murray JJ, Marcus P, Kosinski M, Pendergraft TB, Jhingran P. Assessing the relative contribution of the asthma control test™ and spirometry in predicting asthma control. *Journal of Allergy and Clinical Immunology*. 2004 Feb 1;113(2):S279.
2. Prasad BD, Prasad PK, Sagar Y. A comparative study of machine learning algorithms as expert systems in medical diagnosis (Asthma). In *International Conference on Computer Science and Information Technology* 2011 Jan 2, pp. 570–576, Springer, Berlin, Heidelberg.
3. Prasadl BD, Prasad PE, Sagar Y. An approach to develop expert systems in medical diagnosis using machine learning algorithms (asthma) and a performance study. *International Journal on Soft Computing (IJSC)*. 2011 Feb;2(1):26–33.
4. Tartarisco G, Tonacci A, Minciullo PL, Billeci L, Pioggia G, Incorvaia C, Gangemi S. The soft computing-based approach to investigate allergic diseases: a systematic review. *Clinical and Molecular Allergy*. 2017 Dec;15(1):10.
5. Kocsis O, Arvanitis G, Lalos A, Moustakas K, Sont JK, Honkoop PJ, Chung KF, Bonini M, Usmani OS, Fowler S, Simpson A. Assessing machine learning algorithms for self-management of asthma. In *2017 E-Health and Bioengineering Conference (EHB)* 2017 Jun 22, 571–574, IEEE, Sinaia, Romania.
6. Azizpour Y, Delpisheh A, Montazeri Z, Sayehmiri K, Darabi B. Effect of childhood BMI on asthma: a systematic review and meta-analysis of case-control studies. *BMC Pediatrics*. 2018 Dec;18(1):143.
7. Burgess JA, Matheson MC, Gurrin LC, Byrnes GB, Adams KS, Wharton CL, Giles GG, Jenkins MA, Hopper JL, Abramson MJ, Walters EH. Factors influencing asthma remission: a longitudinal study from childhood to middle age. *Thorax*. 2011 Jun 1;66(6):508–13.
8. Svanes C, Koplin J, Skulstad SM, Johannessen A, Bertelsen RJ, Benediktsdottir B, Bråbäck L, ElieCarsin A, Dharmage S, Dratva J, Forsberg B. Father's environment before conception and asthma risk in his children: a multi-generation analysis of the respiratory health in Northern Europe study. *International Journal of Epidemiology*. 2016 Aug 25;46(1):235–45.
9. Deo RC. Machine learning in medicine. *Circulation*. 2015 Nov 17;132(20):1920–30.
10. Pushpalatha, M. P., and M. R. Pooja. A predictive model for the effective prognosis of Asthma using Asthma severity indicators. In *2017 International Conference on Computer Communication and Informatics (ICCCI)*, IEEE, Coimbatore, India, 2017.
11. Sorkness CA, Schatz M, Li JT, Nathan RA, Murray JJ, Marcus P, Kosinski M, Pendergraft TB, Jhingran P. Assessing the relative contribution of the asthma control test™ and spirometry in predicting asthma control. *Journal of Allergy and Clinical Immunology*. 2004 Feb 1;113(2):S279.
12. Pooja MR, Pushpalatha MP. A neural network approach for risk assessment of asthma disease. *Journal of Health Informatics & Management*, 2018;2(1):1–6.
13. Manoharan SC, Ramakrishnan S. Prediction of forced expiratory volume in pulmonary function test using radial basis neural networks and k-means clustering. *Journal of Medical Systems*. 2009 Oct 1;33(5):347.

14. Manoharan SC, Swaminathan R. Prediction of forced expiratory volume in normal and restrictive respiratory functions using spirometry and self-organizing map. *Journal of Medical Engineering & Technology.* 2009 Jan 1;33(7):538–43.

15. Salau, AO and Jain, S. Feature extraction: A survey of the types, techniques, and applications. In *5th IEEE International Conference on Signal Processing and Communication (ICSC),* Noida, India, 2019, 158–164. Doi: 10.1109/ICSC45622.2019.8938371.

16. Chhabra SK. Clinical application of spirometry in asthma: Why, when and how often? *Lung India: Official Organ of Indian Chest Society.* 2015 Nov;32(6):635.

17. Forno E, Fuhlbrigge A, Soto-Quirós ME, Avila L, Raby BA, Brehm J, Sylvia JM, Weiss ST, Celedón JC. Risk factors and predictive clinical scores for asthma exacerbations in childhood. Chest. 2010 Nov 1;138(5):1156–65.

18. Lachman BS, Pengetnze Y. Improved asthma outcomes from use of predictive modeling as part of a system of care. *Journal of Allergy and Clinical Immunology.* 2017 Feb 1;139(2):AB191.

19. Pooja, MR. On the potential of machine learning to improve disease outcomes. *International Journal of Clinical & Medical Informatics* 2020; 3(1): 4–6.

4 Artificial Intelligence Emergence in Disruptive Technology

J. E. T. Akinsola and M. A. Adeagbo
First Technical University

K. A. Oladapo
Babcock University

S. A. Akinsehinde
The Amateur Polymath

F. O. Onipede
First Technical University

CONTENTS

4.1   Introduction 64
4.2   Artificial Intelligence... 64
4.3   Components of Artificial Intelligence 65
4.4   Types of Artificial Intelligence... 67
    4.4.1   Reactive Machines.. 67
    4.4.2   Limited Memory.. 67
    4.4.3   Theory of Mind 67
    4.4.4   Self-Awareness.. 67
4.5   Artificial Intelligence for Modern Businesses... 67
    4.5.1   Interactive Artificial Intelligence (IAI) 68
    4.5.2   Functional Artificial Intelligence (FAI) 68
    4.5.3   Analytic Artificial Intelligence (AAI)... 69
    4.5.4   Text Artificial Intelligence (TAI) 69
    4.5.5   Visual Artificial Intelligence (VAI).. 69
4.6   Disruptive Technology... 69
    4.6.1   Digital Transformation 70
    4.6.2   Examples of Disruptive Technology... 71
    4.6.3   Impact of Big Data in Disruptive Technology 73
4.7   Artificial Intelligence as a Disruptive Technology 74

DOI: 10.1201/9781003224068-4

    4.7.1   Artificial Intelligence as a Disruptive Technology in Various
        Sectors... 75
           4.7.1.1   Accounting and Finance 75
           4.7.1.2   Marketing.. 77
           4.7.1.3   E-Commerce 79
           4.7.1.4   Contact Centre 81
           4.7.1.5   Telecommunications 81
4.8   Business Benefits of Adopting AI 82
4.9   Conclusions.. 84
References.. 85

4.1 INTRODUCTION

Artificial intelligence (AI) has created a significant paradigm shift in the enterprise when it concerns digital transformation. This shift is the disruption arising from AI, which is called disruptive technology. The fundamental disruptive technology is the Internet. Morgan (2019) opined that the digital transformation has affected the way we live daily because AI is being incorporated into products and services. Machine learning (ML) predictive analytics systems are now being employed in diverse areas (Akinsola et al., 2019). There is now high level of automation with great prediction power using ML algorithms, especially with deep learning and reinforcement learning. The disruptive technology occasioned by AI has affected every sector of the economy. According to Hinmikaiye et al. (2021), there have been dramatic changes from mobile telephony to mobile computing, which is greatly affecting the sustainability of the telecommunication industry. In order to experience more successful outcomes from AI disruption, there is a need for balancing competing social interests by putting greater regulatory activities in place. While AI is an enormous challenge, ML's limited applications affect many parts of our daily living. Without artificial intelligence or AI, it is rather difficult to talk about technology today. To this end, to succeed in this global economy, AI is the major force that any business must take cognizance of.

4.2 ARTIFICIAL INTELLIGENCE

AI centres around machines imitating human thought processes and behaviours and is applicable in most fields of study, for example philosophy, psychology and linguistics. AI has been the driving force of disruptive innovations in technologies such as video games, voice assistant, self-driving cars and fraud detection. But with disruption come issues and opportunities (Khamis, 2019). AI has transformed every aspect of our life, and it requires an intelligent user interface (IUI) based on ML paradigms (Akinsola et al., 2021). To achieve greater results in disruptive technology, human–computer interaction is essential. The evolution of HCI has resulted in many advances leading to technological growth driven by ubiquitous computing (Alao et al., 2019).

The disruptive nature of AI has made it susceptible to privacy and ethical issues, which has brought about a push for more regulations and ethics in AI. One of the major issues negatively affecting the rise of AI is deepfake, a technology that is used to impersonate anybody via fake media such as pictures and videos (Nguyen et al., 2021). Face recognition mismatch is another AI misfortune that has turned into a race-related issue. This has resulted from numerous events of security face recognition systems wrongfully identifying people as miscreants (Allyn, 2020); (Hill, 2020, 2021). Notwithstanding, the future of AI still poses a lot of uncertainty and the volume of opportunities currently created by AI continues to yield exponential growth (Acemoglu & Restrepo, 2020).

4.3 COMPONENTS OF ARTIFICIAL INTELLIGENCE

As a branch of computer science, the major components of AI are as highlighted below and as shown in Figure 4.1.

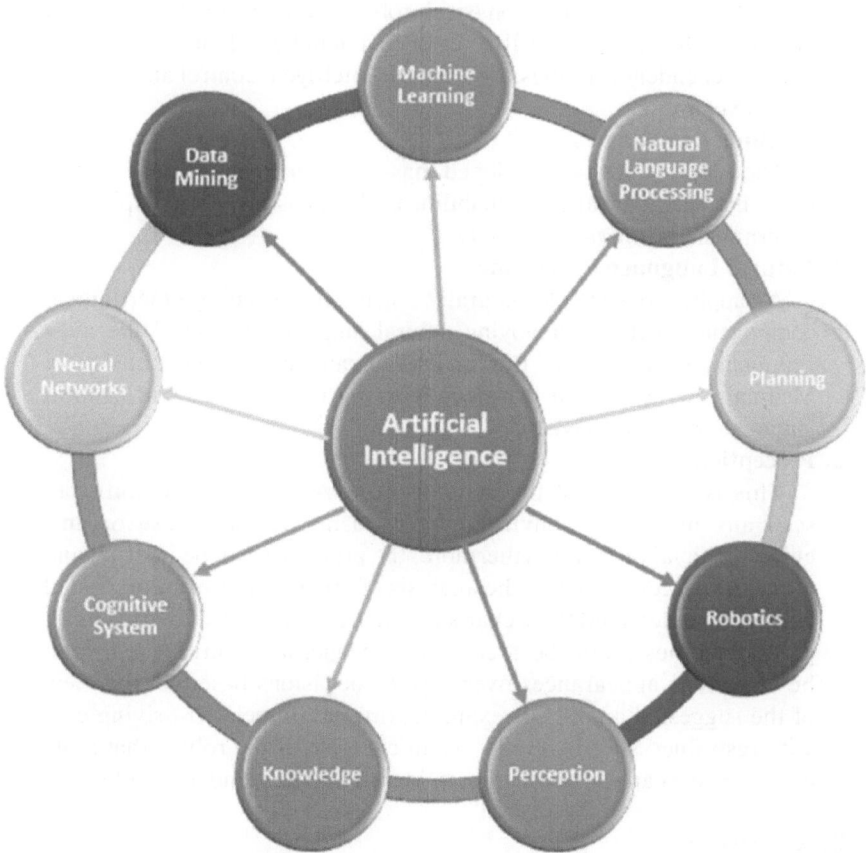

FIGURE 4.1 Components of artificial intelligence.

a. Cognitive System

This deals with how the individuals operate their mind, how an individual behaves, and how an individual's brain processes data and information.

b. Neural Networks

It attempts at building an artificial system using a simplified network of simplified artificial neurons. For instance, time series prediction and brain modelling are typical examples for the development of artificial systems based on simple artificial neuron networks.

c. Planning

AI has the capacity to bring up various general methods for theorem proving, planning and problem solving. This include case-based reasoning, reduction of problem, constraint satisfaction, resolution, adversarial search and state space search.

d. Robotics

The robot handles and moves objects. Robotics is a technological industry dedicated to physical robots. Robots are programmed machines that can generally perform a number of autonomous or semi-independent activities, for example intelligent control and autonomous exploration.

e. Machine Learning

The scope of AI has developed massively since the machine intelligence has adapted learning capabilities, which has generated impacts on government, business and society.

f. Natural Language Processing

AI applications should naturally engage by matching other human components, thereby employing natural language as one of the most significant techniques to this end, for example machine translation. Medmain (2018) opined that deep learning is associated with natural language processing (NLP).

g. Perception

This is an aspect of the component of AI that has the ability to scan any highlighted environment by using various sense organs, either artificial or real. Furthermore, the processes can be internal and allow the perceiver to give the analysis of other scenarios in suggested objects to understand their characteristics and relationship. This analysis sometimes might be a complicated one, and similar items can be of various appearances over various occasions based on the view of the suggested angle. For example, this can impel self-driving cars with restrained speeds. Freddy is an example of the robots that at its earliest stage can use perception to identify objects and assemble various artefacts.

h. Knowledge

This is a vital part of the development of any AI application (Oke, 2008; Haenlein & Kaplan, 2019; Cioffi et al., 2020).

i. Data Mining

This is crucial for the invention of model due to the application of profound learning methods. It is the automation of pattern detection and prediction by utilizing huge amounts of data.

4.4   TYPES OF ARTIFICIAL INTELLIGENCE

AI has generally been grouped into four types, which include reactive machines, limited memory, theory of mind and self-awareness (Johnson, 2020; Tucci, 2020).

4.4.1   REACTIVE MACHINES

This is the first phase of AI which has basic or simple use. It lacks data training or advanced ML models and is not considerate of other states of events (past or future) apart from the immediate. These were the earliest introductions of AI like the IBM Deep Blue, a chess expert system trained with only the rules of chess playing.

4.4.2   LIMITED MEMORY

This is a data-enabled reactive machine that relies heavily on existing data for ML model predictions. The limited memory type undergoes a series of ML model training from new data or feedback data from previous model training. Examples of ML models in this phase are reinforcement learning and long short-term memory (LSTM).

4.4.3   THEORY OF MIND

This is the current development stage of AI where much is still futuristic, but fast approaching. It attempts to embody the human mind concerning thoughts and emotions on AI systems. It is currently being applied for the optimization of driverless cars and AI assistants.

4.4.4   SELF-AWARENESS

This is the epitome of the prospects of AI which is still theory based or fictional to a large extent. It is the phase where an AI system has an autonomy of mind and can function on a par with the human mind if not better. Theory of mind and self-awareness are the focal points of AI's role in disruptive technologies.

4.5   ARTIFICIAL INTELLIGENCE FOR
MODERN BUSINESSES

There are five areas in which AI can be used for modern business. These areas have shaped the way modern businesses are conducted.

4.5.1   INTERACTIVE ARTIFICIAL INTELLIGENCE (IAI)

Interactive analytical systems can be used for the assessment of AI. In order to contextualize, externalize and interpret the knowledge, there is a need to address explainable artificial intelligence (XAI) and explainable cognitive intelligence (XCI). Examples of interactive AI are chatbots and smart personal assistants. Interactive AI is a type of AI in modern business that allows automation of communication without interactivity compromise. Smart personal assistants and chatbots are used to visualize interactive AI because they have capacities that can vary from replying to pre-built questions to understanding the conversation framework. Interactive AI can also be used in improving a company's interior procedures (Bekker, 2021). Figure 4.2 shows the various types of AI.

4.5.2   FUNCTIONAL ARTIFICIAL INTELLIGENCE (FAI)

Functional AI is very similar to analytic AI. Enormous quantities of data are needed for scanning and locating patterns as well as dependencies. However, functional AI jumps into action instead of giving references. For instance, functional AI can identify a machine breakdown pattern in the device data received from a confident machine and a command is generated to turn off this machine because of being the part of the Internet of things (IoT) cloud. Another example is robots being used by Amazon to bring the tables with the items on it to the pickers; during this process, the picking process is being sped up. Functional AI can also improve a company's interior procedures, for example creation of a chatbot to facilitate the corporate process of vacation booking (Bekker, 2021).

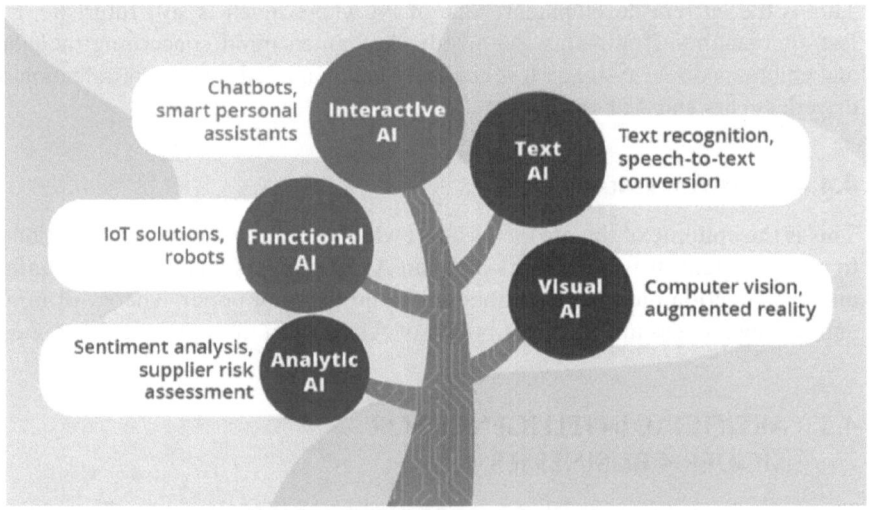

FIGURE 4.2   Types of artificial intelligence (Bekker, 2021).

4.5.3 ANALYTIC ARTIFICIAL INTELLIGENCE (AAI)

It is a type of AI that is motorized with ML (including its most innovative unrestricted learning systems); analytic AI scans loads of data for dependences and outlines to finally produce references or provide a business with visions, thus contributing to data-driven decision-making. Dealer risk assessment and sentiment study are the examples of analytic AI in action (Bekker, 2021).

4.5.4 TEXT ARTIFICIAL INTELLIGENCE (TAI)

Examples of text AI that can be enjoyed by businesses are *speech-to-text conversion*, *content generation* capabilities, *text recognition* and *machine translation*. Text powered by AI can find the document containing the most appropriate answer even if the document doesn't have full keywords contrary to a traditional knowledge base that rests upon a search by keywords. AI is permitted to build semantic maps by keywords and identify synonyms to appreciate the framework of the user's questions through the help of ordinary language dispensation and semantic search (Bekker, 2021).

4.5.5 VISUAL ARTIFICIAL INTELLIGENCE (VAI)

Visual AI enables businesses to classify, identify and categorize objects or change videos and images into visions. An example of visual AI is a computer system that enables a guarantor to evaluate damage based on damaged machine/car or a photo that grades apples based on their size and colour. This type of AI covers *augmented reality* or *computer vision* fields (Bekker, 2021).

4.6 DISRUPTIVE TECHNOLOGY

Disruptive technology invented by Professor Clayton M. Christensen of Harvard Business School is the kind of technology that changes the usual course of action as it concerns a market or an industry. Disruptive technology is a technological know-how that changes the conduct of industries function, consumer or business paradigms. Disruptive technology usually seduces an unconfirmed real application, restricted viewers and concert topics. Some of the expressively disruptive technologies influencing the future are virtual/augmented reality, IoT, e-commerce, AI and blockchain technology.

Disruptive technology replaces a deep-rooted technology or product by generating an innovative market or industry. Current disruptive technology examples include ridesharing apps, online news sites, global positioning system (GPS) and e-commerce; schemes discovery, electricity service, television and automobile have disruptive technologies in their own time. Disruptive technologies are usually invented from young companies and start-ups compared to the foremost firms.

A disruptive technology curves away the schemes it changes because it has potentials that are recognizably higher (Smith, 2020). Types of disruptive technology are blockchain, AI and IoT. Disruptive technology can be made a successful one by making business models to be innovative. The usage of a well-formulated software development life cycle (SDLC) model (Akinsola, Ogunbanwo, et al., 2020) will help when building ML models. This technology is utilized by a network of suppliers, thus making their products accessible to customers and creating a value chain of network that brings about success while making the products affordable to a bigger audience (Corporate Finance Institute (CFI), 2021b).

The five steps established by Christensen and his colleague, Professor Joseph Bower, in the Harvard Business Review for evaluating and cultivating innovations that are naturally disruptive are as follows: determine whether the innovation is sustaining (in other words, an expansion of something that has already been in existence) or disruptive, recognize the original market for the disruptive innovation, evaluate your innovation's potential, generate a sovereign association to follow the disruptive route and retain the disruptive association sovereign.

A disruptive technology replaces an already existing habit, product or process. It usually has greater qualities that are directly noticeable to initial adopters. Newcomers rather than recognized firms are the common bases of disruptive technologies (Smith, 2020).

4.6.1 Digital Transformation

Digital transformation is a technology proposition. There's no hesitation that the impact of technology is extraordinary in terms of things such as unrestricted computing power and unrestricted storage that are becoming a certainty today, as well as opportunities offered by disruptive technologies such as AI, IoT, mixed reality and deep learning. Digital transformation provides a chance for businesses to consider and function like digital companies in the technique and empower their employees. They engage their customers, renovate their products and improve their operations. Some of the essential opinions and observations regarding digital transformation are leadership matter, research and fail fast, initiative change through culture, actual change management, customers' loyalty, products, resources and people, adoption of data culture and think ecosystem in order to ask a pertinent question who is my Uber? (Microsoft, 2016).

Digital transformation and successive typical business revolution have basically changed consumers' behaviours and expectations, disrupting many markets and putting huge pressure on traditional companies.

Three stages of digital transformation identified are digital transformation, digitalization and digitization, and each stage places certain demands on companies' growth strategies, metrics, organization structure and digital resources.

Digital transformation is comprised of change in marketing, policy, supply chains, organization and technology, which makes it multi-disciplinary in nature. The understanding on how companies can gain a managerial guidance for digital transformation and gain cheap benefit by building on certain resources which must be increased by defining what methods they should approve to win, and how the company's inner organization structure must change to support these methods in order to win. The three major factors affecting digital transformation are as follows:

First, an increasing number of accompanying technologies (for example smartphones, speech recognition, broadband Internet, cloud computing, Web 2.0, SEO, cryptocurrencies and online payment systems) have appeared since the advent of the World Wide Web and its worldwide adoption, which has supported the development of e-commerce.

Second, competition is changing vividly; that is, technologies have drastically uttered the competition landscape, shifting sales to moderately young digital firms due to these new digital technologies.

Third, consumers' behaviour is changing as a response to the digital revolution; that is, both online and offline sales are affected by consumers' behaviour through shifting of their purchases to online stores and digital touchpoints, which have a significant role in the customer journey (Verhoef et al., 2021).

4.6.2   EXAMPLES OF DISRUPTIVE TECHNOLOGY

a. Internet-Based Video

Netflix is one of the examples of Web-based video. It is now well known and continues to develop the way people watch television and movies. On-demand viewing has twisted the old dissemination model. Netflix and other similar companies allow viewers to avoid irritating commercials and watch shows on their own time schedule. Other popular online games for online TV are Crackle, Sling TV and Hulu (Harris, 2021). Hulu and HBO are videos streaming platforms, which seem like a noticeable development over Blockbuster and Cable (Moore, 2019).

b. Ridesharing Services

Uber is one of the ridesharing services. It is a fast-growing ridesharing service, and it has become the print child for disruptive innovation. The traditional taxi cab business has abruptly been changed forever by a mobile platform linking consumers who need rides with drivers eager to provide them. Customers no longer need to wave down a car on the street, and a rider with an automatic Uber account does not even need to hand over cash. The trip is usually cheaper because consistent cabs have to charge more to cover the huge upfront asset in a taxi license (Harris, 2021). Ridesharing companies such as Uber and Lyft have become most people's choice for their travels, simply getting around town and commutes which have replaced the taxi industry. These companies practised impressive rises, making it an exciting example of innovative and fast success in tech. Uber wasn't a low-end substitute to an unreachable, expensive and difficult service. It was actually invented in an ordinary market first and then increased total demand and appealed to lower-end sections later. In fact, when Uber first came into existence, it solely used extravagant black cars. Ridesharing services seem to be sustaining innovations that basically progress on the current taxi exercise (Moore, 2019).

c. Virtualization of Reality

It is not just for entertaining or gaming, it could disrupt how people do business. Goldman said, "Gaming and entertainment will drive much of the growth, but car makers, retailers and even interior designers could bank on virtual reality (VR) technology" (Harris, 2021).

d. Augmentation of Reality

It could disrupt mobile phones and in-car navigation systems. Augmented reality technology can be applied in all means of transport. This technology makes cars safer and supports the driver in many daily conditions such as parking. It can also be shifted to other means of transport such as a plane and help during landing by making the floor translucent. Another application is in ships making windowless lodges more attractive to passengers on a trip (Harris, 2021).

e. Cryptocurrency

It could disrupt the payment services systems and banking. Examples are Ethereum, BitGold and Bitcoin (with smartphones offering a new payment system). The most attractive method out of the three methods of non-traditional currencies is BitGold, because it unites the advanced tech of cryptocurrency with human civilization's original ordinary unit of value – gold, by contributing physical call on all BitGold a purchaser owns in the form of 10-gram gold cubes, as well as the ability to load a card and spend gold as you would spend dollars in an inspection account.

BitGold represents a concrete, ready-to-use solution for the use of a structured service. While cryptocurrencies are still a long way from the mainstream, solutions such as BitGold are making progress to link the gap between the borders of tech and our daily financial transactions, not to mention creating a level of reality not present with digital wallet solutions for cryptocurrency storage and transfer (Harris, 2021).

f. Commerce Collaboration

It is a disrupting retail and e-commerce. Airbnb is heating up in the business space as it identifies a huge chance to grow its company margin which is currently at 10% and rising. Airbnb is a company asking the right questions about what consumers have been seeking in the warmth industry for years. And, the answer is elasticity and simplicity. Airbnb is making steady inroads through direct supplier corporations and devoted corporate programs such as Airbnb for Business. Airbnb team have attunement with both the hosts and the guests and understand the real needs of the consumer because it started with air mattresses on the floor (Harris, 2021).

g. 3D Printers

These are disrupting the manufacturing industry. A 3D printer actually produces a thing, not just a copy. These things range from houses, body parts to buttons and everything in between. At first look, it may not appear to be disruptive technology. 3D printing is a preservative process of manufacturing, as opposite to a subtractive process such as sculpting. Blueprints from digital files or scans of a three-dimensional object are used; the images are uploaded to a digital file, reducing the images layer by layer. Resins or liquefied materials are used by the printer to be moulded into the necessary shapes, which was then hardened to form the printed structure (Harris, 2021).

It's immobile and uncertain of how extensive the use of 3D printers will become and whether they could present new business models or feeding ways that disrupt compulsory manufacturers. But as Freddie Dawson writes for Forbes,

The more hopeful we see 3D printers as a significant article that will ultimately be in every house if prices for machines and printing materials continue to drift downwards. Then instead of orderinsg products through online retailers or shops, consumers would be able to download and print their own – either through commercial sites or open source (Moore, 2019).

h. Edition of Genes by Recoding Cancer

It is experimental and considers how it could disrupt cancer treatment. It's not ready for figure creation. In terms of a disruption in how we treat incurable diseases such as cancer, this experience opened a big, bright door to the future (Harris, 2021).

i. Li-Fi, 100X Faster than Wi-Fi

This could be the next disruptive technology to change how we connect and communicate. In a recent TED talk, Haas claimed that household LED light bulbs could easily be converted into Li-Fi transmitters, providing Internet users with more effective influences. "All we need to do is to fit a small microchip to every potential brilliance device, and this would then combine two basic functionalities: illumination and wireless data transmission," he said. It is also worth mentioning that the speed at which these LEDs trace in order to relay data is too fast for the human eye to perceive, so users will not have to worry about annoying flashes in their ambient light. It has been suggested that Li-Fi could provide the answer to increasing incidence crowding as Internet usage continues to rise across the world (Harris, 2021).

4.6.3 IMPACT OF BIG DATA IN DISRUPTIVE TECHNOLOGY

In recent years, there has been vast penetration of Internet-enabled devices across the globe, which has coincided with an influx of users across various online platforms. This significant growth induces a daily exponential growth in data from the digital footprints of humans online. In 2018 alone, 90% of all yearly data were generated over the previous 2 years (Lackey, 2019). This unprecedented surge in data generation has brought about big data analytics. This availability of huge amount of data is what AI is leveraging on using ML algorithms to create the disruptive technology needed for the modern businesses. Different supervised learning algorithms can be implemented to identify the most efficient method based on the dataset, the number of instances and variables involved (Osisanwo et al., 2017). Nonetheless, formal methods for performance assessment contrasted with conventional techniques and have been proven to be profound (Akinsola, Kuyoro, et al., 2020).

4.7   ARTIFICIAL INTELLIGENCE AS A DISRUPTIVE TECHNOLOGY

The disruptive technology that is creating a new path for IT explorers is AI. The importance of a pattern-driven dataset implemented on supervised ML makes AI achieve its disruptiveness (Akinsola et al., 2020).

As a technology, the ambiguity around AI that most IT professionals try to figure out is what it can and cannot do as an innovative technology. AI definitely looks like a real driving force for digital platforms and businesses with the exponential growth in the field of digital marketing, at least for the next 5 years.

AI is one of disruptive technologies that create a rapid evolution from approaching technology to one that surrounds us in our day-to-day lives. AI is being combined into services and products we use every day to improve our lives better, from taking faultless pictures to predicting what we can say next in an email (Trivedi, 2021).

AI is the most disruptive among all the technologies that are driving digital transformation in the creativity. AI is really bringing about New Societal Structures which is altering the way human work. Thus, the conception of AI replacing humans. Therefore, if this is so, then how will humans work for the living? It needs a practical response that is investigational and comprehensive in order to see the effects of revolution and guarantee that no section of society lags behind. AI is in the process of disrupting people's day-to-day jobs because of the elegant automation.

For a business, AI increases the value of proposition by turning features such as visions and analytics to productivity speeds and value addition (Morgan, 2019). AI is helping to develop and bring improvement to many industries such as computer science, finance, education, marketing, healthcare, human resources, heavy industry and aviation. The disruptive technology that companies are willing to invest heavily in when considering the promise it has shown so far is AI. Ace and its partners are helping such companies to be at the lead, applying AI in their existing tech structure to reap the enormous benefits and advantages it brings (Trivedi, 2021). AI applications fall into three comprehensive groups: insight applications, process applications and product applications (Deloitte, 2021).

AI exists in three types, namely:

a. Artificial Narrow Intelligence (ANI)
Robots or similar alternatives can perform a solitary task very well.
b. Artificial General Intelligence (AGI)
AI strives to vie with the capabilities and intelligence of human beings.

    c. Artificial Superintelligence (ASI): Here, the mechanisms of AI are expected to be superior to the intelligence of human in the nearest future (Daqar & Smoudy, 2019).

4.7.1 ARTIFICIAL INTELLIGENCE AS A DISRUPTIVE TECHNOLOGY IN VARIOUS SECTORS

AI has transformed virtually all the sectors with vivid outstanding results. This is made possible with the usage of nimiety of data that are readily available as a result of digital transformation.

4.7.1.1 Accounting and Finance

AI is globally and rapidly transforming financial service industries (Buchanan, 2019). AI is an embodiment of different technologies that use device, data and cloud-based tools in an efficient way by disrupting old ways of doing things, paving and facilitating new ways which bring about digital transformation (Eckert, 2017). It intelligently performs tasks expected of human in different circumstances without significant human interventions and improves its performance when given datasets it can learn from (Girasa, 2020). The adoption of AI to various applications of financial services is rapidly increasing and contributing positively to transforming the sector (Financial Stability Board, 2017). The inclusion of AI has revolutionized the financial industry, bringing about enormous changes and creating numbers of innovations in financial services. Examples of these are intelligent consultant, intelligent lending, monitoring and warning, and intelligent customer service as required (Xie, 2019).

4.7.1.1.1 Applications of Artificial Intelligence in Financial Services

AI can be applied in relation to financial services in the following ways.

    a. Assessments of Credit Quality
       One of the major problems in financial institution is the ability to establish the creditworthiness of borrowers through the use of credit rating (Kagan, 2021). Lending involves evaluation of borrowers' capacity, motivation and protection of lender from losses if the borrower fails to repay the loan (Bazarbash, 2019). ML is designed to help computer learn from the past or present and to anticipate or foretell what will happen in future in unknown situations (Akinsola et al., 2019). AI with the help of ML evaluates loan application of borrowers through the use of digital footprints, predictive analytics and various data points and computational algorithms (ICICI, 2020). Digital footprints allow the prediction of personality traits through the collection and analysis of implicit and explicit information about the borrowers (Azucar et al., 2018). The application of AI in lending determines if the financial entity is healthy and has the possibility to sufficiently generate cash flows for servicing the debt (Corporate Finance Institute, 2021a). AI and ML explore available

data on credit rating factors in financial industry, such as capacity, coverage, condition, capital structure and character to evaluate credit risk.

b. Price and Market Insurance Contracts

The increase in the expectation of customers has brought about a new viewpoint into insurance industry (Butzmann et al., 2017). This rise in expectation is due to the fall in profit pools using traditional perspectives, shift in key player landscape and advancement in technological innovations. Insurance industries are now largely data-driven by nature using data analysis to effectively understand and evaluate the risk (Hussain & Prieto, 2016). It uses high volume of customer data at the disposal of financial institutes for data analysis to aid in decision-making, especially in the area of trading in the capital market. The availability of large volume of data gives room for the application ML for analysis and prediction which eventually metamorphosed into AI. The application of AI in insurance sector is technologically driven along with a key change in the expectation of customers (Matthews, 2020). AI is being applied in and is transforming the insurance industry, especially in the area of marketing, customer service, fraud detection, underwriting and claims. AI is embedded with many potentials to change the unsatisfying and bureaucratic experience of customers in insurance to something better, fast and affordable. With the high volume of data at their disposal, AI can bring about tailor-made and flexible insurance products such as pay-as-you-go insurance, a product that is responsive and automatically adjusts to customer situations such as accidents and health (Shroff, 2019). Among the benefits that AI has brought into insurance industry are saving of time and cost, improvement in customer experience, reduction in fraudulent claim and increase in profitability (Dilmegani, 2021). Some of the current AI application trend in insurance industry are behavioural premium pricing, customer experience & coverage personalization and faster, customized claim settlement (Faggella, 2020).

c. Automated Client Interaction

Among the limitations of customer service using traditional approaches are poor response time, inefficiency and costly management (Ping et al., 2019). As customer relationship is one of the topmost priorities of every business (Resnick, 2021), several organizations are trying to overcome these challenges, which brought about the introduction of AI into customer service. AI is currently becoming a popular word among the experts in the industry and is transforming the services of industrialists, vendors, product developers, marketers, logistic services, to mention but a few (Pandian, 2019). AI together with human agents enhances the productivity and brings perfections to services to customers (Ping et al., 2019). An example of automated client interaction is chatbot software, which is cost-effective, provides easy business transaction with fast response time and addresses prospective customer in a convenient way and preferred medium (Podnar, 2016). Another example is IRIS

ClaimIT that assists debt collection agencies in the recovery of debt and makes their work efficient with the ability to handle high volume of task (TORQ Software, 2021). Also, Pipedrive software in customer relationship management (CRM) has several features such as AI sales assist, pipeline management and Web forms to enhance the work of CRM officers (Riserbato, 2021).

d. Capital Market and Trading

Other areas of application where AI is being applied in the financial industry are analysing the effect of market in trading, optimizing the execution of trading, scarce capital optimization and finding indicators for better returns (Amorin, 2019).

e. Data Security

The AI technologies are also positively impacting both private and public institutions in the area of quality of data assessment, compliance regulation, detection of fraud and surveillance.

4.7.1.2 Marketing

The advancement in technologies have necessitated bringing new dimensions into marketing strategies in order to meet the needs and demand of the teaming customers. As AI is driven by high volume of data, any business that has data-oriented marketing landscapes and the ability to convert its data into AI has a competitive advantage over the other. Some of the contemporary technologies that ensure the effectiveness of AI marketing today are big data, ML and the right solutions (Hariharan & Chandrakhanthan, 2020). As shown in Figure 4.3, these contemporary also known as disruptive technologies are regarded as core elements of AI marketing right from data gathering to powerful solutions (Mcdonald, 2020). The application of disruptive technologies such as IoT through the use of cameras, sensors, and GPS devices allows seamless gathering of unparallel volume of data which give AI high potential to transform the marketing industries. As experts are trying to find out best AI solutions that can address their marketing

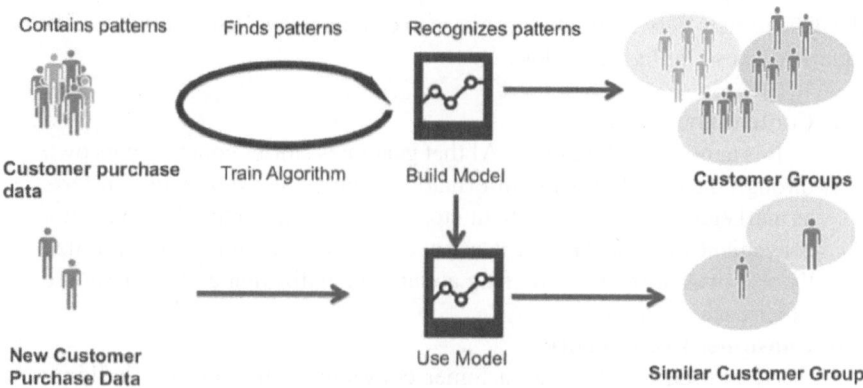

FIGURE 4.3 Core elements of AI marketing (Mcdonald, 2020).

functions (Verma et al., 2020), AI is now currently manifesting in different aspects of marketing such as marketing plan, management of brand and sustainable CRM (Yeğİn, 2020). Based on the foregoing, several indicators show that the interest of experts on AI in marketing from different quarters is continually rising. Among the indicators are the inflows of researchers and educationists into AI in marketing. The report has it that the number of research publications in peer-reviewed journals on AI in marketing has increased significantly, especially from 2017 (Feng et al., 2020). Also, the attempt to introduce AI marketing courses in educational sector makes the understanding of the impact of AI in marketing available to educators, practitioners and students. This introduction will eventually increase the awareness of AI in marketing and make students working in marketing sector acquainted with the necessary skills and reduce the initial training cost for the employers (Elhajjar et al., 2021). Application of AI in marketing implies that target customers are continually and routinely tracked on their next purchasing decisions in order to ensure that enterprises realize efficient, effective and profitable sales (Yeğİn, 2020). AI has recently made an impressive progress in marketing applications through the deep learning field improvement (De Bruyn et al., 2020). The introduction of deep artificial neural network in the majority of applications of AI in the area of business has helped solving predictive tasks that seem to be complex and unsolvable some decades ago. Among the products of complex tasks that have been solved is predictive analytics in marketing that allows the marketer to forecast the future of actions in marketing, impacting performance, to create intuition to enhance leads, to obtain new customers and to attain optimization of price (De Bruyn et al., 2020). Therefore, as advancement in technology usually brings about shift in the structure of business paradigm, so does AI in marketing industries (Kumar et al., 2019). AI has necessitated companies to improve on their data collection strategies, which brings about several investments to facilitate several related marketing tasks such as image recognition, personalization optimization of search engine, profiling and CRM (Mustak et al., 2021).

4.7.1.2.1 *Applications of Artificial Intelligence in Marketing*

AI can be used to deliver greater promises to customers and users alike in relation to marketing as highlighted below.

a. Content Generation

It is another application of AI that generates stories content on its own through the use of existing information sources. It requires little efforts at marketers end to bring about stories on the topic the user loves in a convenient manner. Hence, it is necessary for marketers to maximize the use of data in order to bring about personalization and improve the experience of customer.

b. Consumer Convenience

AI is used to improve customer convenience through the building of emotional connect. For instance, the application of semantic tools into social media called sentiment uses NLP, which is one of the AI components. This tool helps in gaining insight into the customers' emotion, which assists in understanding the feeling of customers about their brand and product (Devang et al., 2019).

c. Lead Nurturing

AI solutions are playing a vital role to marketers by making the method of attracting, finding and nurturing new leads efficient (Agrawal, 2021). An example of AI-driven software in lead nurturing is Conversica. It is a software driven by AI to reach, qualify, engage and follow up the leads in an automatic way through two-way exchange of email. Conversica software is regarded as an automated sales assisting software.

d. Data Governance

The increase in data collection gives AI the potential to facilitate creating better insights into the high volume of information through a deeper observation of customers' insights and effectively building customer interaction.

4.7.1.3 E-Commerce

E-commerce is playing an important role in improving the revenue and increasing customer base, as well as creating jobs in information technology industry (Girdher, 2019). As AI now turns out to be an essential part of every human endeavour, its introduction into e-commerce has over the years further boosted the performance of e-commerce sector. The result of the survey conducted by Anuj et al. (2018) shows that the increase in sales on e-commerce is a result of the drastic increase in the application of ML and AI in e-commerce sector. Following the data pattern of this result up to 2021 further predicted the sales to be approximately $4878 billion in 2021 (Tapan & Monica, 2019). As e-commerce gains more popularity, the leading brands in e-commerce sector are increasingly investing more funds into research to discover how best AI can give them a competitive advantage and improve customer loyalty (Girdher, 2019). AI allows real-time investigation of business with the view to improve its efficiency and enhance country security and safety (Mohammad, 2020). E-commerce sector harnesses the potentials embedded in AI by creating customer base, conducting research in real time, getting acquainted with customer needs and bringing about solutions that will further improve the sector (Soni, 2020). AI is mostly used in e-commerce to offer better understanding of buyers and generating new leads, as well as improving the experience of user. The major components of AI used in e-commerce sector are data mining, NLP and ML, as shown in Figure 4.4. AI is currently being applied in e-commerce in the area of CRM, customer service, sale assistant to achieve the sales goals, chatbots, to mention but a few.

4.7.1.3.1 Applications of AI in E-commerce

The following are the areas of application of AI in e-commerce.

a. Chatbots and AI Assistant

Chatbot is a sophisticated AI software that mimics intelligent being and exchanges conversation with people in a natural language through

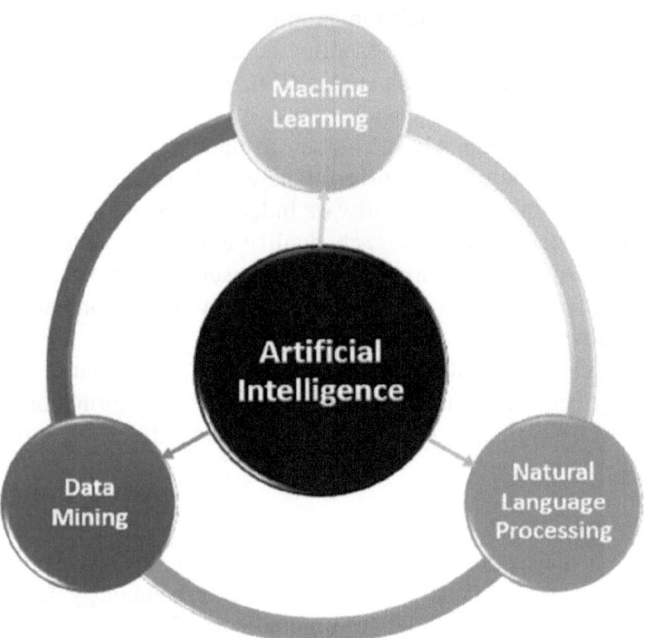

FIGURE 4.4   Components of AI in e-commerce.

sites, messaging apps and mobile apps (Girdher, 2019). Chatbot utilizes AI to ensure 24*7 support for customers on e-commerce website and provide response to clients based on their enquiries (Tapan & Monica, 2019).

b. Customer Relationship Management

Employees use CRM systems to access information about the primary contact of sales lead and work history, as well as the profiles on the social network and detailed information about their experience (Girdher, 2019). AI in CRM helps in tackling the evaluation and making reasonable suggestions regarding the prospect of customers based on the information gathered by the device about the person. Therefore, with the introduction of AI, high volume of data can be transferred and used to observe the trend of customer choices and buying, as well as factors responsible for their decision in buying. This will eventually allow proper engagement in a secured way (Soni, 2020). CRM through the use of AI avails users the ability to forecast and predict lead score, make recommendations and search natural language.

c. Recommendation Engines

E-commerce through the introduction of ML and AI can easily analyse the behaviours of customers on e-commerce websites and provide product recommendations to customers. As exemplified in Amazon and Flipkart websites, the data on search pattern are trained and appropriate algorithms are applied to predict products that will be of interest to customers (Tapan & Monica, 2019).

d. Product Content Management and Improved Sales Goals

AI assists e-commerce websites in providing an enhanced customer experience to all clients in the area of product design and cataloguing and offers e-commerce companies a clear perception to ensure seamless customer journey process and improved sales.

4.7.1.4   Contact Centre

The activities of AI in contact centres are in two spheres: at first, it can effortlessly offer the correct data as needed at the appropriate time with self-service alternatives, thus reducing the need for a call to support a customer. Also, it has the potential to offer the representative of customer service additional data to assist in complicated issues that the self-service cannot resolve. Most establishments have majorly been using some AI applications in contact centre for a longer time. A number of contact centres have introduced more advanced AI solutions such as language processing AIs and chatbots with the mindset that full automation using digital assistance is still very far away. Jeffs (2018) put forward a necessity for AI in an organization with the use of quantitative analysis of data. It showed that 80% of chief executive officers (CEOs) had the intention that the customer experience was great, while it was just 8% that agreed with the thought. The study was on 362 corporations, and this indicates that there is a massive gap in delivering a superior experience to the corporations' customers. It may be due to the lack of getting to know what the customers really want. This is an aspect where AI adoption is highly needed. IBM (2018) indicated that AI can improve contact centre, thus increasing customer experience through insight (AI can assist in clearly understanding the customer needs and provide most appropriate channel combinations to engage additional customers); customer communications (customers can connect and interact with one another, thus yielding a higher customer experience); and automation (for working smartly and obtaining better results).

In a study by Klein et al. (2017), concerning the foremost progressive retail executives regarding the end product of AI applications for contact centre transformation, the use of quantitative analysis of data suggested that AI power-driven contact centre is the new reality of retailers. Gori (2018) emphasized that 87% of establishments need to put additional efforts for the provision of a steady contact centre experience. Despite the fact that AI cannot solve every issue, there is no deficiency of the positive influence of AI in the delivery of consistent and quality contact centre.

4.7.1.5   Telecommunications

The telecommunications industry has been a fertile field of application for AI, and a variety of applications for the support of telecommunication have been identified in the literature. One of the major issues in telecommunications is the allocation of resources at the backbone and reconfiguration of network. AI can provide a dynamic dimensioning of the backbone. Neural network was used to perform and predict reconfiguration based on the volume of traffic and direction (Morales et al., 2017).

Also, AI can support the routing in IP networks and the setting up of connections. The proposed AI applications such as genetic algorithms, swarm intelligence and neural networks can assist in selecting the optimal network paths (Kyriakopoulos et al., 2016). It can also be used for the optimization of parameters for optical transmission in the telecoms networks, which take account of laser amplitude and phase noise (Zibar et al., 2015) and transmission quality measurement (Mata et al., 2017).

It is very important to have the ability to quickly identify network failures and predict the need for maintenance. Zhang et al. (2016) proposed an AI technique to be used in diagnosing and predicting optical access for network maintenance. Additionally, there is provision for AI-based self-healing characteristics for automatic restart of a server in case of an issue (Sendra et al., 2017). The new AI-based applications as provided by telecommunications industry and partners could for that reason take a look at designs to detect, predict and localize irregularities in the network and also in a straight line take steps for mitigation (Guibao et al., 2017).

With the above analysis, it is evident that AI applications transform the way telecommunications operate, improve and make available great service to their customers and provide guidance majorly on preventive maintenance, network optimization, robotic process automation (RPA) and virtual assistants.

4.7.1.5.1   Applications of AI in Telecommunications

AI can be applied in the following areas as regards telecommunications.

    a. AI Enhances the Service Level for Customer
        Essential technologies used by telecom operators have improved the customer service with the assistance of AI, such as natural language understanding and speech recognition.

    b. AI Enhances the Intellective Level of Network
        The techniques of AI endow the telecom operators with the competence in analysing, judging and predicting huge data such as promoting the planning of network development, operation, construction and optimization.

    c. AI Expands the Business Areas to Many Industries
        With the assistance of AI, telecom operators do not need to be restrained to outdated network communication services and can enlarge their business capabilities to various vertical commerce. This is a very big occasion to broaden the digital transformation (Qi et al., 2007).

4.8   BUSINESS BENEFITS OF ADOPTING AI

There are several benefits that can be derived from adopting AI. Some of them are highlighted below.

a. Quick Business Decision-Making

An entrepreneur must have the ability to manage a successful business by making fast decisions because there will always be competitors that can make the same decisions before you. The most important reasons why an entrepreneur must have the requisite skills to make fast decisions are ability to respond fast to competitor's activities, quick response to customer's needs and fast response to competitor's action. There are two types of decisions that we make: personal decisions, that is decisions that are associated with our own life and our family, and business decisions, that is decisions that are associated with business life. A quality decision depends on the form of information, analytical skills of the entrepreneur and the optimal amount of information (Sutevski, 2021).

b. Improvement in Operational Efficiencies and Productivity

Productivity can be increased by 40% or more by using AI through decision-making, automation and cyber security, which can help free up valuable time for employees. AI is capable of collecting huge amount data that are highly essential for businesses and their productivity. It reduces employees' time by collecting data automatically, which is formerly done manually. AI programs are capable of understanding and implementing data correctly, crawl information collected quickly and understand the patterns, which is simply difficult for humans. AI brings about an increase in production and productivity and makes a company superior when it comes to gaining insights into initiating marketing decisions and products. From the data and insights collected, AI enables companies to make future predictions on opportunities and references and which actions to take. AI is used for customer service chats because it understands natural language, and automated chatbots are developed to answer customers' questions in a natural way and a realistic manner. This helps companies to have more free time and be more productive while allowing customers to still get an accurate and personal interaction. It enables companies to spend less money on customer service employees and invest more in engineering, marketing or sales. Businesses should find all the ways AI can help them avoid being left behind or get ahead of others (Carter, 2018).

c. Improvement in Services and Quality

AI provides the ability to deal with new circumstances, data and situations that have not been initially predicted. Autonomous system is achieved by constant performance measurement and repetition of tasks (Jade, 2021). AI reduces the amount of employee expenses; that is, the development of robots to execute some tasks which are performed by employees leads to a reduction in the number of employees, which will ultimately leads to saving up company resources for productive pursuits due to the reduction in employee expenditure budget. There will be minimization of errors when there is less distractions. Employees can cause some damages that are sometimes irreparable due to distraction that can be emotional and physical situations. Creation of work using AI enables environment for filling the gaps of inaccuracies caused by predictable human errors and for implementing tasks by strictly following definite rules and quality management. AI enables good decision-making; that is, no matter how little, there is always a possibility of data inaccuracies when data collection is in the hand of employees and this can be identified by the employer by using AI in data collection and business insights. AI provides good delivery service to customers. The relationship and provision of excellent customer service is an integral part of a business. Customer experience can be enhanced by using AI for better customer service delivery (Kabadaian, 2018).

d. Reduction in Technology Cost

Analysing a particular buyer's purchase behaviour can be done through actionable sales intelligence. Smoothing of the entire process and plans is possible using actionable sales intelligence. With the reduction in the cost of complex business practices, more transparency can be created. Single customer engagement can strengthen the future perspective of business (QUYTECH, 2019). Some employees can be replaced partially or fully using a software powered by AI. Their salaries will be excluded in the budget because of the automation of their tasks by the software. The use of AI is especially important if your company started with a low fund. Crises can be reduced using AI, and such savings can be crucial when your business is facing a hard time (Haponik, 2019).

e. Improvement in Time to Market

AI in marketing concerns the collection of data, customer insights, anticipation of consumer buying move and adoption of automatic marketing decisions. In marketing where the speed is of importance, AI is generally employed. In fact, AI can increase the marketing return on investment (ROI). AI allows sellers to fully comprehend the behaviours, activities and indicators of their customers. They may thus focus on the proper strategy in a time-saving and successful manner towards the right customer. Gmail and Google Docs that utilize Smart Compose to read what is being typed, and then comprehend and propose what to type next; this is a perfect example of employing AI for marketing.

4.9   CONCLUSIONS

The effect of AI on business in general and economy in particular cannot be overemphasized as a disruptive technology. Every sector and business has witnessed one disruptive technology or the other, but the most prominent disruptive technology is AI. The emergence of AI as a disruptive technology not only helps organization increase their bottom line, but has propelled superlative research in the field of ML using algorithms that focus on deep learning and reinforcement learning. The Internet as a classic example of disruptive innovation laid the foundation upon which other forms of disruptive technologies are built. Disruptive technology is transforming work and organizations. It is important for businesses to embrace digital transformation towards an increasingly technology future. Digital transformation has altered the equilibrium of the market in recent years by changing the basic concept of business which must be more digital, technical and intelligent nowadays. As the era of AI is sure to face data issues, transformations in the business models and security risks, a business model based on AI is needed to help make a business competitive. Advancements in AI will also require a range of human abilities which will take time to develop, but today's AI trends appear highly unlikely that this will take time to occur.

REFERENCES

Acemoglu, D., & Restrepo, P. (2020). Robots and jobs: Evidence from us labor markets. *Journal of Political Economy*, *128*(6), 2188–2244. Doi: 10.1086/705716.

Agrawal, A. (2021). *The Power of AI in Lead Generation*. Startupgrind. https://www.startupgrind.com/blog/the-power-of-ai-in-lead-generation/.

Akinsola, J. E. T., Adeagbo, M. A., & Awoseyi, A. A. (2019). Breast cancer predictive analytics using supervised machine learning techniques. *International Journal of Advanced Trends in Computer Science and Engineering*, *8*(6), 3095–3104. Doi: 10.30534/ijatcse/2019/70862019.

Akinsola, J. E. T., Akinseinde, S., Kalesanwo, O., Adeagbo, M., Oladapo, K., Awoseyi, A., & Kasali, F. (2021). Application of Artificial Intelligence in user interfaces design for cyber security threat modeling. In R. Heimgärtner (Ed.), *Intelligent User Interfaces* (pp. 1–28). IntechOpen. Doi: 10.5772/intechopen.96534.

Akinsola, J. E. T., Awodele, O., Idowu, S. A., & Kuyoro, S. O. (2020). SQL injection attacks predictive analytics using supervised machine learning techniques. *International Journal of Computer Applications Technology and Research*, *9*(4), 139–149. Doi: 10.7753/ijcatr0904.1004.

Akinsola, J. E. T., Awodele, O., Kuyoro, S. O., & Kasali, F. A. (2019). Performance evaluation of supervised machine learning algorithms using multi-criteria decision-making techniques. *International Conference on Information Technology in Education and Development (ITED)*, 17–34. https://ir.tech-u.edu.ng/416/1/Performance Evaluation of Supervised Machine Learning Algorithms Using Multi-Criteria Decision Making %28MCDM%29 Techniques ITED.pdf.

Akinsola, J. E. T., Kuyoro, A., Adeagbo, M. A., & Awoseyi, A. A. (2020). Performance evaluation of software using formal methods. *Global Journal of Computer Science and Technology: C Software & Data Engineering*, *20*(1). https://computerresearch.org/index.php/computer/article/view/1930/1914.

Akinsola, J. E. T., Ogunbanwo, A. S., Okesola, O. J., Odun-Ayo, I. J., Ayegbusi, F. D., & Adebiyi, A. A. (2020). Comparative analysis of software development life cycle models (SDLC). *Intelligent Algorithms in Software Engineering. CSOC 2020. Advances in Intelligent Systems and Computing*, *1*, 310–322. Doi: 10.1007/978-3-030-51965-0_27.

Alao, O. D., Joshua, J. V, & Akinsola, J. E. T. (2019). Human computer interaction (HCI) and smart home applications. *IUP Journal of Information Technology*, *15*(3), 7–21. https://search.proquest.com/openview/70e74bf39099ec671c013b7bf9d9258a/1?pq-origsite=gscholar&cbl=2029987.

Allyn, B. (2020). *Facial Recognition Leads to False Arrest of Black Man in Detroit*. NPR. https://www.npr.org/2020/06/24/882683463/the-computer-got-it-wrong-how-facial-recognition-led-to-a-false-arrest-in-michig.

Amorin, G. (2019). *Artificial Intelligence for Everyone - Workshop - The Future Of Finance Summit 2019.* . The Asian Banker. http://www.theasianbanker. com/future-of-finance-summit-2019/artificial-intelligence-everyone-workshop/.

Anuj, K., Fayaz, F., & Kapoor, M. N. (2018). Impact of E-commerce in Indian economy. *Journal of Business and Management, 20*(5), 59–71. Doi: 10.9790/487X-2005065971.

Azucar, D., Marengo, D., & Settanni, M. (2018). Predicting the Big 5 personality traits from digital footprints on social media: A meta-analysis. *Personality and Individual Differences, 124*(September 2017), 150–159. Doi: 10.1016/j.paid.2017.12.018.

Bazarbash, M. (2019). *FinTech in Financial Inclusion Machine Learning Applications in Assessing Credit Risk (M. Bazarbash (ed.)).* International Monetary Fund. https://www.imf.org/en/Publications/WP/Issues/2019/05/17/FinTech-in-Financial-Inclusion Machine-Learning-Applications-in-Assessing-Credit-Risk-46883.

Bekker, A. (2021). *5 Types of AI to Propel Your Business.* ScienceSoft. https://www. scnsoft.com/blog/artificial-intelligence-types.

Buchanan, B. G. (2019). *Artificial intelligence in finance_ New landscaping report from The Alan Turing Institute.* https://harisportal.hanken.fi/sv/publications/artificial-intelligence-in-finance.

Butzmann, L., Daweke, E., Geimer, J., Kolev, N., & Stiller, M. (2017). From mystery to mastery: Unlocking the business value of Artificial Intelligence in the insurance industry. In *Deloitte Digital (Issue 11). Deloitte,* 1–45. https://www2.deloitte.com/ content/dam/Deloitte/ru/Documents/financial-services/artificial-intelligence-in-insurance.pdf.

Carter, S. (2018). *How Artificial Intelligence Increases Business Productivity - Business 2 Community.* Business 2 Community. https://www.business2community.com/tech-gadgets/how-artificial-intelligence-increases-business-productivity-02059942.

Cioffi, R., Travaglioni, M., Piscitelli, G., Petrillo, A., & De Felice, F. (2020). Artificial Intelligence and machine learning applications in smart production: Progress, trends, and directions. *Sustainability (Switzerland), 12*(2). Doi: 10.3390/su12020492.

Corporate Finance Institute. (2021a). *Credit Analysis Process - Overview and Evaluation Stages.* CFI Education. https://corporatefinanceinstitute.com/resources/knowledge/ credit/credit-analysis-process/.

Corporate Finance Institute (CFI). (2021b). *Disruptive Technology - Overview, Examples, Success Factors.* CFI Education Inc. https://corporatefinanceinstitute. com/resources/knowledge/other/disruptive-technology/.

Daqar, M. A. M. A., & Smoudy, A. K. A. (2019). The role of Artificial Intelligence on enhancing customer experience. *International Review of Management and Marketing, 9*(4), 22–31. Doi: 10.32479/irmm.8166.

De Bruyn, A., Viswanathan, V., Beh, Y. S., Brock, J. K. U., & von Wangenheim, F. (2020). Artificial Intelligence and marketing: Pitfalls and opportunities. *Journal of Interactive Marketing, 51,* 91–105. Doi: 10.1016/j.intmar.2020.04.007.

Deloitte. (2021). *Artificial Intelligence Disruption | Deloitte US.* Deloitte US. https:// www2.deloitte.com/us/en/pages/technology-media-and-telecommunications/ articles/artificial-intelligence-disruption.html.

Devang, V., Chintan, S., Gunjan, T., & Krupa, R. (2019). Applications of Artificial Intelligence in marketing. *Annals of Dunarea de Jos University of Galati. Fascicle I. Economics and Applied Informatics, 25*(1), 28–36. Doi: 10.35219/eai158404094.

Dilmegani, C. (2021). *9 AI Insurance Applications / Use Cases in 2021 : In-depth guide.* AI Multiples. https://research.aimultiple.com/ai-insurance/.

Eckert, D. (2017). *The Essential Eight Technologies Board byte: Artificial Intelligence.* PwC Global. https://www.pwc.com.au/pdf/essential-8-emerging-technologies-artificial-intelligence.pdf.

Elhajjar, S., Karam, S., & Borna, S. (2021). Artificial Intelligence in marketing education programs. *Marketing Education Review, 31*(1), 2–13. Doi: 10.1080/10528008.2020.1835492.

Faggella, D. (2020). *Artificial Intelligence in Insurance - Three Trends That Matter.* Emerj. https://emerj.com/ai-sector-overviews/artificial-intelligence-in-insurance-trends/.

Feng, C. M., Park, A., Pitt, L., Kietzmann, J., & Northey, G. (2020). Artificial Intelligence in marketing: A bibliographic perspective. *Australasian Marketing Journal.* Doi: 10.1016/j.ausmj.2020.07.006.

Financial Stability Board. (2017). *Artificial Intelligence and machine learning in financial services market developments and financial stability implications.* Financial Stability Board. http://www.fsb.org/2017/11/artificial-intelligence-and-machine-learning-in-financial-service/.

Girasa, R. (2020). Artificial Intelligence as a disruptive technology: Economic transformation and government regulation. In R. Ballard (Ed.), *Springer Nature Switzerland AG.* Palgrave Macmillan. Doi:10.1007/978-3-030-35975-1.

Girdher, S. (2019). Role of Artificial Intelligence in transforming e-commerce sector. *Research Review InternationalInternational Journal of Multidisciplinary, 4*(6), 282–284.

Gori, A. (2018). *5 Customer Experience Trends for 2018.* Zendesk. https://www.zendesk.co.uk/blog/5-customer-experience-trends-for-2018/.

Guibao, X., Yubo, M., & Jialiang, L. (2017). The impact of Artificial Intelligence on communication networks and services. *ITU Journal: ICT Discoveries, 1*(1), 33–38.

Haenlein, M., & Kaplan, A. (2019). A brief history of Artificial Intelligence: On the past, present, and future of Artificial Intelligence. *California Management Review, 61*(4), 5–14. Doi: 10.1177/0008125619864925.

Haponik, A. (2019). *How does AI Reduce Operating Costs and Improve Efficiency?* | Addepto. Addepto. https://addepto.com/reduce-operating-costs-and-improve-efficiency-using-ai/.

Hariharan, B., & Chandrakhanthan, J. (2020). Impact of Artificial Intelligence in marketing. *Test Engineering and Management, 83*, 104–109. Doi: 10.13140/RG.2.2.17946.36802.

Harris, E. (2021). *9 Amazing Examples of Disruptive Technology (Inspired by the WTIA).* Resultist Consulting. https://www.resultist.com/blog/9-amazing-examples-of-disruptive-technology-inspired-by-the-wtia.

Hill, K. (2020). Flawed facial recognition leads to arrest and jail for New Jersey man. *The New York Times.*

Hill, K. (2021). Wrongfully accused by an algorithm. *The New York Times.*

Hinmikaiye, J. O., Awodele, O., & Akinsola, J. E. T. (2021). Disruptive technology and regulatory response: The Nigerian perspective. *Computer Engineering and Intelligent Systems, 12*(1), 42–47. Doi: 10.7176/ceis/12-1-06.

Hussain, K., & Prieto, E. (2016). Big data in the finance and insurance sectors. In J. M. Cavanillas (Ed.), *New Horizons for a Data-Driven Economy: A Roadmap for Usage and Exploitation of Big Data in Europe* (Issue December, pp. 1–303). Doi: 10.1007/978-3-319-21569-3.

IBM. (2018). Trusting AI. In M. Rodrigues (Ed.), *IBM Research AI.* IEEE. Doi: 10.1109/JSTSP.2018.2865887.

ICICI. (2020). *Artificial Intelligence in Banking for Lending & Loan Assessment: How Does it Work?* ICICI Blog. ICICI Bank. https://www.icicibank.com/blogs/personal-loan/artificial-intelligence-in-loan-assessment-how-does-it-work.page.

Jade. (2021). *AI Marketing: What, Why & How to Use Artificial Intelligence in Marketing – Mageplaza.* Mega Plaza. https://www.mageplaza.com/blog/ai-marketing-what-why-how.html.

Jeffs, V. (2018). *Artificial Intelligence and Improving the Customer Experience.* https://www.pega.com/system/files/resources/2019-09/ai-and-improving-cx-en.pdf.

Johnson, J. (2020). *4 Types of Artificial Intelligence – BMC Software | Blogs.* BMC.

Kabadaian, H. (2018). *5 Ways Artificial Intelligence Can Improve Your Business Right Now*. Entrepreneur Media Inc. https://www.entrepreneur.com/article/317937.

Kagan, J. (2021). *Credit Rating*. Investopedia.Com. https://www.investopedia.com/terms/i/insurance-company-credit-rating.asp.

Khamis, A. (2019). *AI and Disruptive Innovation. AI Will Have a Major Impact on The... | by Alaa Khamis | Towards Data Science*. Towards Data Science. https://towardsdatascience.com/ai-and-disruptive-innovation-393ee89eb5dd.

Klein, R., Parsons, D., Sonsev, V., & Peetermann, L. (2017). *How AI Technology Will Transform Customer Engagement*. Linc. https://www.letslinc.com/wp-content/uploads/2017/07/Linc_Brand-Garage_Customer-Service-and-AI-Report.pdf.

Kumar, V., Rajan, B., Venkatesan, R., & Lecinski, J. (2019). Understanding the role of Artificial Intelligence in personalized engagement marketing. *California Management Review, 61*(4), 135–155. Doi: 10.1177/0008125619859317.

Kyriakopoulos, C. A., Papadimitriou, G. I., Nicopolitidis, P., & Varvarigos, E. (2016). Energy-efficient lightpath establishment in backbone optical networks based on ant colony optimization. *Journal of Lightwave Technology, 34*(23), 5534–5541. Doi: 10.1109/JLT.2016.2623678.

Lackey, D. (2019). *How Much Data Do We Create Every Day? The Mind-Blowing Stats Everyone Should Read – Content For marketers*. Blazon. https://blazon.online/how-much-data-do-we-create-every-day-the-mind-blowing-stats-everyone-should-read/.

Mata, J., De Miguel, I., Durán, R. J., Aguado, J. C., Merayo, N., Ruiz, L., Fernández, P., Lorenzo, R. M., & Abril, E. J. (2017). A SVM approach for lightpath QoT estimation in optical transport networks. *Proceedings -2017 IEEE International Conference on Big Data, Big Data 2017, 2018-Janua*, 4795–4797. Doi: 10.1109/BigData.2017.8258545.

Matthews, D. (2020). *Artificial Intelligence*. National Association of Insurance Commissioners. https://content.naic.org/cipr_topics/topic_artificial_intelligence.html.

Mcdonald, C. (2020). *Demystifying AI, Machine Learning and Deep Learning*. HPE Developer. https://developer.hpe.com/blog/demystifying-ai-machine-learning-and-deep-learning/.

Medmain. (2018). *My First Steps into the World of A.I.* Medmain. https://medium.com/@Medmain/my-first-steps-into-the-world-of-ai-d7d591b2fe22.

Microsoft. (2016). *Digital Transformation: Seven Steps to Success*. In *Microsoft*. https://info.microsoft.com/rs/157-GQE-382/images/Digitaltransformation-seven steps to success.v2.pdf?aliId=520174389.

Mohammad, S. M. (2020). Artificial Intelligence in information technology. *Jm*, 1–15.

Moore, R. (2019). *11 Disruptive Innovation Examples (And Why Uber and Tesla Don't Make the Cut) | OpenView*. OPenView. https://openviewpartners.com/blog/11-disruptive-innovation-examples-and-why-uber-and-tesla-dont-make-the-cut/#.YLQoh6hKjMX.

Morales, F., Ruiz, M., Gifre, L., Contreras, L. M., López, V., & Velasco, L. (2017). Virtual network topology adaptability based on data analytics for traffic prediction. *Journal of Optical Communications and Networking, 9*(1), 35–45. Doi: 10.1364/JOCN.9.000A35.

Morgan, H. (2019). *How AI Can be One of the Disruptive Technology in History*. Medium Towards Data Science. https://towardsdatascience.com/how-ai-can-be-one-of-the-disruptive-technology-in-history-1dc3f7d38cfa.

Mustak, M., Salminen, J., Plé, L., & Wirtz, J. (2021). Artificial Intelligence in marketing: Topic modeling, scientometric analysis, and research agenda. *Journal of Business Research*, 389–404. Doi: 10.1016/j.jbusres.2020.10.044.

Nguyen, T. T., Viet, Q., Nguyen, H., Nguyen, C. M., Nguyen, D., Nguyen, T., Nahavandi, S., Nguyen, T. T., Nguyen, D. T., Nguyen, Q. V. H., Nguyen, D., & Nahavandi, S. (2021). Deep Learning for deepfakes creation and detection: A survey. *Cornell University, 11573*(3), 1–16. https://arxiv.org/abs/1909.11573.

Oke, S. A. (2008). A literature review on Artificial Intelligence. *International Journal of Information and Management Sciences*, *19*(4), 535–570.

Osisanwo, F. Y., Akinsola, J. E. T., Awodele, O., Hinmikaiye, J. O., Olakanmi, O., & Akinjobi, J. (2017). Supervised machine learning algorithms : Classification and comparison. *International Journal of Computer Trends and Technology*, *48*(3), 128–138.

Pandian, D. A. P. (2019). Artificial Intelligence application in smart warehousing environment for automated logistics. *Journal of Artificial Intelligence and Capsule Networks*, *1*(2), 63–72. Doi: 10.36548/jaicn.2019.2.002.

Ping, N. L., Hussin, A. R. B. C., & Ali, N. B. M. (2019). Constructs for Artificial Intelligence customer service in E-commerce. *International Conference on Research and Innovation in Information Systems, ICRIIS*, December-2, 19–24. Doi: 10.1109/IC RIIS48246.2019.9073486.

Podnar, K. (2016). *Automating Your Customer Interactions : Get Ready for Chatbots*. Content Marketing Institute. https://contentmarketinginstitute. com/2016/11/automating-customer-interactions-chatbots/.

Qi, J., Wu, F., Li, L., & Shu, H. (2007). Artificial Intelligence applications in the telecommunications industry. *Expert Systems*, *24*(4), 271–291. Doi: 10.1111/j.1468-0394.20 07.00433.x.

QUYTECH. (2019). *How Artificial Intelligence Reduces the Cost of Doing Business*. QUYTECH. https://www.quytech.com/blog/how-artificial-intelligence-reduces-the-cost-of-doing-business/.

Resnick, N. (2021). *4 Things to Automate In 2020 for Better Customer Relationships*. HubSpot. https://blog.hubspot.com/service/automate-customer-relationships.

Riserbato, R. (2021). *The 16 Best Client Management Software for Any- Sized Team*. HubSpot. https://blog.hubspot.com/sales/client-management-software.

Sendra, S., Rego, A., Lloret, J., Jimenez, J. M., & Romero, O. (2017). Including Artificial Intelligence in a routing protocol using Software Defined Networks. *2017 IEEE International Conference on Communications Workshops, ICC Workshops 2017*, SCPA, 670–674. Doi: 10.1109/ICCW.2017.7962735.

Shroff, R. (2019). *How are Insurance Companies Implementing Artificial Intelligence (AI)?* Towards Data Science. https://towardsdatascience.com/how-are-insurance-companies-implementing-artificial-intelligence-ai-aaf845fce6a7

Smith, T. (2020). *Disruptive Technology Definition*. Investopedia Dotdash Publishing. https://www.investopedia.com/terms/d/disruptive-technology.asp.

Soni, V. D. (2020). Emerging roles of Artificial Intelligence in ecommerce. *International Journal of Trend in Scientific Research and Development*, *4*(5), 223–225.

Sutevski, D. (2021). *Fast Decision-Making Process VS Quality Decisions*. Entrepreneurship in a Box. https://www.entrepreneurshipinabox.com/1084/importance-quick-decisions/.

Tapan, K., & Monica, T. (2019). The colossal impact of Artificial Intelligence in e- commerce : Statistics and facts. *International Research Journal of Engineering and Technology*, *6*(5), 570–572.

TORQ Software. (2021). *Iris Debt Collection System*. http://www.torqsoftware. com/products/114-iris-system.

Trivedi, A. (2021). *Artificial Intelligence: The Disruptive Technology of The Year*. AceInfoway. https://www.aceinfoway.com/blog/artificial-intelligence-disruptive-technology

Tucci, L. (2020). *What is Artificial Intelligence (AI)?* TechTarget. https://searchenterpriseai. techtarget.com/definition/AI-Artificial-Intelligence.

Verhoef, P. C., Broekhuizen, T., Bart, Y., Bhattacharya, A., Qi Dong, J., Fabian, N., & Haenlein, M. (2021). Digital transformation: A multidisciplinary reflection and research agenda. *Journal of Business Research*, *122*, 889–901. Doi: 10.1016/j.jbusres.2019.09.022.

Verma, S., Sharma, R., Deb, S., & Maitra, D. (2020). Artificial Intelligence in market-
    ing: Systematic review and future research direction. *International Journal of
    Information Management Data Insights*, 1–8. Doi: 10.1016/j.jjimei.2020.100002.
Xie, M. (2019). Development of Artificial Intelligence and effects on financial system.
    Journal of Physics: Conference Series, *1187*(3), 1–6. Doi: 10.1088/1742-6596/118
    7/3/032084.
Yeğİn, T. (2020). The place and future of Artificial Intelligence in marketing strategies.
    Ekev Akademi Dergis, *24*(81), 489–506.
Zhang, X., Hou, W., Guo, L., Wang, S., Sun, Y., & Yang, X. (2016). Failure recovery
    solutions using cognitive mechanisms for software defined optical networks.
    *ICOCN 2016-2016 15th International Conference on Optical Communications and
    Networks, September*. Doi: 10.1109/ICOCN.2016.7875600.
Zibar, D., De Carvalho, L. H. H., Piels, M., Doberstein, A., Diniz, J., Nebendahl, B.,
    Franciscangelis, C., Estaran, J., Haisch, H., Gonzalez, N. G., De Oliveira, J. C. R.
    F., & Monroy, I. T. (2015). Application of machine learning techniques for ampli-
    tude and phase noise characterization. *Journal of Lightwave Technology*, *33*(7),
    1333–1343. Doi: 10.1109/JLT.2015.2394808.

5 An Optimal Diabetic Features-Based Intelligent System to Predict Diabetic Retinal Disease

M. Shanmuga Eswari and S. Balamurali
Kalasalingam Academy of Research and Education

CONTENTS

5.1   Introduction 91
5.2   Experimental Methods 92
    5.2.1   Dataset Description 93
    5.2.2   Preprocessing of Data.. 93
    5.2.3   Dataset Splitting 95
5.3   Machine Learning Classification Approach.................................... 96
    5.3.1   Kernel-Based SVMs 96
    5.3.2   Linear Model 97
    5.3.3   Boosted Regression.. 97
    5.3.4   K-nearest Neighbor (KNN) 98
    5.3.5   CART (Classification and Regression Tree) 98
    5.3.6   Ensemble-Based Algorithms 99
        5.3.6.1   Random Forest Ensemble Machine
                  Learning Algorithm... 99
        5.3.6.2   AdaBoost Random Forest Ensemble Learner................. 99
        5.3.6.3   Gradient Boost Random Forest Ensemble Learner 99
5.4   Results and Impact.. 100
5.5   Conclusions... 104
Acknowledgement 105
References.. 105

5.1   INTRODUCTION

There will be nearly 439 million diabetic patients worldwide in the year of 2030 as per the proclamation report given by the world health organization [1]. If a person

DOI: 10.1201/9781003224068-5

has had a many years of high glucose and blood pressure, they may be concerned that the pressure on their retina raised [2]. Heart diseases together with high lipid levels lead to retinal damage [3]. Leaser et al. [4] meta-analyzed the details of the patients who had diabetic retinopathy. One of the most evil optical obstacle was diabetic retinopathy (DR) [5].Glaucoma is the next foremost cause of vision loss, which increases the intraocular pressure (IOP) and damages the ocular nerve head [6]. Diabetic macular edema (DME) is another retinal illness that can lead to DR [7]. The retinal glaucoma detection based on SVM with discrete wavelet transforms obtained an accuracy of 0.95 [8]. Iyer et al. [9] used association mining classifier to detect the diabetic mellitus [9]. Dissimilar machine learning techniques are used to predict the illness [10]. Ophthalmologists are using different patterns and machine learning techniques to recognize retinal disorders [11,12].

Tsao et al. [13] applied different mining techniques to predict the DR among type 2 diabetes patients. Patil and Malpe [14] used neural networks and naïve Bayes classification to predict DR.

The frequency of DR has been determined by population-based studies all across the world [15]. Artificial intelligence (AI) was used to assist in the automatic detection of eye diseases [16]. The study by Ameena and Ashadevi [17] explained the diabetic risk among female patients [18]. To forecast the risk of gestational diabetic mellitus, artificial intelligence was implanted in electronic health data [19]. A multi-illness prediction model was created by Jerlin and Eswaran [20] to attain the best classification result. Eswari and Karkuzhali [21] reviewed various segmentation and classification techniques for detecting glaucoma. Eswari and Balamurali [22] proposed Bayesian optimization SVM to detect glaucoma among diabetic patients with an accuracy of 97.4%. Shankar et al. [23] proposed a multi-kernel approach to classify thyroid diseases with optimal features. Kumar and Rajasekaran [24] used Internet of things in combination with Raspberry Pi to collect healthcare data from patients to monitoring the patients health very effectively.

From the above literature review, a new system has been proposed to achieve the following objectives:

- to find the probability of getting diabetic retinal disease among diabetic people
- to establish an expert general prediction system
- to distinguish optimal features from numerous features
- to save time by early detection
- to boost an accuracy using ensembles.

5.2 EXPERIMENTAL METHODS

An experimental configuration is made use of Windows 7 as a platform and the R studio programming language as an implementation language. R is a free and open-source statistics program. The real-time information is acquired from a number of local hospitals. A new dataset is produced based on the chronic dataset [20], as well.

5.2.1 Dataset Description

The new actual diabetes dataset has 28 parameters and 2,500 tuples at first. Age, glucose, blood pressure, hypertension, hemoglobin, diabetic mellitus, diabetic duration, red blood cells, IoP, and CDR were among the variables in the dataset. The dataset was divided into 70% and 30% for training purpose from the total number of 2,400, and the remaining 100 healthcare data entries were reserved for prediction purpose. In this, diabetic retinal disease predicted column as defined as "1" and "0" for chance and no chance of retinal disease. Random forest is implemented for feature selection.

5.2.2 Preprocessing of Data

Preprocessing is an initial stage to understanding of data and knowledge of data, which demands to meet the requirements of the problem. Processes based on new data created, used, managed, and measured as part of operational procedures to gain a better understanding.

Data preprocessing is a data mining approach that entails converting raw data into a format that can be understood. Real-world data are frequently inadequate, inconsistent, and lacking in specific behaviors or trends to include numerous inaccuracies. Data preprocessing is a simple, but effective way of resolving such challenges. Generally, preprocessing involves the following steps:

- Data consolidation that entails gathering, selecting, and integrating information on diabetic patients' data.
- Data cleaning that entails substituting values and removing duplicates from the new dataset.
- Normalization and discretization of data and construction of attributes in the dataset.
- Data reduction that consists of reducing dimension based on principle components.

The collected information is cleaned and preprocessed before entering into machine learning techniques. Now, the above cleaned dataset is used for further processing. Here, row values that are missing are eliminated from the existing dataset. The mean values of the attribute value are used to restore hypertension and blood glucose levels (Figure 5.1). In this, feature is defined as a measurable individual property or attribute. Features can be categorical and numerical. Categorical features are expressed by a set of values or boolean set. Numerical features have a sequence of values or an integer value. The present dataset has both types of features.

The process of cleaning and altering raw data prior to processing and analysis is known as data preparation. It's a crucial stage before processing that often includes reformatting data, making data changes, and integrating datasets to enrich data. Data preparation ensures data accuracy, resulting in accurate

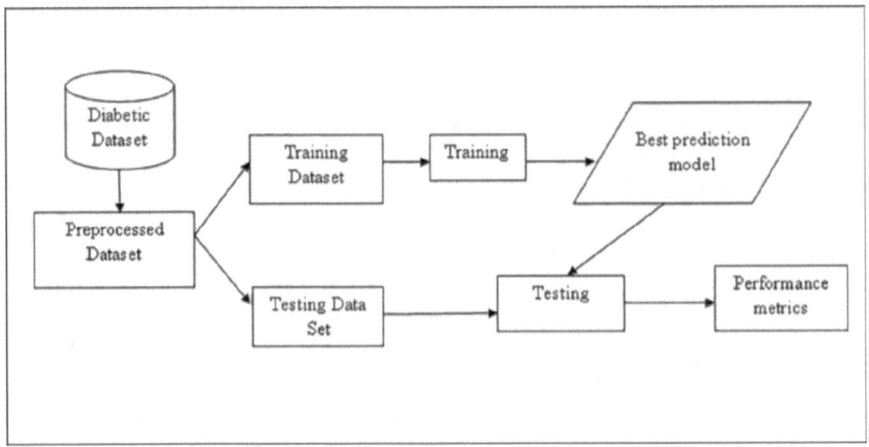

FIGURE 5.1 Workflow of the proposed model.

insights. It's likely that insights will be off owing to garbage data, an ignored calibration issue, or an easily corrected mismatch across datasets if data preparation isn't done.

Preparation also allows you to deal with problems more swiftly and efficiently because you will already have the solutions on hand and ready to utilize. Anomalies or outliers are unexpected values that frequently appear in a value distribution, especially when working with data from unknown sources with insufficient data validation measures. Data must occasionally be removed and saved in a different format or place. Consult domain experts or combine data from various sources to solve improperly formatted data. Inconsistent values and categorical variables aren't standardized. When merging data from several sources, and frequently encounter differences in variables. Finding and standardizing all variations will considerably improve the model's accuracy.

Feature enrichment, or expanding the features in proposed data, frequently necessitates combining datasets from many sources. When there are no straightforward or accurate columns to match the datasets, joining files from separate systems might be difficult. This necessitates the capacity to execute fuzzy matching, which might also be accomplished by merging numerous columns.

Even if all necessary data are accessible, data preparation techniques such as feature engineering may be required to develop additional material that will result in more accurate and relevant models. In applied statistics and machine learning, data visualization is a critical ability. This can helps in spotting patterns, corrupt data, outliers, and much more, when exploring and getting to know a dataset. The graphical representation of information aids in the comprehension of data by summarizing and presenting a large quantity of facts in a simple and easy-to-understand style, and aids in the clear and effective communication of information.

When working on a machine learning task, people spend a majority of work time dealing with data quality issues because data are often gathered from many sources that are not really accurate and they are also in different forms. There could be issues owing to human error, measurement instrument limits, or defects in the data collection process. When the dataset's dimensionality is minimal, descriptive statistics performs better. This is due to the decomposition of non-essential characteristics and ambient noise. Models based on lower-dimensional data are easier to comprehend, and data visualization is easy.

Model evaluation is an important step in the creation of a model. It aids in the selection of the best model to represent our data, as well as the prediction of how well the chosen model will perform in the future. Both methods use a test set (not visible to the model) to evaluate model performance in order to avoid over-fitting. Accuracy, precision, and recall are the three basic measures used to evaluate a classification model. The percentage of correct predictions for the test data is known as accuracy. It's simple to figure out by dividing the number of correct predictions by the total number of projections.

Model selection is the process of choosing one final machine learning model for a training dataset from a pool of candidate machine learning models. Model selection is a technique that can be applied to a variety of models such as logistic regression, SVM, Knn, ensemble, and neural network. To this objective, people must compare the relative performance of different models. As a result, the loss function and the metric that represents it become critical in determining the best and least over-fitted models.

Model presentation is the final step in which the selected model will be trained to the stakeholder prior to deployment in the system (Table 5.1).

5.2.3 DATASET SPLITTING

The suitable datasets are divided into two datasets, namely training and test sets. The training data are a set of data that a system uses to learn how to exploit technologies such as neural networks to learn and produce complex outputs. A training set, training dataset, or learning set is a term used to describe the training data. A test set is a tertiary dataset used to examine a machine learning program after it has been trained on an original training dataset. To avoid over-fitting, there are numerous ways to generate and use the validation set. The holdout method, random subsampling, k-fold cross-validation, and leave-one-out cross-validation

TABLE 5.1
Listing of Few Parameters

Features	Chance	No chance
BP	>130	≤130
BS	>120	≤120
Diabetic years	>10	≤10
Patient_age	>40	≤40
DM	1	0

are some of the different ways. The holdout approach is the first way to employ a validation set. Over-fitting occurs when a statistical model or machine learning method captures the noise in the data. Under-fitting occurs when a model or algorithm seems to have a low variance, but a large bias and typically the outcome of an incredibly simplistic model. The learning dataset usually contains 70% of the data for finest prediction. Of the remaining dataset, 15% is used for checking and 15% for validation to make decision support model.

5.3   MACHINE LEARNING CLASSIFICATION APPROACH

Classification in statistics is defined as identifying categories to which the things based on observations. Machine learning classifiers are very supportive to do the tasks automatically, saving time and cost. The algorithm or the principles that machines employ to classify data are referred to as a classifier. The final outcome of proposed classifier's machine learning is a classification model. The classifier is used to train the model, which then sorts your data. Unsupervised and supervised classifiers are also available. Unsupervised machine learning classifiers are fed only unlabeled datasets and must categorize them based on pattern recognition, data structures, and anomalies. Training datasets are provided to supervised and semi-supervised classifiers, which teach them how to categorize data into predefined categories. Few technical terminologies are discussed below.

A learning example is a data point from a learning set, which is used to solve a predictive modeling problem. The terms "training instance" and "training example" are sometimes used interchangeably.

The terms hypothesis and model are frequently interchanged in the field of machine learning. They can have distinct connotations in other fields, such as the hypothesis being the scientist's "informed guess" and the framework being the personification that presumes with the intention of theory proposed.

A learner predictor is a premise or the process of assigning categorical discrete value class labels to data points. A hypothesis, on the other hand, is not really the same as a classifier.

Our goal is to find or approximate the target function. Therefore, the learning algorithm is a series of instructions that tries to model the target function using our training dataset. The set of possible hypotheses that a learning algorithm can create in order to model an unknown target function by creating the final hypothesis is referred to as a hypothesis space.

Machine learning methods are utilized to find a good machine fit model for retinal illness prediction, and the terms listed below are frequently used.

5.3.1   KERNEL-BASED SVMs

The SVM is a well-known machine learning algorithm. Kernel SVM deals with high-dimensional linear discriminant mathematical function space which is more effective in pattern reorganization and machine learning techniques. There are seven kernels available, and few are implemented to achieve the model. Most of the time, the model fits perfectly in this way, but in real data, the margin might not be so small and the discretized Euclidean distances are less accurate, so it should change the space of the data used. In this algorithm, there are three parameters available, namely C, kernel, and degree. The radial basis values are based on the distance of the origin and gamma values. An Increasing of gamma values obtained the over fixation of the model. The system attained an accuracy of 0.9645 based on the sigma tuning parameter 0.08841.

Steps
- upload preprocessed dataset
- define labels
- select different kernels
- predict the result
- evaluate the performance.

5.3.2 LINEAR MODEL

The generalized linear model (GLM) is used to execute logistic regression. A preset number of assessments are done to get the better result and outputs, response variables not based on preset. In statistics, it is represented as error distribution model, which allows response variables as a linear value. The following steps are implemented:

- upload the dataset
- check factors and continuous variables
- apply feature engineering
- perform statistical analysis on training and test datasets
- test with new data
- perform model fitting

5.3.3 BOOSTED REGRESSION

Boosted regression on logistic pertains to distinguishing strong learners from weak learners. This is one of the improved ensemble techniques. Here, fragile learners are successively qualified to obtain suitable antecedent. This iterative process follows by taking lessons learned from the weak learners and incrementally adding those learners to the strong learners. The weak learners are generally ignored for ensemble learning. The compositional learning model does not rely on strong learners to improve the weak learners. Instead, the weak learners are kept from the environment. Thus, the combined learning models get better results than either one alone. The following steps are incorporated to select the various distributions on every step:

- assign equal weight to each observation to base learners
- higher prediction error occurred in first base learning, then concentrated to select next base algorithms

- do this repeatedly to achieve good accuracy
- at last, it combines all weak learners and creates a strong learner to get good prediction of the model

Eventually, this method increases the prediction rate and highly spotlights the misclassified error rates.

5.3.4 K-NEAREST NEIGHBOR (KNN)

KNN is a straight forward procedure used for classification and regression problem solving. This is a non-parametric lazy learner which means not having any best guess, and not learned from history of training set. The working condition is based on similarity distance measure on supervised lazy learners. It stores up all data and classifies new similarity-based data. The accuracy is based on the highest Euclidean distance value for k. The below-listed steps are implemented on diabetic dataset to obtain results:

- assigning number of k neighbors
- calculation of Euclidean distance
- counting data points in each category
- assigning new data points based on the maximum number of neighbors
- creating model.

This method is more suitable for larger datasets and also fits for noisy dataset and has a high computational cost for distance calculation.

5.3.5 CART (CLASSIFICATION AND REGRESSION TREE)

This is depicted to forecast the model values on target variables. Every tree is built as a two-node tree. The root node has all learned models and leaf nodes with an objective-dependent variable. This will do repeated splitting until the best result is attained. Here, tree construction is based on top-down approach and depicted in the following steps:

- all training leaves are started with its root
- based on gain ratio and other impurity methods, splitting is done
- perform recursive partitioning until there is no node for further separation.

In this, impurity is defined as uncertainty of degree and the heuristic method of measuring is known as Gini index. This is defined as information gaining of the selected instances and good for larger segregation and iteration(i) as 1 to "n" number of attributes to yield fine classification (Equation 5.1):

$$\text{Gini index}(Gi) = 1 - \sum (pi)^2 \qquad (5.1)$$

Here, pi is the probability of record R to belong to the class of c_i as a binary split.

5.3.6   Ensemble-Based Algorithms

5.3.6.1   Random Forest Ensemble Machine Learning Algorithm

This provides an enhancement more on bagged trees by small non-correlates tree. At the time of splitting of tree, random prediction of sample is selected from the entire set of predictors. That means averaging ofless correlated trees is taken for prediction. An ensemble learning of random tree hyperparameter tuning is performed by selecting random predictors from each division. This learning produces a less bias and less variance model. There are two ensembles: bagging and boosting. Bagging is defined as the process of an individual model training a subset in random in parallel. Boosting is opposite to bagging, which is done as sequential execution individual model and learns from the prior fault. Random forest uses the bagging method and creates a decision tree individually and obtained the result of 0.973, by the following steps:

- sample selection in a random manner
- construction of every sample of decision tree
- prediction on maximum voting of sample.

5.3.6.2   AdaBoost Random Forest Ensemble Learner

The Boosting ensemble is creates a decision tree based on prior mistakes by inculcating on misclassified data weights. This method is executed as follows:

- decision tree training
- weighted error rate of first decision tree
- calculate ensemble decision tree weight

$$\text{Tree weight} = \text{tree learning rate} * \log\left(\frac{1-e}{e}\right) \qquad (5.2)$$

- weight update to the misclassified points
- repeat the process
- model prediction.

From the above steps, the final decision is according to the power of higher weight of the tree.

5.3.6.3   Gradient Boost Random Forest Ensemble Learner

This is also a boosting algorithm,instead of weights ,it takes residual error directly. It generates new prediction through appending of all predicted trees. The steps below are coined to get the prediction:

- training of decision tree
- applying decision tree for prediction
- calculation of residual error of the tree and adding it as new

- repeating the above steps until the maximum number of tree sets is reached
- prediction of the final model.

5.4 RESULTS AND IMPACT

Figure 5.2 furnishes the baseline instance range in the diabetic health dataset. The properties of the anticipated model created and dependent variables are prioritized in terms of their correlation. For a long period of time,it has been considered that diabetes mellitus increases the risk of diabetic retinal disorders such as DR, glaucoma, and DME. Blood sugar, age, red blood cell, IOP, and CDR are further factors to consider. Data selection is the significant process to select the data for an effective processing. Here, feature selections are made through ranking correlations and it depicts the importance of features.

Table 5.2 illustrates an evaluation of 6 machine learning methods without ensemble. At this juncture, novel diabetic dataset is trained with CART, XGBoost, SVM, logistic regression, random forest, and KNN machine learning algorithms. In this case, accuracies are scaling measures to select the fit model. After the performance evaluation, the finest fit is random forest. This fit learner is supported to make a conclusion by the experts.

Table 5.3 provides the performance accuracies of various random forest ensemble classifiers. In order to increase the performance values, this proposed system introduced ensembles in the form of boosting and bagging. When compared to boosting ensembles, bagged random forest ensemble gives better performance measures. General random forest obtains an accuracy of 0.962, and added

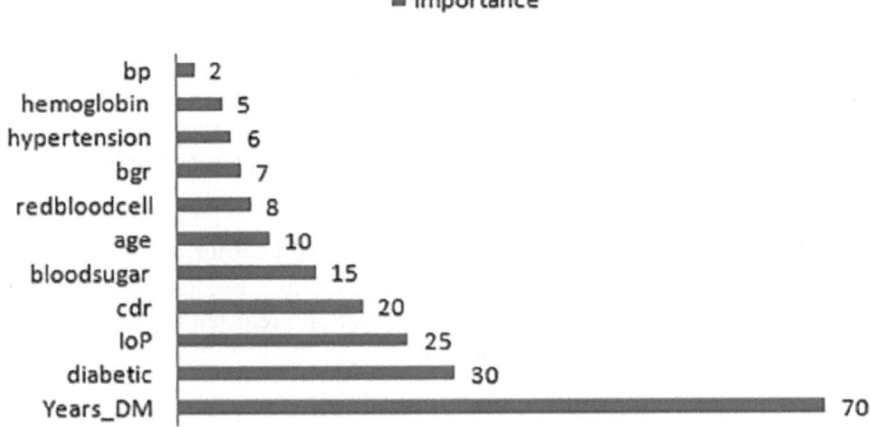

FIGURE 5.2   Feature importance.

TABLE 5.2

Comparison of ML Algorithms Without Ensemble

ML Training Algorithm	Training Accuracy
Random forest	0.962
CART	0.9631
XGBoost	0.9712
Kernel SVM	0.9578
LR (logistic regression)	0.9610
KNN	0.8941

TABLE 5.3

Comparison of Best Fit Random Forest with Various Ensembles

Ensemble Classifier	Accuracy
Ensemble random forest	0.9722
AdaBoost random forest	0.9645
Gradient boost random forest	0.9623

up with bagged mechanism, it produced an improved accuracy of 0.012, which means 0.9722.

Figure 5.3 depicts the performance measures of different machine learners. In this, KNN has the very least performance measures due to its distance values in the data points. So, this is not a best fit model. Again, this dataset is incorporated in to CART, XGBoost, kernel SVM, and logistic regression. These methods obtained moderate level of accuracies, which are not up to the mark for real-time validation. But random forest obtained a good accuracy level to classify and predict the diabetic retinal diseases among diabetic people.But this is not enough to be a best fit model, so it is necessary to adopt any ensemble techniques,optimization, or metaheuristic algorithms. In this proposed system, random forest is ensembled with bagging and boosting algorithms to improve the performances (Figure 5.4). When comparing all the random forest ensembles, bagging produces the best result, but adaptive and gradient boosting obtained good results compared to that of normal random forest.

Table 5.4 summarizes the point of training and real-time test datasets with and without ensemble random forest. There are 25 real-time patient data inputs, and accuracies of 0.967 and 0.99 are obtained with random forest classifier with and without ensemble, respectively (Figure 5.5).

The uncertainty or confusion matrix of the learning dataset is estimated with the confidence level of 0.95, achieved the rate of misclassification of 0.22, and

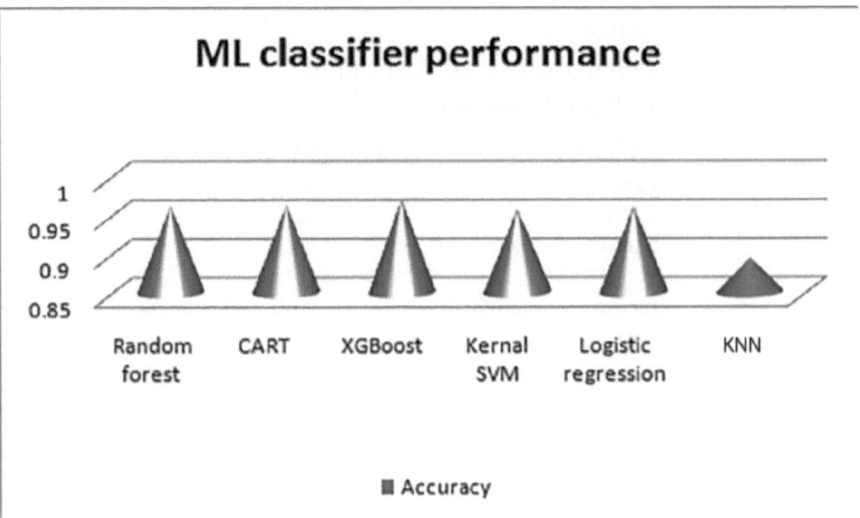

FIGURE 5.3   Performance comparison of various machine learners.

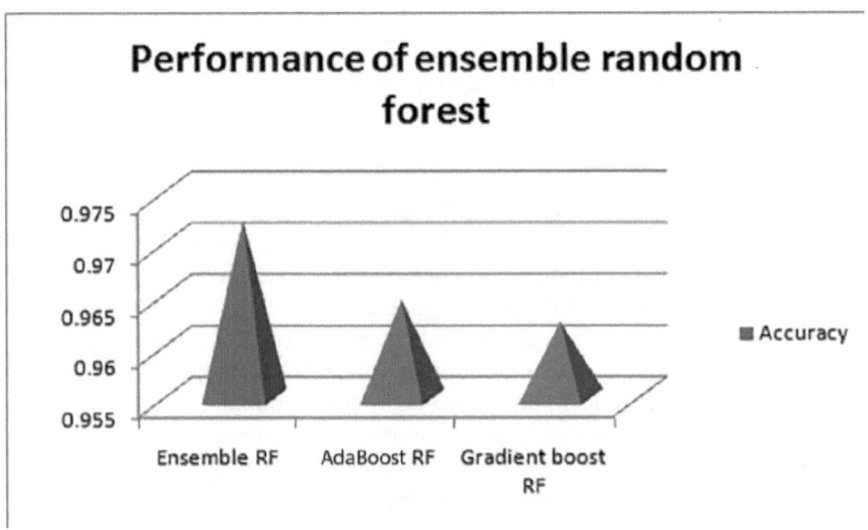

FIGURE 5.4   Performance comparison of various ensemble random forests.

TABLE 5.4

Summarization of Classifiers with Real-Time Test Dataset

Classifier	Training	Testing with Real-Time Data
Without ensemble	0.96	0.9679
With ensemble	0.9722	0.99

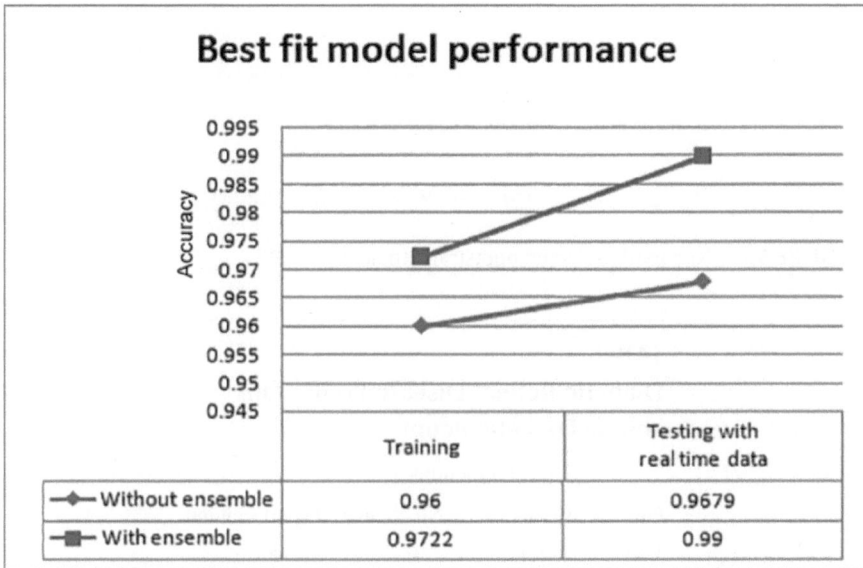

FIGURE 5.5   Performance comparison of best fit model.

with a classification rate on probability of 0.0176 and non-probability of 0.209, from this, an ensemble random forest model is the finest fit model for foretell the diabetic retinal disease found on the performance competence.

Figure 5.6 is explaining the random selected predictor features of bootstrap learner stuck in 0.98 and 0.99, and attained the result accurateness of 0.94, with 0.93 of sensitivity, 0.957 as specificity. Analytical measures explain the measures and their values (Table 5.3). The area under ROC curve sprawled between 0.98 and 1.

Here, 25 real-time patient records are provided to ensemble random forest model to predict the diabetes-based retinal disease. Here, few predictions are listed for discussion (Table 5.5). The Pid21 has very modest probability of retinal disease, because the person has normal blood sugar, normal blood pressure, abnormal IOP, normal CDR, and no DM. But the Pid12 has more risk of retinal disease. The patient needs an instant attention from an expert or

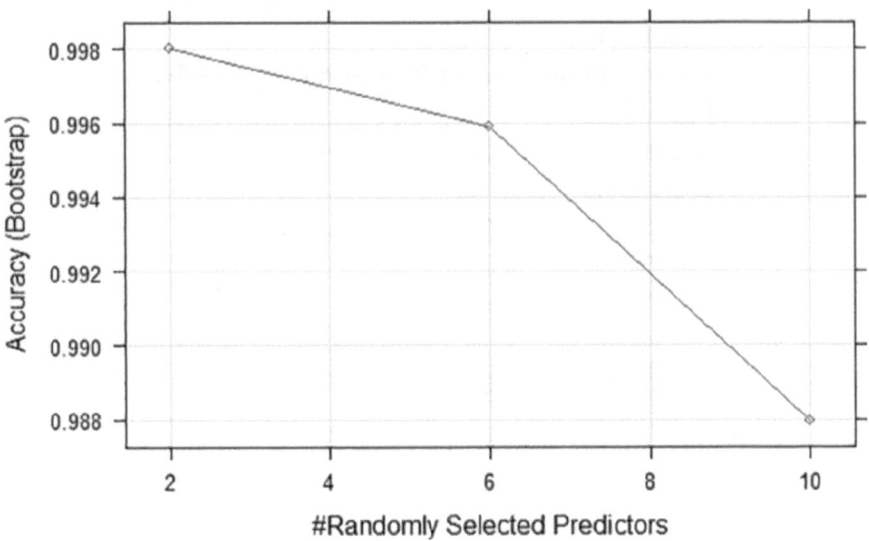

FIGURE 5.6   Randomly selected bootstrap efficiency.

TABLE 5.5
Diabetic Retinal Disease Predictions among Diabetic People

	Predictions		
Pid	Chance	No chance	Expert opinion
12	1.000	0.000	Correct
03	0.874	0.126	Correct
21	0.105	0.985	Moderate

ophthalmologist due to the high-risk correlation factor. Also the Pid03 has a fifteen-year-long history of DM,along with high blood sugar and hypertension, so that patient also requires expert's implications to avoid being visionless in future.

5.5   CONCLUSIONS

A clinically prepared new dataset was used for accurate data analysis. Data are preprocessed, and appropriate features for advanced processing are chosen. Among SVM, logistic, boosted logistic, KNN, and CART, the random forest model had an excellent accuracy of 0.96, while ensemble random forest had an accuracy of 0.97 with an area under curve of 0.9, when compared the result with an AdaBoost random forest and gradient boost random forest.

In propinquity, to prospect will use numerous feature mixture to predict diabetic vision defect in an efficient way. The use of multiple classifiers instead of binary classifiers will advance the experts' assessment swiftly, to find DR, glaucoma, and DME in a trouble-free manner. This model will be handy to the frontline healthcare workers and ophthalmologists, as well as to diabetic patients, to detect diabetic retinal disease too early at the time of general check-up itself. In future, the training dataset will be extended with few more features related to pandemic conditions and adding up eye fungal disease factors. However, this system could act as a telemedicine system in situations such as pandemic and epidemic.

ACKNOWLEDGEMENT

The first author would like to thank Kalasalingam Academy of Research and Education for offering fellowship to carry out the research work.

REFERENCES

1. J. E. Shaw, R. A. Sicree, and P. Z. Zimmet, "Global estimates of the prevalence of diabetes for 2010 and 2030," *Diabetes Research and Clinical Practice*, vol. 87, no. 1, (2010), pp. 4–14.
2. P. Romero-Aroca, R. Navarro-Gil, A. Valls-Mateu, R. Sagarra-Alamo, A. Moreno-Ribas, and N. Soler, "Differences in incidence of diabetic retinopathy between type 1 and 2 diabetes mellitus: a nine-year follow-up study," *British Journal of Ophthalmology*, vol. 101, no. 10, (2017), pp. 1346–1351.
3. G. S. Wander, M. Bansal, and R. R. Kasliwal, "Prediction and early detection of cardiovascular disease in South Asians with diabetes mellitus," *Diabetes & Metabolic Syndrome: Clinical Research & Reviews*, vol. 14, no. 4, (2020), pp. 385–393.
4. J.L. Leasher, R.R.A. Bourne, S.R. Flaxman, J.R. Jonas, J. Keeffe, K. Naidoo, K. Pesudoves, H. Price, R.A. White, T.Y. Wong, S. Resnikoff, and H.R. Taylor, "Global estimates on the number of people blind or visually impaired by diabetic retinopathy: a meta-analysis from 1990 to 2010," *Diabetes Care*, vol. 39, no. 9, (2016), pp. 1643–1649.
5. C-H. Tan, B.M. Kyaw, H. Smith, C.-S. Tan, and L. Car Tudor, "Use of smart phones to detect diabetic retinopathy: scoping review and meta-analysis of diagnostic test accuracy studies. - PubMed – NCBI," *Journal of Medical Internet Research*, vol. 22, no. 5, (2020), pp. e16658.
6. S. Jain and A.O. Salau, "Detection of glaucoma using two dimensional tensor empirical wavelet transform,"*SN Applied Sciences*, vol. 1, no. 11, pp. 1–8. Doi: 10.1007/s42452-019-1467-3.
7. X. Zou, X. Zhao, Y. Yang, and N. Li, "Learning-based visual saliency model for detecting diabetic macular edema in retinal image,"*Computational Intelligence and Neuroscience*, vol. 2016, pp. 1–10.
8. M. R. K. Mookiah, U. Rajendra Acharya, C. M. Lim, A. Petznick, and J. S. Suri, "Data mining technique for automated diagnosis of glaucoma using higher order spectra and wavelet energy features," *Knowledge-Based Systems*, vol. 33, (2012), pp. 73–82.
9. A. Iyer, S. Jeyalatha, and R. Sumbaly, "Diagnosis of diabetes using classification mining techniques," *International Journal of Data Mining & Knowledge Management Process*, vol. 5, no. 1,(2015), pp. 01–14.

10. A. Lee, P. Taylor, J. Kalpathy-Cramer, and A. Tufail, "Machine learning has arrived!" *Ophthalmology*, vol. 124, no. 12, (2017), pp. 1726–1728.
11. M. Caixinha and S. Nunes, "Machine learning techniques in clinical vision sciences," *Current Eye Research*, vol. 42, no. 1, (2017), pp. 1–15.
12. C. Bowd and M. H. Goldbaum, "Machine learning classifiers in glaucoma," *Optometry and Vision Science*, vol. 85, no. 6, (2008), pp. 396–405.
13. H.-Y. Tsao, P.-Y. Chan, and E. C.-Y. Su, "Predicting diabetic retinopathy and identifying interpretable biomedical features using machine learning algorithms," *BMC Bioinformatics*, vol. 19, no. Suppl 9, (2018), pp. 111–121.
14. S. S. Patil and K. Malpe, "Implementation of diabetic retinopathy prediction system using data mining," in *IEEE 3rd International Conference on Computing Methodologies and Communication (ICCMC)*, Erode, Tamilnadu, (2019), Mar. 1206–1210.
15. M. Zhao and Y. Jiang, "Great expectations and challenges of Artificial Intelligence in the screening of diabetic retinopathy," *Eye*, vol. 34, (2019), pp. 418–419.
16. V. Bellemo, G. Lim, T. HyungtaekRim, G.S.W. Tan, C.Y. Cheung, S. Sadda, M. He, A. Tufail, M.L. Lee, W. Hsu, and D.S.W. Ting, "Artificial Intelligence screening for diabetic retinopathy: the real-world emerging application," *Current Diabetes Reports*, vol. 19, no. 9, (2019), pp. 1–12.
17. R. R. Ameena and B. Ashadevi, "Chapter 6- Predictive analysis of diabetic women patients using R," in *Systems Simulation and Modeling for Cloud Computing and Big Data Applications*, J. D. Peter and S. L. Fernandes, Eds. Academic Press, Elsevier, (2020), pp. 99–113.
18. B.A. Mateen, A. L. David, and S. Denaxas, "Electronic health records to predict gestational diabetes risk,"*Trends in Pharmacological Sciences*, vol. 41, no. 5, (2020), pp. 301–304.
19. K. Harimoorthy and M. Thangavelu, "Multi-disease prediction model using improved SVM-radial bias technique in healthcare monitoring system," *Journal of Ambient Intelligence and Humanized Computing*, vol. 12, no. 5,(2020), pp. 3715–3723.
20. M.S. Eswari and S. Karkuzhali, "Survey on segmentation and classification methods for diagnosis of glaucoma," in *Proceedings of 2020 IEEE International Conference on Computer Communication and Informatics (ICCCI)*, Coimbatore, India, (2020), pp.1–6.
21. M.S. Eswari and S. Balamurali," An intelligent machine learning support system for glaucoma prediction among diabetic patients," in *Proceedings of 2021 IEEE International Conference on Advance Computing and Innovative Technologies in Engineering (ICACITE)*, Greater Noida, India (2021), pp. 447–449.
22. K. Shankar, S.K. Lakshmanaprabu, D. Gupta, A. Maseleno, and V.H.C. De Albuquerque, (2020). "Optimal feature-based multi-kernel SVM approach for thyroid disease classification." *The Journal of Supercomputing*, vol. 76, no. 2, (2020), pp. 1128–1143.
23. R. Kumar and M.P. Rajasekaran, "An IoT based patient monitoring system using raspberry Pi," in *Proceedings of 2016 IEEE International Conference on Computing Technologies and Intelligent Data Engineering*, Kovilpatti, India (2016), pp.1–4.
24. L. Jerlin and P. Eswaran, "UCI machine learning repository: Chronic_Kidney_Disease Data Set," (2015), https://archive.ics.uci.edu/ml/datasets/chronic_kidney_disease.

6 Cross-Recurrence Quantification Analysis for Distinguishing Emotions Induced by Indian Classical Music

M. Sushrutha Bharadwaj
Dayananda Sagar College of Engineering
Vellore Institute of Technology

V. G. Sangam
Dayananda Sagar College of Engineering

Shantala Hegde
National Institute of Mental Health and Neurosciences

Anand Prem Rajan
Vellore Institute of Technology

CONTENTS

6.1 Introduction 108
6.2 Music, Emotion and Cognition 108
6.3 Materials and Methods 111
    6.3.1 Signal Acquisition.. 111
    6.3.2 Music Stimulus 112
    6.3.3 Pre-processing of EEG Signals... 112
6.4 Phase Space Plots 112
6.5 Cross-Recurrence Plots.. 114
6.6 Cross-Recurrence Quantification Analysis 114
6.7 Results and Discussion 116
6.8 Conclusions... 118
Acknowledgments... 119
References.. 119

DOI: 10.1201/9781003224068-6

6.1 INTRODUCTION

The human brain is the central part of the nervous system and is extremely complex in carrying out its functions. All the actions and reactions of the body are monitored and regulated by the brain. The brain controls the functions and different actions of the body by acquiring information from the sense organs, analyzing the data received and responding to the information suitably. Brain is a complex nonlinear system where all information is perceived, and nonlinear processing of signals obtained from the brain can be used to assess the behavior of the person. Signals from the brain can be recorded using modalities such as electroencephalography (EEG), computed tomography (CT), positron emission tomography (PET), magnetic resonance imaging (MRI) and functional magnetic resonance imaging (*fMRI*), which give valuable information about the structural and functional changes in the brain due to several activities (D'Elia and Madaffari 2012).

The behavior of a person is based on the changes that are induced in the brain, and these changes are a result of the signals perceived through the sense organs induced by external stimuli. Out of these senses, hearing or the sound has the capability to induce powerful emotions and imaginations in the brain. Among all the random sounds, structured music is known to induce strong emotional experience in the listener (Hegde 2010). Musical experience is accompanied by changes in electrophysiology and autonomic responses of the body. Music has the capability to induce various emotions consistently across subjects. These emotions can be manifested as pleasant and unpleasant emotions or physical sensations such as tears from the eyes and chills down the spine (Bharadwaj et al. 2019). It has been known from ages that listening to music can introduce a soothing or relaxing effect on the brain, and of late, this property has been exploited to be used in treating brain disorders. A critical study of the effects of music on a person can yield valuable information that can be helpful clinically (Biswas et al. 2016; Rao, Hegde, and Nagendra 2015; Hegde et al. 2016). Many signal analysis measures have been used to analyze the information recorded from the brain during music processing using EEG signals. This work discusses the effects of Indian Classical Music (ICM) on happy and sad emotions using a nonlinear analysis technique—the Cross Recurrence Quantification Analysis (CRQA).

6.2 MUSIC, EMOTION AND COGNITION

Music is ubiquitous. Directly or indirectly, every person indulges in some sort of musical behavior. It may be singing or playing an instrument or listening to music during performance of daily chores. Many people are able to criticize and differentiate between good and bad music, expert and amateur performer, specific instrument or a singer and compositions and are able to judge various aspects of music such as timbre, tonal quality etc. (Hegde 2010). In order to perform these musical functions, one's brain has to engage in multiple cognitive processes such as processing musical information, recollecting from memory, attention, making decisions and learning.

Music and emotions share an intimate and inseparable relation. The studies involving music and its effect are not a new initiative altogether since the effect of music has been known from times immemorial (Bhatkhande 1934). It would be ideal to study the effect of music on the listeners rather than examining the characteristics of music itself since every musical form will have its own unique features that have a distinct effect on the listener (Gulati et al. 2020). For example, in the ICM, the artist tries to express the emotional flavor called the 'rasa' associated with the musical phrase or the 'raga' (Jairazbhoy 1971).

The mechanisms of the brain underlying the cognitive processes that are accompanied with music are listening to music, performing music, composing and improvising, memorizing and recalling music, aesthetics, emotion and mental imagery related to music. Broadly, the effects of music on a person can be emotional, cognitive, motor or social. Studies have been conducted to understand the musical behaviors such as listening, performing, remembering, composing, movement and dancing among musically trained and untrained individuals and the relation between music and language and between music and emotion (Cox 2016).

Musical emotions are the various emotional states experienced by the listener due to music. For example, as per Putkinen et al. (2020), affect defines the valence states associated with music such as mood, emotion and preference. Mood is a lower level feature when compared to emotion and features less intense positive or negative valence of music. Mood can sustain within a person from a few hours to several days after listening to music (Putkinen et al. 2020; Juslin and Västfjäll 2008).

Many researches have shown that emotions produced from music are able to alter all the subcomponents of emotions. Qualitative methods have revealed that music listening has been able to induce various emotions in the listener (Balkwill, Thompson, and Matsunaga 2004; Mohan and Thomas 2020; Balkwill and Thompson 1999). Similar to facial expressions, people can identify and distinguish happy and sad emotions through music from other cultures also. Valence and arousal aspect of musical emotion experiences have a significant influence on the autonomic responses such as the heart rate, respiration and temperature (Lang, Bradley, and Cuthbert 1998). Music is known to induce expressive changes and can influence action tendencies among listeners. Emotions induced by listening to music are highly dependent on past events that have occurred in the listener's life (Hallam, Cross, and Thaut 2016). Therefore, since music is known to influence every action of the listener, it would be interesting to study how music induces emotion and its underlying brain correlates.

Since ICM is monophonic or quasi-monophonic, and it has no separate notation for representation, the processing of the ragas of ICM may require a completely different cognitive engagement (Chordia and Rae 2004; Deva and Virmani 1980). The traditional ICM systems are taught orally, and the student is expected to remember the compositions and ragas instead of referring to the notations, which helps in the improvement of verbal memory and auditory skills (Kraus and Chandrasekaran 2010; Chan, Ho, and Cheung 1998). The semantics of ICM are completely different from the western or other systems of music. It may also be noted that usually music systems within ICM are associated with religion and spirituality (van der Merwe and Habron 2015).

The rendering of the ragas are subject to the creativity of the artist as the artist has to perform within a strict complex framework of the raga system and is not supposed to deviate from it. Even though every particular raga is known to induce a specific primary emotion (Bhatkhande 1934; Jairazbhoy 1971), it can also elicit multiple secondary emotions. A particular emotion may be induced or highlighted by a trained musician by making variations in the musical structures such as specific notes, accents, slurs, gamakas or taans varying in tempo, etc. (Hegde et al. 2012). It has to be noted that every single note or swara and its presentation can convey a particular emotion (Mathur et al. 2015).

ICM is primarily based also on the rhythm or the meter to which it is set, which is a very complex system when compared to other systems of music in the world (Clayton 2008). Such differences in the musical systems make these systems perform differently on the cognitive functions of the listener, and it is expected that the involvement of the underlying neural networks would also differ to a great extent while processing different musical systems.

In order to study the effects of music and the brain correlates, the EEG signals have to be analyzed using suitable signal processing techniques. There are different signal processing techniques available for analyzing EEG signals such as the time domain, frequency domain, time–frequency and nonlinear. While conventionally linear methods have been used for the analysis of EEG signals, since the brain is inherently nonlinear, the scope of linear methods is limited for analyzing nonlinear systems. In 1985, Rapp and Bobloyantz published the theory related to nonlinear dynamic systems called the chaos theory, which has been used to analyze EEG signals thereafter (Jacob and Gopakumar 2018; Rey and Guillemant 1997). Nonlinear parameters such as the Lyapunov exponent and correlation dimension, and entropies such as approximate entropy and sample entropy have thus been used to analyze EEG signals (Stam 2005; Acharya et al. 2013).

Biomedical signals are represented as time series, which is a series of data samples indexed in the time order. Time series is a data sequence recorded at successive equally spaced points in time. Analysis of time series comprises extraction of useful statistics and other features of the data (Galka 2000). There are several tools and techniques used to analyze time series. The methods used to analyze time series may be divided into distinct types based on the domain, parameters, linearity or the number of variables used for analysis. Since brain is a nonlinear dynamical system, nonlinear and chaotic analysis tools would be more useful to analyze the time series and the underlying system (Diks 1999).

Many studies before have focused on analyzing brain activity through the EEG frequency bands or spectral analysis of brain activity maps. Since previous knowledge and experience alters brain activity, brain demonstrates a chaotic behavior. Therefore, it would be more advantageous to apply nonlinear methods for time series analysis since the complexities in the EEG signal that cannot be comprehended using linear methods may be discovered (Acharya et al. 2012).

One such technique is the cross-recurrence plot and quantification analysis method, which is an extension of the recurrence plot and quantification analysis technique for analyzing nonlinear systems. There have been very less studies involving cross-recurrence plots and quantification analysis methods, more so with ICM. Therefore, this work aims at analyzing the emotions induced by listening to excerpts of ICM using the cross-recurrence analysis method.

6.3   MATERIALS AND METHODS

The primary objective of this work is to analyze the EEG signals acquired while listening to excerpts from Hindustani classical music using CRQA technique and classify the emotions induced by the music in the brain into happy and sad emotions.

6.3.1   SIGNAL ACQUISITION

The EEG signals were acquired at the Music Cognition Laboratory, National Institute of Mental Health and Neurosciences (NIMHANS), Bengaluru. The signals were acquired from the Neuroscan acquisition system (Neuroscan, Inc., USA) using a 32-channel electrode cap. The 32 electrodes were placed based on the globally accepted standard ten–twenty system of placement of electrodes (Jasper 1958; Klem et al. 1999). The placement and positions of the 32 electrodes are shown in Figure 6.1 obtained from the EEG acquisition system.

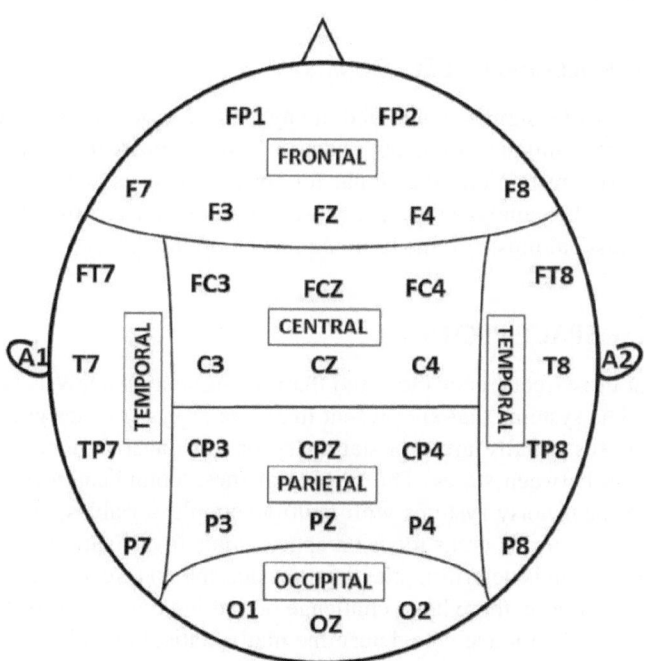

FIGURE 6.1   Placement of electrodes on the scalp for EEG recording.

6.3.2 Music Stimulus

The music stimuli used in this work consisted of excerpts of ragas (structured notes) from Hindustani classical music, a form of classical music in India. These ragas were classified as ragas capable of inducing happy or sad emotions in the person listening to them based on the classical music theory (Bhatkhande 1934). Each raga was presented in a slow (Alap) and a fast (Jod-Jhala) phase on the flute. Three ragas were known to elicit happy emotion, and three were known to elicit sad emotion. The excerpts were played by the same artist. The tonic note for all the excerpts was the same. The subjects were informed to feel the emotion that each phase elicits in them while listening. The participants were made to listen to each raga for an average duration of 129 ± 16 seconds. The signals were recorded during eyes closed rest condition also. A tanpura sound equivalent to the tonic note was also presented as stimulus. The number of participants recruited for the study were 20 (M:F = 10:10). Since it was hypothesized based on the previous literature that music should induce calmness in the brain of the listener, the signals acquired with music condition were compared with eyes closed rest conditions. Behavioral ratings were also collected on a five-point Likert scale a week prior to the recording from the participants. The raga excerpt was presented and followed by a silence for 15 seconds. A white noise was presented in between the musical stimuli to cancel out the memory of the previous raga followed by a silence for 15 seconds before presenting the next raga excerpt.

6.3.3 Pre-processing of EEG Signals

The acquired EEG signals contained many artifacts and noise components. Pre-processing techniques were applied using the Neuroscan 4.5 (Neuroscan, Inc., USA) software to clean the signal for further analysis. The pre-processed EEG signal was then analyzed using CRQA technique to determine the effects of Hindustani classical music on the brain and its emotions.

6.4 PHASE SPACE PLOTS

The basis for cross-recurrence plots and their quantification analysis is the phase space plots. The systems that are present in nature or systems designed for engineering purposes usually are non-stationary or are quasi-stationary, owing to their transition between states. The origins of these complicated processes are complex nonlinear noisy systems with multiple couple variables, and they resist analysis. Many scientific areas focus on approximate investigations of these processes. If the system is deterministic, then it is possible to forecast future states of the systems. Therefore, there lies a challenge of making dynamical systems deterministic. The usual practice is to determine mathematical models that mimic the actual systems and processes so that the problems further can be solved using these models. The initial steps in understanding and analyzing a process is to measure the observations of the state and analyze these data (Kantz and Schreiber 2003; Takens 1981).

The state of a system can be described by its d state variables: $x_1(t)$, $x_2(t)$, $x_3(t) \ldots x_d(t)$.

A vector $x(t)$ is formed at time 't' by the 'd' state variables in a space of dimension 'd', known as the phase space. A phase space trajectory or orbit is formed due to the limited-time progression of the phase space vectors. The dynamics or the attractor of the system is well explained by the time evolution of the trajectory. By mathematically integrating the equation of the system, the state of the system at a particular time can be known. If the mathematical integration part has to be avoided, then a graphical representation and visualization of the trajectory can be helpful in determining the system's state. The shape of the trajectory is an important indicator about whether the system is periodic or chaotic based on its typical phase space signatures.

All possible variables of the states are normally not obtained from a real process, i.e., either all variables are unknown, or a few of them cannot be measured. In most cases, the only available observation is a single observation or variable $u(t)$. Since different components of the system couple together with each other, it implies that vital information about the complete system's dynamics can be obtained using a single component or variable also. This also means that a single time series or a single observation obtained from the system can be used to reconstruct a phase space trajectory that is equivalent to the original phase space trajectory with the same topological structures (Packard et al. 1980; Takens 1981). This forms the first step in nonlinear time series analysis, and the trajectory is known as a phase space plot where, in an m-dimensional Euclidean space denoted by R^m, a one-dimensional time series is mapped. The original unknown attractor's topological properties are expected to be retained by an attractor that is formed by the trajectories that join the different data appoints in the state space. The time delay method is more often applied for reconstructing a trajectory $\dot{x}(t)$, where $\dot{x}_i = (u_i, u_{i+\tau}, \ldots u_{i+(m-1)\tau})^T$, '$m$' is the embedding dimension, and 'τ' is the time delay. For $m \geq 2d + 1$, the topological structures of the original trajectory are definitely preserved, where 'd' is the dimension of the attractor (Lancaster et al. 2018; Takens 1981).

The embedding dimension 'm' and the delay parameter 'τ' have to be chosen carefully. There are many methods that are available to find the minimum embedding dimension that is sufficient for the study. In this work, the false nearest neighbors approach has been used to determine the embedding dimension. This method works on the concept that the number of phase space points that will be projected to the neighborhood of an already existing phase space point, even though they are not actual neighbors, increases by reducing the embedding dimension. These points are known as *false nearest neighbors (FNNs)*. The numbers of FNNs can be used as a function of the embedding dimension to calculate the smallest embedding dimension. The dimension where the FNNs are not found or vanish is considered as the embedding dimension for determining the phase space plots (Henriques et al. 2020).

The phase space plot can be obtained by representing the time series $x(k)$ on the X-axis and the time-delayed series $x(k + t)$ on the Y-axis. The time delay has to be chosen appropriately so as to remove linear dependencies that are observed as a result of random errors and lower precision of measurement between vectors x_i. In the current work, the mutual information method is used to calculate the suitable delay (Fraser and Swinney 1986; Grassberger and Procaccia 1983; Zou et al. 2019). Usually, for experimental data, the dimensionality of the attractor and thus the dimension are not known. An appropriate delay is assumed. Based on previous studies, the delay for an EEG is chosen to be either 8 or 10 (Vlachas et al. 2018; Takens 1981).

6.5 CROSS-RECURRENCE PLOTS

The dynamics of a system that are represented by two time series simultaneously can be compared using the cross-recurrence plots that are very much analogous to recurrence plots. A same phase space simultaneously embeds both time series. The test for closeness of each point in the first trajectory $\vec{x}_i\,(i = 1...N)$ with each point in the second trajectory $\vec{y}_j\,(i = 1...M)$ results in an $N{\times}M$ array $CR_{i,j} = \theta\left(\varepsilon - \left\|\vec{x}_i - \vec{y}_j\right\|\right)$ known as the cross-recurrence plot, where ε is a predefined threshold and $\vec{x}_i$ and $\vec{y}_j$ are phase space trajectories in an m-dimensional phase space. Worthy information is revealed about the inter-relations between systems or time series by mere visual inspection of cross-recurrence plots. Identical phase space behavior in both time series is reflected through lengthy diagonal patterns. Very obviously, the diagonal line will turn completely black if there is no difference between both the systems. Distortion of the diagonal line indicates that there is time delay or compression in the trajectories (Marwan, Thiel, and Nowaczyk 2002). Along the main diagonal, the density of recurrence points goes on increasing as the similarity between systems increases until the mail diagonal becomes black.

6.6 CROSS-RECURRENCE QUANTIFICATION ANALYSIS

Comparing and finding relations between different time series is an important task in multivariate analysis of data that can give vital insights into the underlying systems that are often non-stationary and inherently complex. More often, long data series aren't available from these systems and thus linear methods are not suitable and sufficient for complete analysis of these systems. The recurrence plots were introduced specifically to overcome the shortcomings due to shorter lengths of data and non-stationary behavior of the systems (Casdagli 1997; Eckmann, Kamphorst, and Ruelle 1987), followed by a quantitative analysis of these plots which can be used to detect transitions in complex systems (Trulla et al. 1996; Marwan et al. 2002; Zbilut and Webber 1992). Two processes that are recorded in a single time series or that are simultaneously recorded can be analyzed and compared for their behaviors using cross-recurrence plots, which are an extension of the recurrence plots method (Zbilut, Giuliani, and Webber 1998; Marwan, Thiel, and Nowaczyk 2002; Marwan and Kurths 2002). The phase space trajectories of two different processes may be compared in a single phase space using the cross-recurrence plots.

Next, a few parameters are defined for quantifying cross-recurrence plots and are collectively called the CRQA parameters. Similarity measures in RPs are based on the discontinuous main diagonal. For CRPs, the initial RQA measures are slightly modified and are computed as functions of the distance from the main diagonal for every diagonal that is parallel to the main diagonal (Marwan and Kurths 2002).

Hence, similarity in dynamics can be determined based on a particular delay, and for each parallel diagonal, the distributions of the diagonal line length $P_t(l)$ are analyzed. The diagonal line number is indicated by the index $t \in [-T...T]$ where the main diagonal is represented by $t=0$, the diagonals above the main diagonal by $t>0$, and the diagonals below the main diagonal by $t<0$, which represent positive and negative time delays, respectively (Marwan and Kurths 2002; Marwan, Thiel, and Nowaczyk 2002; Lira-Palma et al. 2018).

The probability that states are identical in both time series with a specified delay is represented by the recurrence rate defined mathematically as represented in (6.1).

$$\text{RR}(t) = \frac{1}{N-t} \sum_{l=1}^{N-t} l\, P_t(l) \tag{6.1}$$

A higher rate recurrence indicates higher density of recurrence points in a diagonal which is usually seen in systems with their trajectories returning to the same phase space regions.

Similarly, analogous to the recurrence parameters, the determinism is defined as in (6.2).

$$\text{DET}(t) = \frac{\sum_{l=l_{\min}}^{N-t} l\, P_t(l)}{\sum_{l=1}^{N-t} l\, P_t(l)} \tag{6.2}$$

Determinism is the proportion of recurrence points forming long diagonal structures of all recurrence points. Longer diagonals are a result of deterministic systems, while none or shorter diagonals are produced by stochastic and fluctuating systems. In case if both the systems are deterministic and have identical phase space behavior, then there is a more number of longer diagonals than the shorter ones.

The average diagonal line length is defined using (6.3).

$$L(t) = \frac{\sum_{l=l_{\min}}^{N-t} l\, P_t(l)}{\sum_{l=l_{\min}}^{N-t} P_t(l)} \tag{6.3}$$

The duration of similarity in the dynamics of time series or systems under study is represented by the average diagonal line length. A higher similarity between systems implies longer diagonals. If the chance of same states occurring in both systems is higher, it is indicated by a higher RR. If similar dynamics in both systems sustain for a longer time, they are indicated by higher values of determinism and average diagonal line length. Determinism and average diagonal line length are highly sensitive to noisy and fluctuating data, but RR measures the similarities in spite of the data being noisy. Since all these measures are statistical measures, their validity will increase if the size of the CRP increases. In the same way, all other measures may also be defined (Marwan, Thiel, and Nowaczyk 2002; Shockley 2005).

The CRP tool is advantageous over other measures as different time delays are used for computation of parameters and lags can be identified and causal links can be proposed. Positive and negative relations can be compared and separated by analyzing another time series with opposite polarity. For convention purposes and to identify these measures, a '+' index and a '−' index are added for positive and negative linkages, respectively, e.g., RR_+ and RR_-. The nonlinear similarities, even though the time series is short, non-stationary and noisy, which are typical for physiological systems, can be effectively determined using this method.

The cross-recurrence plot and the CRQA measures have been implemented using the *Cross Recurrence Plot Toolbox for MATLAB (http://tocsy.pik-potsdam. de/CRPtoolbox)* developed at the Potsdam Institute for Climate Impact Research, Germany (Marwan et al. 2007).

Cross-recurrence plots have been plotted for signals from the same region, different hemispheres and across regions. Similar to recurrence plots, since cross-recurrence plots are visual observations, quantification parameters are derived from cross-recurrence plots. The following parameters were derived from cross-recurrence quantification techniques: recurrence time, determinism, average and longest diagonal line length, entropy, laminarity, trapping time, length of the longest vertical line, recurrence times, trapping time, recurrence period density entropy, clustering coefficient and transitivity. All these are implemented using inbuilt functions in the CRP toolbox. The dimension and delay are chosen as 8 and 6, respectively. Correlations between parameters obtained from different music stimuli are calculated, and the correlation and average values are statistically analyzed using ANOVA measures.

6.7 RESULTS AND DISCUSSION

Various parameters were determined using the CRQA methods from the EEG signals recorded simultaneously from different parts of the brain. Out of all the parameters, determinism, trapping time and second recurrence time showed significant differences for happy and sad musical stimuli. Table 6.1 shows the average determinism values for happy slow, sad slow and eyes closed rest conditions. Table 6.2 shows the average determinism values for happy slow, sad slow and eyes closed rest conditions in frontal, temporal and parietal regions. Figure 6.2 shows the average trapping time for happy fast, sad slow and eyes closed rest conditions, and Table 6.3 shows the average RT2 values for happy fast, sad slow and eyes

TABLE 6.1
Avg. Determinism Values across Left and Right Hemispheres for Happy Slow, Sad Slow and Eyes Closed Rest Conditions

Avg. Determinism	Happy Slow	Sad Slow	Eyes Closed Rest
Left hemisphere	0.5±0.03 (b)	0.61±0.07	0.35±0.02
Right hemisphere	0.42±0.1	0.68±0.04 (c)	0.45±0.03

Correlation coefficient between happy slow and eyes closed rest conditions in the left hemispheres is **0.827**.

TABLE 6.2
Avg. Determinism Values for Happy Slow, Sad Slow and Eyes Closed Rest Conditions in Frontal, Temporal and Parietal Regions

Avg. Determinism	Happy Slow	Sad Slow	Eyes Closed Rest
Frontal region	0.65±0.02	0.62±0.06	0.45±0.01
Temporal region	0.64±0.04 (c)	0.54±0.05	0.47±0.03
Parietal	0.84±0.03 (c)	0.72±0.02	0.65±0.03

Correlation coefficient between happy slow and eyes closed rest conditions is **0.754 in temporal and 0.62 in parietal regions**.

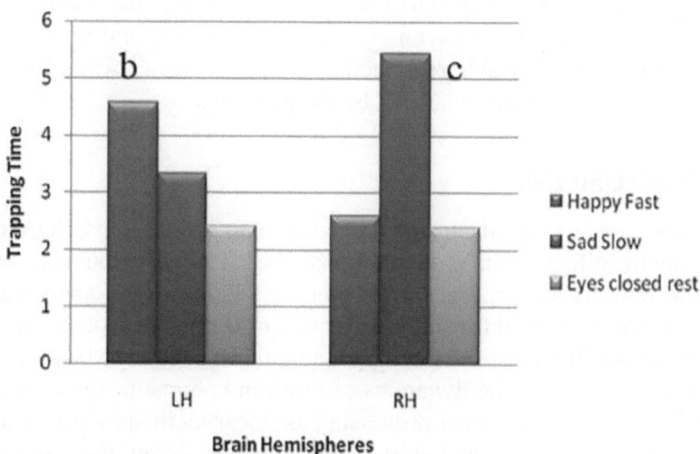

FIGURE 6.2 Trapping time for happy fast, sad slow and eyes closed rest conditions between hemispheres.

TABLE 6.3

Avg. RT2 Values for Happy Fast, Sad Slow and Eyes Closed Rest Conditions

Avg. Second Recurrence Time	Happy Fast	Sad Slow	Eyes Closed Rest
Left frontal & temporal	25±2.6 (b)	58±2	35±1.3
Right frontal & temporal	42±4.3	72±3.5 (a)	38±2.1
Left frontal region	28±2.1 (a)	32±6.3	31±0.7
Right frontal region	34±3.2	49±3.3	38±2.8

closed rest conditions. Clearly, significant differences can be seen between the parameters for happy and sad music stimuli and for eyes closed rest conditions. Statistical significances are denoted in three levels in tables 6.1 to 6.3 where (a) represents α-0.05, (b) represents $\alpha=0.01$ and (c) represents $\alpha=0.001$.

CRQA analyzes the recurrence of same states in the signals induced in different electrodes due to music listening. High correlation was observed between determinism due to happy slow and eyes closed rest conditions in the left hemisphere, indicating that the happy slow music can induce calmness in the listener's brain. Determinism was found to be significantly varying between left and right hemispheres for happy and sad slow music, respectively.

A higher determinism indicates a state of drowsiness or calmness that was induced due to slow tempo music. Trapping time was found to be significantly high for happy fast and sad slow music between electrodes in right and left hemispheres, respectively. Between left frontal and temporal regions, RT2 was found to significantly reduce for happy fast stimulus, indicating more similar states within the region and involvement of temporal region for music processing. A significant increase in RT2 between right frontal and temporal regions for sad slow music shows the probable recollection of sad memories while listening to sad music, which was confirmed orally by the participants.

6.8 CONCLUSIONS

This work focused on a nonlinear signal processing technique, the CRQA method, for analyzing the effects of ICM on emotions. The effect of various raga excerpts from Hindustani classical music on the emotions induced was analyzed. Time series parameters were used for analysis because time domain processing is faster and reduces loss of information. Nonlinear method was found to be more advantageous in understanding the dynamics of brain in contrast to the conventional linear methods during emotional processing as linear methods were found to be poor in differentiating chaos and noise in the complex system. Participants rated ragas with major pitches with tempo as having positive valence and raga excerpts with minor pitches with no tempo as having negative valence. Variation in tempo influenced and enhanced the intensity of the experience of positive emotions, which are evident from the nonlinear parameters. All these differences were observed at 10 s after the onset of stimuli and indicate the time taken by classical music to induce emotions. It also infers that the recording time could be reduced by a great extent while implementing on patients with cognitive disorders. The effects of music remained unchanged and reached saturation after 60 s of music listening. The current study focused on musically untrained subjects. Since various confounding factors affect music processing in musically trained subjects, it would be wiser to analyze them through advanced connectivity studies. Since the dynamics of the brain will vary for abnormal subjects, differences in these parameters can be used to detect various cognitive disorders. These findings may be helpful in further understanding emotional processing via music in trained and untrained individuals. Deeper understanding may help us in studying emotion processing in general, which is a complex brain process. Further studies may examine how music is often used in emotion regulation and induction of mood in participants. Also, since these features are varying specifically with individual electrodes, the effects of music can be analyzed through specific electrodes instead of the whole brain maps, thus reducing the number of electrodes used for recording.

ACKNOWLEDGMENTS

This work was financially supported by the Department of Science and Technology (DST), Government of India, for the research project under the SERC FAST Track Scheme (SR/FT/LS-058/2008 dated 15 July 2009). The principal investigator of the project was Dr. Shantala Hegde who is also one of the authors of this book chapter.

REFERENCES

Acharya, U.R., Molinari, F., Sree, S.V., Chattopadhyay, S., Ng, K.H. and Suri, J.S. 2012. "Automated diagnosis of epileptic EEG using entropies." *Biomedical Signal Processing and Control* 7 (4). Elsevier Ltd: 401–8. Doi: 10.1016/j.bspc.2011.07.007.

Acharya, U.R., Sree, S.V., Swapna, G., Martis, R.J. and Suri, J.S. 2013. "Automated EEG analysis of epilepsy: a review." *Knowledge-Based Systems* 45 (June): 147–65. Doi: 10.1016/j.knosys.2013.02.014.

Balkwill, L.-L. and Thompson, W.F. 1999. "A cross-cultural investigation of the perception of emotion in music: psychophysical and cultural cues." *Music Perception* 17 (1): 43–64. doi:10.2307/40285811.

Balkwill, L.L., Thompson, W.F. and Matsunaga, R.I.E. 2004. "Recognition of emotion in Japanese, Western, and Hindustani music by Japanese listeners 1." *Japanese Psychological Research* 46 (4): 337–49. Doi:10.1111/j.1468-5584.2004.00265.x.

Bhatkhande, V. 1934. "Hindustani Sangeet Paddhati." *Sangeet Karyalaya.*

Biswas, A., Hegde, S., Jhunjhunwala, K. and Pal, P.K. 2016. "Two sides of the same coin: impairment in perception of temporal components of rhythm and cognitive functions in Parkinson's disease." *Basal Ganglia* 6 (1). Elsevier GmbH.: 63–70. doi:10.1016/j.baga.2015.12.001.

Casdagli, M.C. 1997. "Recurrence plots revisited." *Physica D: Nonlinear Phenomena* 108 (1–2): 12–44. doi:10.1016/S0167-2789(97)82003-9.

Chan, A.S., Ho, Y.-C. and Cheung, M.-C. 1998. "Music training improves verbal memory." *Nature* 396 (6707): 128–128. doi:10.1038/24075.

Chordia, P. and Rae, A. 2004. "Understanding emotion in raag: an empirical study of listener responses." In *Computer Music Modeling and Retrieval. Sense of Sounds*, 110–24. Berlin, Heidelberg: Springer Berlin Heidelberg. doi:10.1007/978-3-540–85035-9_7.

Clayton, M. 2008. *Time in Indian Music: Rhythm, Metre, and Form in North Indian Rāg Performance*. Oxford University Press, USA.

Cox, A. 2016. *Music and Embodied Cognition: Listening, Moving, Feeling, and Thinking*. Indiana University Press, USA.

D'Elia, M. and A. Madaffari, eds. 2012. *Medical Imaging: Procedures, Techniques & Applications*. Nova Science Publishers Inc, USA.

Deva, B.C. and Virmani, K.G. 1980. "A study in the psychological response to 'Raga-S.'" *Journal of the Indian Musicological Society; Baroda* 11 (1): 33.

Diks, C. 1999. *Nonlinear Time Series Analysis: Methods and Applications*. Edited by H.A.M. Tong. World Scientific Publishing Co Pte Ltd, Singapore

Eckmann, J.-P., Oliffson Kamphorst, S. and Ruelle, D. 1987. "Recurrence plots of dynamical systems." *Europhysics Letters (EPL)* 4 (9): 973–77. doi:10.1209/0295-5075/4/9/004.

Fraser, A.M. and Swinney, H.L. 1986. "Independent coordinates for strange attractors from mutual information." *Physical Review A* 33 (2): 1134–40. doi:10.1103/PhysRevA.33.1134.

Galka, A. 2000. *No Topics in Nonlinear Time Series Analysis, With Implications for EEG Analysis*. World Scientific Publishing Co Pte Ltd, Singapore

Grassberger, P. and Procaccia, I. 1983. "Characterization of strange attractors." *Physical Review Letters* 50 (5): 346–49. doi:10.1103/PhysRevLett.50.346.

Gulati, A., Joshi, B., Jain, C. and Shukla, J. 2020. "It's not what they play, it's what you hear: understanding perceived vs. induced emotions in Hindustani classical music." In *Companion Publication of the 2020 International Conference on Multimodal Interaction*, 42–46. New York, NY: ACM. doi:10.1145/3395035.3425246.

Hallam, S., Cross, I. and Thaut, M. 2016. *The Oxford Handbook of Music Psychology*. Oxford University Press, London.

Hegde, S. 2010. "Music and emotion." *Journal of ITC Sangeet Research Academy* 24: 16–27.

Hegde, S., Aucouturier, J.-J., Ramanujam, B., and Bigand, E. 2012. "Variations in emotional experience during phases of elaboration of North Indian raga performance." In *12th International Conference on Music Perception and Cognition and the 8th Triennial Conference of the European Society for the Cognitive Sciences of Music*, edited by Cambouropoulus, E., Tsougras, C., Mavromatis, P., and Pastiads, K. 412–13.

Hegde, S., Bharath, R.D. Rao, M.B., Shiva, K. Arimappamagan, A., Sinha, S., Rajeswaran, J., and Satishchandra, P. 2016. "Preservation of cognitive and musical abilities of a musician following surgery for chronic drug-resistant temporal lobe epilepsy: a case report." *Neurocase* 22 (6): 512–17. doi:10.1080/13554794.2016.1198815.

Henriques, T., Ribeiro, M., Teixeira, A., Castro, L., Antunes, L., and Costa-Santos, C. 2020. "Nonlinear methods most applied to heart-rate time series: a review." *Entropy* 22 (3): 309. doi:10.3390/e22030309.

Jacob, J.E. and Gopakumar, K. 2018. "A review of chaotic analysis of EEG in neurological diseases." In *2018 International CET Conference on Control, Communication, and Computing (IC4)*, 181–86. IEEE. doi:10.1109/CETIC4.2018.8530960.

Jairazbhoy, N.A. 1971. "The Rags of North Indian Music," 222. Faber and Faber Limited, Great Britain

Jasper, H.H. 1958. "The ten-twenty electrode system of the international federation." *Electroencephalography and Clinical Neurophysiology* 10: 371–75.

Juslin, P.N., and Västfjäll, D. 2008. "Emotional responses to music: The need to consider underlying mechanisms." *Behavioral and Brain Sciences* 31 (5): 559–75. doi:10.1017/S0140525X08005293.

Kantz, H., and Schreiber, T. 2003. *Nonlinear Time Series Analysis*. Cambridge University Press. doi:10.1017/CBO9780511755798.

Klem, G.H., Lüders, H.O., Jasper, H.H., and Elger, C. 1999. "The ten-twenty electrode system of the international federation. the international federation of clinical neurophysiology." *Electroencephalography and Clinical Neurophysiology. Supplement* 52: 3–6.

Kraus, N., and Chandrasekaran, B. 2010. "Music training for the development of auditory skills." *Nature Reviews Neuroscience* 11 (8): 599–605. doi:10.1038/nrn2882.

Lancaster, G., Iatsenko, D., Pidde, A., Ticcinelli, V. and Stefanovska, A. 2018. "Surrogate data for hypothesis testing of physical systems." *Physics Reports* 748. Elsevier B.V.: 1–60. doi:10.1016/j.physrep.2018.06.001.

Lang, P.J., Bradley, M.M. and Cuthbert, B.N. 1998. "Emotion and motivation: measuring affective perception." *Journal of Clinical Neurophysiology* 15 (5): 397–408. doi:10.1097/00004691-199809000-00004.

Lira-Palma, D., González-Rosales, K., Castillo, R.D., Spencer, R. and Fresno, A. 2018. "Categorical cross-recurrence quantification analysis applied to communicative interaction during ainsworth's strange situation." *Complexity* 2018. doi:10.1155/2018/4547029.

Marwan, N., Carmenromano, M., Thiel, M. and Kurths, J. 2007. "Recurrence plots for the analysis of complex systems." *Physics Reports* 438 (5–6): 237–329. doi:10.1016/j.physrep.2006.11.001.

Marwan, N., Thiel, M. and Nowaczyk, N.R. 2002. "Cross recurrence plot based synchronization of time series." *Nonlinear Processes in Geophysics* 9 (3/4): 325–31. doi:10.5194/npg-9-325-2002.

Marwan, N. and Kurths, J. 2002. "Nonlinear analysis of bivariate data with cross recurrence plots." *Physics Letters, Section A: General, Atomic and Solid State Physics* 302 (5–6): 299–307. doi:10.1016/S0375-9601(02)01170-2.

Marwan, N., Wessel, N., Meyerfeldt, U., Schirdewan, A. and Kurths, J. 2002. "Recurrence-plot-based measures of complexity and their application to heart-rate-variability data." *Physical Review E - Statistical Physics, Plasmas, Fluids, and Related Interdisciplinary Topics* 66 (2). doi:10.1103/PhysRevE.66.026702.

Mathur, A., Vijayakumar, S.H., Chakrabarti, B. and Singh, N.C. 2015. "Emotional responses to hindustani raga music: the role of musical structure." *Frontiers in Psychology* 6 (April). doi:10.3389/fpsyg.2015.00513.

Merwe, L.v.d., and Habron, J. 2015. "A conceptual model of spirituality in music education." *Journal of Research in Music Education* 63 (1): 47–69. doi:10.1177/0022429415575314.

Mohan, A., and Thomas, E. 2020. "Effect of background music and the cultural preference to music on adolescents' task performance." *International Journal of Adolescence and Youth* 25 (1): 562–73. doi:10.1080/02673843.2019.1689368.

Packard, N.H., Crutchfield, J.P., Farmer, J.D. and Shaw, R.S. 1980. "Geometry from a time series." *Physical Review Letters* 45 (9): 712–16. doi:10.1103/PhysRevLett.45.712.

Putkinen, V., Nazari-Farsani, S., Seppälä, K., Karjalainen, T., Sun, L., Karlsson, H.K., Hudson, M., Heikkilä, T.T., Hirvonen, J. and Nummenmaa, L. 2020. "Decoding music-evoked emotions in the auditory and motor cortex." *Cerebral Cortex*, December. doi:10.1093/cercor/bhaa373.

Rao, T.I., Hegde, S., and Nagendra, H.R. 2015. "Music based intervention to target stress and cognitive dysfunction in Type 2 diabetes." In *Proceedings of the Ninth Triennial Conference of the European Society for the Cognitive Sciences of Music*, edited by Ginsborg, S., Lamont, J., Phillips, A., and Bramley, M, 17–22 August 2015, Manchester, UK.

Rey, M. and Guillemant, P. 1997. "Apport Des Mathématiques Non-Linéaires (Théorie Du Chaos) à l'analyse de l'EEG." *Neurophysiologie Clinique/Clinical Neurophysiology* 27 (5): 406–28. doi:10.1016/S0987-7053(97)88807-7.

Shockley, K. 2005. "Cross recurrence quantification of interpersonal postural activity." In *Tutorials in Contemporary Nonlinear Methods for the Behavioral Sciences*, edited by M.A. Riley and G.C. Van Orden, 142–177.

Stam, C.J. 2005. "Nonlinear dynamical analysis of EEG and MEG: Review of an emerging field." *Clinical Neurophysiology* 116 (10): 2266–2301. Doi: 10.1016/j.clinph.2005.06.011.

Sushrutha Bharadwaj, M., Hegde, S., Dutt, D.N. and Rajan, A.P. 2019. "Nonlinear signal processing method detects emotional changes induced by Indian classical music." *International Journal of Engineering and Advanced Technology* 9 (1): 6200–6206. doi:10.35940/ijeat.A1853.109119.

Takens, F. 1981. "Detecting strange attractors in turbulence." 366–81. doi:10.1007/BFb0091924.

Trulla, L.L., Giuliani, A., Zbilut, J.P. and Webber, C.L. 1996. "Recurrence quantification analysis of the logistic equation with transients." *Physics Letters A* 223 (4): 255–60. doi:10.1016/S0375-9601(96)00741-4.

Vlachas, P.R.H., Byeon, W., Wan, Z.Y., Sapsis, T.P. and Koumoutsakos, P. 2018. "Data-driven forecasting of high-dimensional chaotic systems with long-short term memory networks." *ArXiv*.

Zbilut, J.P., Giuliani, A. and Webber, C.L. 1998. "Detecting deterministic signals in exceptionally noisy environments using cross-recurrence quantification." *Physics Letters A* 246 (1–2): 122–28. doi:10.1016/S0375-9601(98)00457-5.

Zbilut, J.P. and Webber, C.L. 1992. "Embeddings and delays as derived from quantification of recurrence plots." *Physics Letters A* 171 (3–4): 199–203. doi:10.1016/0375-9601(92)90426-M.

Zou, Y., Donner, R.V., Marwan, N., Donges, J.F. and Kurths, J. 2019. "Complex network approaches to nonlinear time series analysis." *Physics Reports* 787. Elsevier B.V.: 1–97. doi:10.1016/j.physrep.2018.10.005.

7 Pattern Recognition and Classification of Remotely Sensed Satellite Imagery

Pramit Pandit
Bidhan Chandra Krishi Viswavidyalaya

K. S. Kiran and Bishvajit Bakshi
University of Agricultural Sciences, Bengaluru

CONTENTS

7.1   Introduction 123
7.2   Methodologies 125
    7.2.1   Classification Techniques 125
        7.2.1.1   MLP Neural Network 125
        7.2.1.2   K-SOM Neural Network 126
        7.2.1.3   Maximum Likelihood Classification Algorithm 126
        7.2.1.4   Mahalanobis Distance Classification Algorithm 127
        7.2.1.5   Spectral Correlation Mapper Classification Algorithm ... 127
    7.2.2   Assessment of Classification Accuracy 128
7.3   Empirical Illustrations 129
    7.3.1   Data and Implementation... 129
7.4   Discussion.. 136
7.5   Conclusions.. 136
Acknowledgements.. 137
References.. 137

7.1   INTRODUCTION

Satellite imageries are a rich source for and play a significant role in generating geographical information. These imageries provide quantitative and qualitative information that substantially reduces the complexity of field work and study time. Technological advancements in remote sensing have boosted the availability of a vast amount of remote sensing data with a high spatial resolution (Mboga et al., 2019). As a result, the ability to produce detailed and accurate land

DOI: 10.1201/9781003224068-7

cover maps has improved. Land cover maps can be used for a variety of purposes, including urban planning, demographic modelling, and socio-economic studies. However, the enormous amount of generated data and challenging fabric in developing countries have necessitated the application of efficient pattern recognition and classification algorithms. In response, several classification algorithms have already been developed to accomplish such tasks (Ediriwickrema and Khorram, 1997; Tuia et al., 2011). The most notable ones among these approaches are based on artificial neural networks (ANNs) (Weller et al., 2006; Dam et al., 2008). In this aspect, ANNs are intended as pattern recognition tools to imitate the brain's neural storing and analytical functions. Neural network-based approaches have a unique edge over the traditional statistical classification algorithms because these are non-parametric and hence do not need to assume any prior input data distribution (Liu et al., 2013). In addition, ANNs can model the non-linear relationship between the inputs and the expected outputs with a faster generalisation ability (Islam and Morimoto, 2017).

In several studies, ANNs are applied as competitive classification strategies to statistical classification algorithms such as maximum likelihood classification algorithm (MLCA), Mahalanobis distance classification algorithm (MDCA), spectral correlation mapper classification algorithm, etc. (SCMCA) (Bischof et al., 1992; Foody, 1995; Benediktsson and Sveinsson, 1997; Foody and Arora, 1997). Erbek et al. (2004) made a comparative assessment between ANN and MLCA for land use activities using Landsat TM data. The multilayer perceptron (MLP) neural network has out-yielded all the competing techniques in terms of overall accuracy and kappa value. Yuan et al. (2009) constructed an automated neural network classification system comprising of an unsupervised Kohonen's self-organising map (K-SOM) neural network module and a backpropagation-based MLP neural network module. The SOM network module is trained using two separate algorithms: the conventional SOM and an improved SOM learning technique that includes simulated annealing. The MLP neural network has been proven to perform much better than both the SOM networks in the parlance of categorisation. Tan et al. (2010) utilised the Landsat data to examine the urban expansion as well as to determine the land use/land cover (LU/LC) changes in Penang Island, Malaysia. Their study revealed that among the employed supervised classification techniques, the MLCA yielded better outcomes and offered a much more precise classification. Yang et al. (2011) analysed high-resolution SPOT 5 satellite imagery for the purpose of crop monitoring and assessment. In their study, both MLCA and support vector machine (SVM) have outperformed all other classification algorithms on the basis of kappa analysis. Srivastava et al. (2012) used Landsat satellite imageries to investigate the LU/LC change in the part of Walnut Creek, Iowa. Outcomes emanated from their study revealed the superiority of ANN over the kernel-based SVMs (linear, radial based, sigmoid, and polynomial) and MLCA. The literature review clearly suggests that notwithstanding the ubiquity and immense power of the neural network-based models, their empirical classification performances with respect to various statistical classification algorithms have rather been inconclusive. Hence, an attempt has been made in this chapter to investigate the applicability of neural network classifiers for pattern recognition and classification of remotely sensed satellite imagery in comparison with the traditional statistical classifiers.

The rest of the chapter proceeds in the following way. The details of the image classification methodologies employed in this study are described in Section 7.2. Sections 7.3 and 7.4 present and discuss the empirical findings, respectively. Section 7.5, finally, concludes the chapter.

7.2   METHODOLOGIES

7.2.1   CLASSIFICATION TECHNIQUES

7.2.1.1   MLP Neural Network

In remote sensing, the MLP is perhaps the most commonly implemented type of neural network (Chauhan et al., 2010; Zare et al., 2013). The determination of the number and size of hidden layers in MLP is usually carried out by a series of trial runs, while various heuristics can be employed to aid in the network architecture selection (Heidari et al., 2019; Pham et al., 2019). In principle, the more complicated the task, the larger the required network is in terms of nodes and layers. A weighted connection connects each unit in a layer to every unit in the subsequent layer(s). These units perform some fundamental tasks. A hidden node, for instance, computes the weighted total of all its inputs and feeds it through its transformation function to obtain the output, which is then propagated to the units of the subsequent layer. The input fed to a unit is computed as:

$$\text{net}_h = \sum_{a=1}^{N} w_{ha} O_a \qquad (7.1)$$

where O_a is the magnitude of the output from the unit a in the previous layer and w_{ha} represents the interconnection between unit a and unit h. The activation function then transforms this net input $\left(\text{net}_h\right)$ to generate an output. The most common activation function is the sigmoidal function specified as:

$$O_h = \left(1 / (1 + \exp\left(-\lambda \text{net}_h\right))\right) \qquad (7.2)$$

where λ represents the gain parameter frequently set to 1. Although each network unit performs quite a simple analysis, the network altogether has the potential to address complicated issues. This task is accomplished by assigning weights to the units that allow the network to precisely identify the class membership based on the data fed to it. The magnitude of each weighted connection is decided through an iterative training procedure such as backpropagation (Vamsidhar et al., 2010). The magnitude is set to random starting points at the beginning of training. The training set is subsequently passed through the network. As in the training sample, the expected output is known for each case, the prediction error can be calculated for the network. The obtained error, in the next step, is passed backward through the network with the magnitude of the weights connecting the units adjusted in relation to the error magnitude. The weight adjustment is usually carried out by applying the function:

$$\Delta w_{ha}(f) = -\eta \delta_h O_a + \alpha \Delta w_{ha}(f-1) \tag{7.3}$$

where f is the iteration number, δ_h denotes the computed error, and η and α represent the learning rate and momentum, respectively. Methodologically, the whole set is first subdivided into the training and the testing sets. The training set is used for network training while retaining the test set for validation purposes. Choosing an appropriate point to cease the network's training is a challenging, but crucial task in order to avoid under- and over-fitting of the network. This task is typically accomplished by using a test set. In this way, this reduces the reliance on potentially deceptive training or learning error and provides a guide to the network's generalisation capability.

7.2.1.2   K-SOM Neural Network

The K-SOM neural network inherits all fundamental characteristics of a neural network. Constructed as an unsupervised clustering technique, the K-SOM yields a one- or two-dimensional association map (Kohonen, 1982). The ability of K-SOM networks to analyse complicated multivariate data from natural systems has also been discovered. During the training process, the input layer of the K-SOM network receives multiple inputs and sends these either to an output or to a one- or two-dimensional competitive layer. The term competitive, in this parlance, implies that during the learning process, the algorithm seeks the best match between the input signal and all nodes in the output grid. That is, all neurons in the K-SOM compete for the input signal and only the winning node associates with it. The winning node is the one in the competitive layer whose weight vector is closest to this input vector in terms of the Euclidean distance. It is worth noticing that during the training procedure, only the weight vectors of the winning and its neighbouring nodes get updated. The K-SOM network is distinguished from other competitive networks by its spatial neighbourhood property (Kurdthongmee, 2011). The K-SOM is specifically well suited to the classification tasks in which obtaining class labels for training patterns is either impractical or incredibly expensive. Moreover, it has a functional edge over the supervised classification algorithms in terms of minimal interaction period. Conversely, it offers limited control over the resulting classes (Thomson et al., 1998).

7.2.1.3   Maximum Likelihood Classification Algorithm

Each pixel in the MLCA is assumed to follow a multivariate normal distribution (MND). Let $\mu_1, \mu_2, \ldots, \mu_m$ and $\Sigma_1, \Sigma_2, \ldots, \Sigma_m$ denote the population mean vectors and population variance–covariance matrices for m classes, respectively. If the observation vector X_r at pixel r belongs to class c and is distributed as an MND (μ_c, Σ_c), then

$$P_{rc} = \left[\frac{1}{2\pi}\right]^{p/2} \left|\sum\right|^{-1/2} \exp(X_r - \mu_c)' \sum_c^{-1} (X_r - \mu_c) \qquad (7.4)$$

Given the likelihood of pixel r belonging to class c,

$$\ln P_{rc} = \frac{p}{2} \ln[\frac{1}{2\pi}] - \frac{1}{2} \ln|\Sigma| - \frac{1}{2}(X_r - \mu_c)' \Sigma_c^{-1}(X_r - \mu_c) \qquad (7.5)$$

Now, the MLCA assigns pixel r to class c iff $\ln P_{rc} \geq \ln P_{rq} \; \forall q = 1, 2, \ldots, m$ classes; $q \neq c$. Since μ_c and Σ_c are unknown, the sample estimates are derived using the training set and utilised subsequently. The whole image classification process is performed on a pixel-by-pixel basis (Erbek et al., 2004). Based on the likelihood outlined above, each pixel is allocated to one of the mutually exclusive groups until no pixel is left unclassified.

7.2.1.4   Mahalanobis Distance Classification Algorithm

In most of the multivariate analyses, variables (typically two) are specified in Euclidean space using a coordinate system. However, it is difficult to represent and measure the variables in planar coordinates when the number of variables exceeds two. The Mahalanobis distance (MD) enters the picture at this point. It considers the mean (sometimes also mentioned as the centroid) of the multivariate data as the reference. It computes the distance between the two variables in relation to the centroid. As a result, the bigger the MD, the farther the variable is from the centroid (Zhang et al., 2011; Gallego et al., 2013).

Similar to MLCA, each pixel in this algorithm is also assumed to follow a MND. Following the usual notations, the MD for pixel r and the class c is represented by:

$$MD_{rc} = \sqrt{(X_r - \mu_c)' \Sigma_c^{-1} (X_r - \mu_c)} \qquad (7.6)$$

where μ_c and Σ_c are obtained and utilised subsequently in the similar fashion.

7.2.1.5   Spectral Correlation Mapper Classification Algorithm

The classification methods for feature spectra rely on comparing the spectral image to a reference spectrum. The comparative evaluation is carried out using a similarity criterion. The spectral angle mapper (SAM) is one of the most widely employed classification algorithms as it compares the spectra by suppressing the effects of shade and emphasising the target reflectance properties (Zhang and Li, 2014; Cho et al., 2010). However, the major bottleneck of the SAM is that it is impossible to discriminate between negative and positive correlations as this algorithm takes into account only the absolute value (Petropoulos et al., 2010; Kumar et al., 2015). Moreover, it is indifferent to shading since it quantifies only the vector direction, not the magnitude. To overcome these lacunas, a further modification has been proposed: the SCMCA (de Carvalho and Meneses, 2000). It eliminates these true inconsistencies by norming each vector to the vector mean. It is basically a variant of the Pearsonian correlation coefficient that retains the SAM property of minimising the shading effect yielding in improved accuracy. If X and Y represent the image spectrum and the reference spectrum, respectively, and $\overline{X}, \overline{Y}$ are their respective mean, then the spectrum correlation coefficient is formulated as:

$$R = \frac{\sum (X - \overline{X})(Y - \overline{Y})}{\sum (X - \overline{X})^2 \sum (Y - \overline{Y})^2} \tag{7.7}$$

The SCM ranges between –1 and 1, whereas the range of cos(SAM) is from 0 to 1.

7.2.2 Assessment of Classification Accuracy

The accuracy assessment is an integral feature of the classification exercise. The correctness of image classification in remote sensing applications implies the extent to which the classifications match with the reference spatial data. To further illustrate various measures of classification accuracy, providing the idea of confusion or error matrix is a prerequisite. Table 7.1 provides the tabular representation of a confusion matrix.

In Table 7.1, y_{ij} denotes the number of pixels classified as category 'i' which actually belong to category 'j' in the reference data, N is the size of the test data set, m is the number of classes, and $y_i.$ and $y_{.j}$ represent the marginal totals of the corresponding rows and columns. Clearly, the diagonal and off-diagonal elements represent the correct classification and misclassification, respectively. The individual class accuracies are assessed in terms of user's accuracy, producer's accuracy, and F-score. The user's accuracy for a particular class is defined as the proportion of correctly classified pixels of that class to the total classified pixels of that class. On the other hand, the producer's accuracy is computed for a particular class by dividing the number of correctly classified pixels of that class by the total number of pixels of its reference class.

TABLE 7.1
A Confusion Matrix

Classification Categories	Reference Categories						Classified Totals
	1	**2**	$\cdots$	**j**	$\cdots$	**m**	
1	y_{11}	y_{12}	$\cdots$	y_{1j}	$\cdots$	y_{1m}	$y_1.$
2	y_{11}	y_{12}	$\cdots$	y_{2j}	$\cdots$	y_{2m}	$y_2.$
$\vdots$	$\vdots$	$\vdots$	$\cdots$	$\vdots$	$\cdots$	$\vdots$	$\vdots$
i	y_{i1}	y_{i2}	$\cdots$	y_{ij}	$\cdots$	y_{im}	$y_i.$
$\vdots$	$\vdots$	$\vdots$	$\cdots$	$\vdots$	$\cdots$	$\vdots$	$\vdots$
m	y_{m1}	y_{m2}	$\cdots$	y_{mj}	$\cdots$	y_{mm}	$y_m.$
Reference totals	$y_{.1}$	$y_{.2}$	$\cdots$	$y_{.j}$	$\cdots$	$y_{.m}$	N

$$\text{User's accuracy } (\%) = \frac{y_{ii}}{y_{i.}} \times 100\% \tag{7.8}$$

$$\text{Producer's accuracy } (\%) = \frac{y_{ii}}{y_{.j}} \times 100\% \tag{7.9}$$

where y_{ii} denotes the number of correctly classified pixels; $y_{i.}$ and $y_{.j}$ represent the marginal totals of classification and reference categories, respectively. F-measure, being the harmonic mean of these measures, provides a way to capture both the properties into a single measure.

However, to measure the overall classification accuracy of a classifier, the overall accuracy and kappa coefficient have been considered. Overall accuracy is a simpler and intuitive way to assess the agreement by considering only the diagonal elements of the confusion matrix and the size of the test data set.

$$\text{Overall accuracy } (\%) = \frac{\sum_{i=1}^{m} y_{ii}}{N} \times 100\% \tag{7.10}$$

where the symbols have the same meaning as above. Kappa coefficient, on the other hand, aims to control the chance agreement by including all the marginal distributions of the confusion matrix and is expressed as:

$$\text{Kappa coefficient } (\hat{\kappa}) = \frac{p_0 - p_e}{1 - p_e} \tag{7.11}$$

where $p_0 = \dfrac{\sum_{i=1}^{m} y_{ii}}{N^2}$, $p_e = \dfrac{\sum_{i=j=1}^{m} y_{i.} y_{.j}}{N^2}$ and the symbols have the same meaning as above.

7.3 EMPIRICAL ILLUSTRATIONS

7.3.1 DATA AND IMPLEMENTATION

The study utilises the imagery of Shettikere Hobli, Chikkanayakanahalli Taluk, Tumkur District, Karnataka, obtained on 14 November 2015 through RESOURCESAT-1 (LISS-IV sensor with a ground sampling distance of 5.8 m). Figure 7.1 represents the location map of the area under study. The geometrically corrected imagery is obtained from the Karnataka State Remote Sensing Applications Centre (KSRSAC), Bengaluru. To extract the digital image of the study region, its geographical map has been overlaid on it. The imagery is recorded in three spectral bands, viz. green $(0.52-0.59 \ \mu m)$, red $(0.62-0.68 \ \mu m)$, and

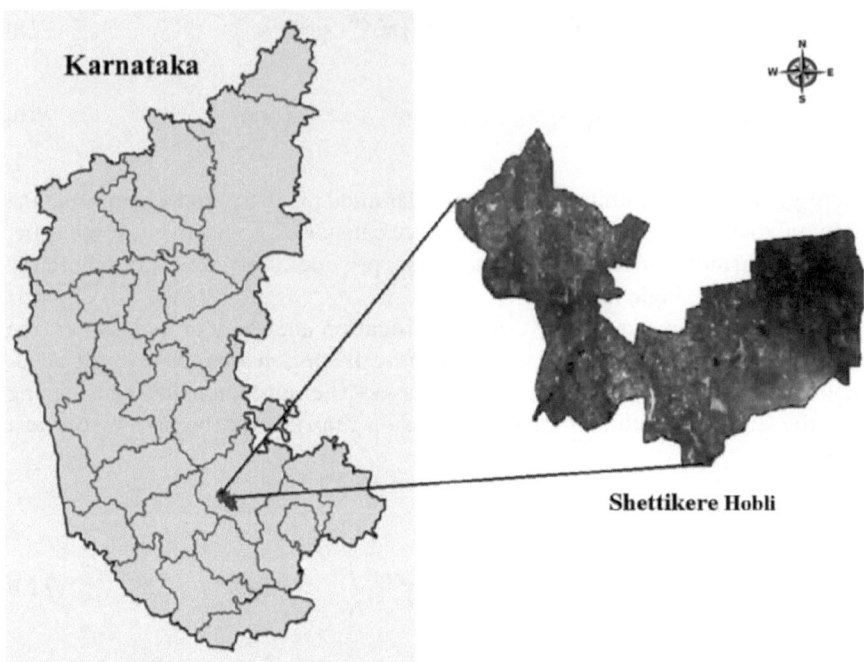

FIGURE 7.1 Location map of the area under study.

TABLE 7.2
Class Descriptions

Class Number	Class Description
1	Horticultural crops
2	Agricultural crops
3	Forest
4	Grazing land
5	Waste land
6	Water bodies
7	Roads
8	Built-ups

near-infrared ($0.77 - 0.86$ μm). ERDAS IMAGINE 9.0 has been used for the purpose of feature extraction. The ground truth data have been obtained by surveying the study area. The topo sheets are then utilised for the selection of training areas for each category. A test data set of 600 pixels has been retained for assessing the classification accuracy. The categories are chosen purposefully and specified carefully to execute digital image categorisation in an effective manner. Table 7.2 provides the LU/LC classification scheme adopted with eight categories.

TABLE 7.3

Assessment of Overall Classification Accuracy

Measure	MLP	K-SOM	MLCA	MDCA	SCMCA
Overall accuracy	95.17	82.50	74.17	66.67	51.67
Kappa coefficient	0.94	0.79	0.69	0.61	0.43

TABLE 7.4

Confusion Matrix for the MLP Neural Network

Classification Categories	Reference Categories								Classified Totals	User's Classified Accuracy (%)
	1	2	3	4	5	6	7	8		
1	93	0	0	0	0	0	2	0	95	97.89
2	7	138	2	1	0	0	0	0	148	93.24
3	0	2	48	0	1	0	1	0	52	92.31
4	0	1	0	32	0	1	0	1	35	91.43
5	0	3	0	1	68	0	0	2	74	91.89
6	0	1	0	0	1	59	1	0	62	95.16
7	0	0	0	0	0	0	61	0	61	100.00
8	0	0	0	1	0	0	0	72	73	98.63
Reference totals	100	145	50	35	70	60	65	75	600	
Producer's accuracy (%)	93.00	95.17	96.00	91.43	97.14	98.33	93.85	96.00		
F-score	0.95	0.94	0.94	0.91	0.94	0.97	0.97	0.97		

For all the algorithms employed in this study, the same training and testing sets have been used for the sake of the direct comparability of the analysis. The assessment of overall classification accuracy, as presented in Table 7.3, clearly reveals the superiority of the neural network classifiers over the traditional statistical classifiers. The classification using the MLP neural network has the highest overall accuracy (95.17%) and kappa coefficient (0.94). It is also evident from the remotely sensed data set that the eight classes have displayed substantially higher separability. The corresponding confusion matrices, as presented in Tables 7.4–7.8, respectively, enable us for a more extensive examination of the five classification techniques (MLP neural network, K-SOM neural network, MLCA, MDCA, and SCMCA). The respective classified maps are provided in Figures 7.2–7.6.

TABLE 7.5
Confusion Matrix for the K-SOM Neural Network

	Reference Categories									User's
Classification									Classified	Accuracy
Categories	1	2	3	4	5	6	7	8	Totals	(%)
1	85	0	0	0	0	0	4	0	89	95.51
2	15	105	4	3	4	0	0	0	131	80.15
3	0	11	43	0	5	0	2	0	61	70.49
4	0	3	3	27	0	4	0	4	41	65.85
5	0	11	0	3	58	0	0	7	79	73.42
6	0	15	0	0	3	56	2	0	76	73.68
7	0	0	0	0	0	0	57	0	57	100.00
8	0	0	0	2	0	0	0	64	66	96.97
Reference totals	100	145	50	35	70	60	65	75	600	
Producer's accuracy (%)	85.00	72.41	86.00	77.14	82.86	93.33	87.69	85.33		
F-score	0.90	0.76	0.77	0.71	0.78	0.82	0.93	0.91		

TABLE 7.6
Confusion Matrix for the MLCA

	Reference Categories									User's
Classification									Classified	Accuracy
Categories	1	2	3	4	5	6	7	8	Totals	(%)
1	80	0	0	0	0	0	10	0	90	88.89
2	20	90	5	5	5	0	0	0	125	72.00
3	0	10	40	0	5	0	5	0	60	66.67
4	0	5	5	20	0	5	0	5	40	50.00
5	0	25	0	5	55	0	0	10	95	57.89
6	0	15	0	0	5	55	5	0	80	68.75
7	0	0	0	0	0	0	45	0	45	100.00
8	0	0	0	5	0	0	0	60	65	92.31
Reference totals	100	145	50	35	70	60	65	75	600	
Producer's accuracy (%)	80.00	62.07	80.00	57.14	78.57	91.67	69.23	80.00		
F-score	0.84	0.67	0.73	0.53	0.67	0.78	0.82	0.86		

TABLE 7.7
Confusion Matrix for the MDCA

Classification Categories	Reference Categories								Classified Totals	User's Accuracy (%)
	1	2	3	4	5	6	7	8		
1	60	0	0	0	0	0	5	0	65	92.31
2	30	80	5	0	0	0	0	0	115	69.56
3	0	5	30	0	0	0	0	0	35	85.71
4	0	5	0	20	0	0	0	0	25	80.00
5	5	25	5	5	55	0	0	10	105	52.38
6	5	30	10	10	15	60	10	20	160	37.50
7	0	0	0	0	0	0	50	0	50	100.00
8	0	0	0	0	0	0	0	45	45	100.00
Reference totals	100	145	50	35	70	60	65	75	600	
Producer's accuracy (%)	60.00	55.17	60.00	57.14	78.57	100.00	76.92	60.00		
F-score	0.73	0.62	0.70	0.67	0.63	0.54	0.87	0.75		

TABLE 7.8
Confusion Matrix for the SCMCA

Classification Categories	Reference Categories								Classified Totals	User's Accuracy (%)
	1	2	3	4	5	6	7	8		
1	65	25	5	5	5	0	15	0	120	54.17
2	5	80	10	20	0	0	0	40	155	51.61
3	10	0	30	0	0	0	0	0	40	75.00
4	5	5	0	5	0	0	5	5	25	20.00
5	5	15	0	0	50	0	10	0	80	62.50
6	0	0	0	0	0	15	0	0	15	100.00
7	10	15	0	0	15	45	35	0	120	29.17
8	0	5	5	5	0	0	0	30	45	66.67
Reference totals	100	145	50	35	70	60	65	75	600	
Producer's accuracy (%)	65.00	55.17	60.00	14.28	71.43	25.00	53.85	40.00		
F-score	0.59	0.53	0.67	0.17	0.67	0.40	0.38	0.50		

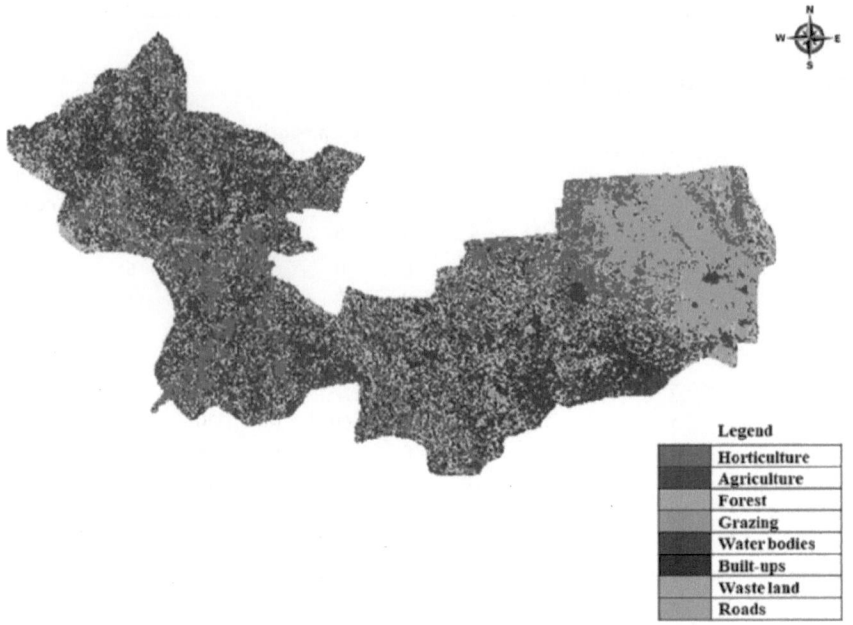

FIGURE 7.2   Classified maps of the MLP neural network.

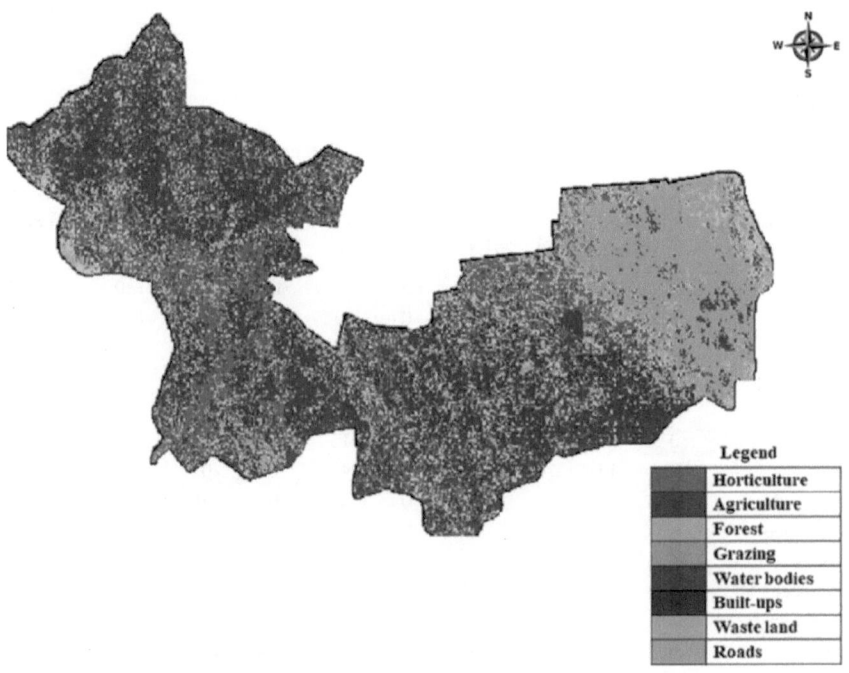

FIGURE 7.3   Classified maps of the K-SOM neural network.

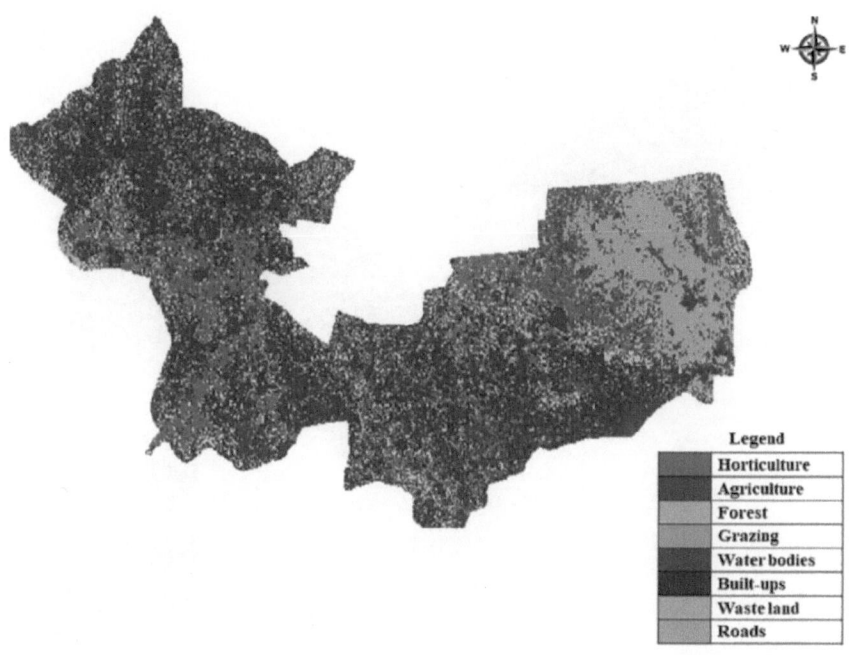

FIGURE 7.4   Classified maps of the MLCA.

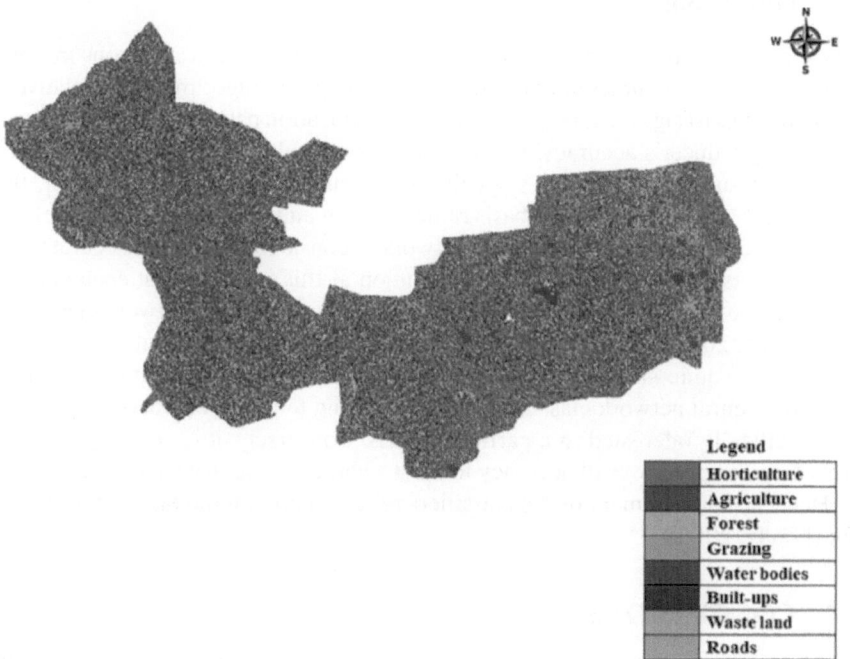

FIGURE 7.5   Classified maps of the MDCA.

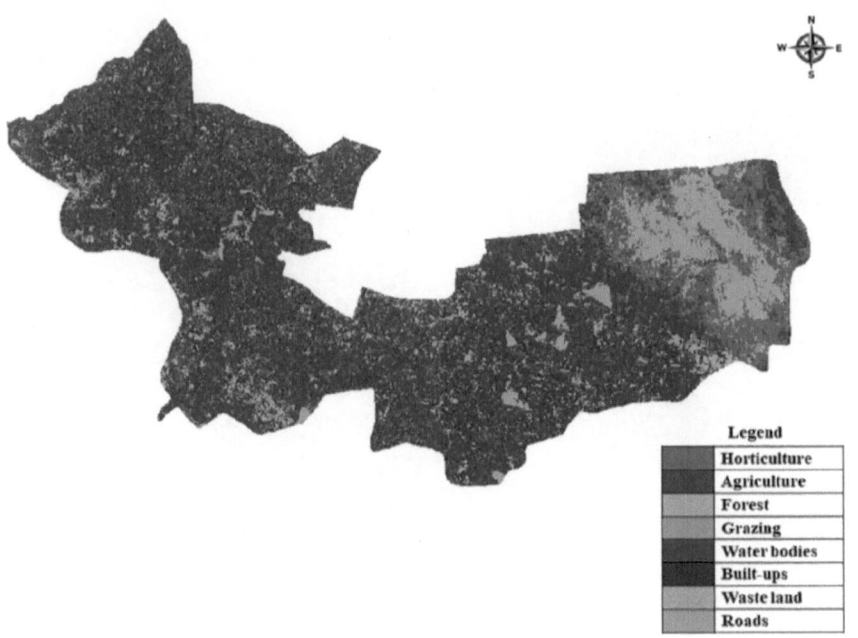

Legend

	Horticulture
	Agriculture
	Forest
	Grazing
	Water bodies
	Built-ups
	Waste land
	Roads

FIGURE 7.6   Classified maps of the SCMCA.

7.4   DISCUSSION

It is apparent from the confusion matrices that even though the variations in over-all classification accuracy among the five classification algorithms are relatively small, there exist significant disparities in class allocation patterns in terms of user's accuracy, producer's accuracy, and F-score. As a result, the classification quality may differ subject to the user's specific requirements. In general, however, the emphasis is given on an overall basis rather than on an individual basis. Therefore, in the current study, the MLP neural network is considered the most accurate and effective classifier. It is noteworthy to mention at this juncture that each class is assigned a reasonably consistent number of cases by the neural network classifiers (Yuan et al., 2009; Petropoulos et al., 2010). Moreover, the individual class accuracies are also quite similar, which is often desirable in classification tasks. These make the neural network classifiers more appealing to the researchers unless they are specifically interested in a particular class. Conversely, it is conceivable for a classifier with poor overall accuracy to most accurately classify a particular class.

Hence, the assessment of the classifiers relies greatly on the task to be carried out (Foody, 2006).

7.5   CONCLUSIONS

ANNs are considered immensely powerful computational tools in divergent domains. Consequently, upsurging research interest has been evident in the implementation of neural networks for the classification and regression problems, where different statistical tools have traditionally been employed. In this context, this study has evaluated the suitability of neural network classifiers for pattern recognition and classification of remotely sensed satellite imagery in comparison with the traditional statistical classifiers. Outcomes emanated from the investigation clearly revealed that the neural network classifiers have clear edges over the conventional statistical classifiers and the classification using the MLP neural network has the highest overall classification accuracy. It is envisaged that the spectrum of neural network applications will expand in the future to fully unearth the potential of neural computing in the domain of remote sensing.

ACKNOWLEDGEMENTS

The authors sincerely acknowledge the Karnataka State Remote Sensing Applications Centre (KSRSAC), Bengaluru, and the University of Agricultural Sciences, Bengaluru, for providing the remote sensing imagery and research facilities, respectively, required for this investigation.

REFERENCES

Benediktsson, J. A., & Sveinsson, J. R. (1997). Feature extraction for multisource data classification with artificial neural networks. *International Journal of Remote Sensing, 18*, 727–740.

Bischof, H., Schneider, W., & Pinz, A. J. (1992). Multispectral classification of landsat images using neural networks. *IEEE Transactions on Geoscience and Remote Sensing, 30*, 482–490.

Chauhan, S., Sharma, M., Arora, M. K., & Gupta, N. K. (2010). Landslide susceptibility zonation through ratings derived from artificial neural network. *International Journal of Applied Earth Observation and Geoinformation, 12*(5), 340–350.

Cho, M. A., Debba, P., Mathieu, R., Naidoo, L., Van Aardt, J. A. N., & Asner, G. P. (2010). Improving discrimination of savanna tree species through a multiple-endmember spectral angle mapper approach: Canopy-level analysis. *IEEE Transactions on Geoscience and Remote Sensing, 48*(11), 4133–4142.

Dam, H. H., Abbass, H. A., Lokan, C., & Yao, X. (2008). Neural-based learning classifier systems. *IEEE Transactions on Knowledge and Data Engineering, 20*, 26–39.

De Carvalho, O. A., & Meneses, P. R. (2000, February). Spectral correlation mapper (SCM): an improvement on the spectral angle mapper (SAM). In *Summaries of the 9th JPL Airborne Earth Science Workshop,* (pp. 1–18). JPL Publication, Pasadena, CA.

Ediriwickrema, J., & Khorram, S. (1997). Hierarchical maximum-likelihood classification for improved accuracies. *IEEE Trans. Geosci. Remote Sens., 35*, 810–816.

Erbek, F. S., Özkan, C., & Taberner, M. (2004). Comparison of maximum likelihood classification method with supervised artificial neural network algorithms for land use activities. *International Journal of Remote Sensing, 25*(9), 1733–1748.

Foody, G. M. (1995). Land-cover classification by an artificial neural-network with ancillary information. *International Journal of Geographical Information Science, 9*, 527–542.

Foody, G. M. (2006). Pattern recognition and classification of remotely sensed images by artificial neural networks. In *Ecological Informatics* (pp. 459–477). Springer, Berlin, Heidelberg.

Foody, G. M., & Arora, M. K. (1997). An evaluation of some factors affecting the accuracy of classification by an artificial neural network. *International Journal of Remote Sensing, 18*, 799–810.

Gallego, G., Cuevas, C., Mohedano, R., & Garcia, N. (2013). On the Mahalanobis distance classification criterion for multidimensional normal distributions. *IEEE Transactions on Signal Processing, 61*(17), 4387–4396.

Heidari, A. A., Faris, H., Aljarah, I., & Mirjalili, S. (2019). An efficient hybrid multilayer perceptron neural network with grasshopper optimization. *Soft Computing, 23*(17), 7941–7958.

Islam, M. P., & Morimoto, T. (2017). Non-linear autoregressive neural network approach for inside air temperature prediction of a pillar cooler. *International Journal of Green Energy, 14*(2), 141–149.

Kohonen, T. (1982). Self-organizing formation of topologically correct feature maps. *Biological Cybernetics, 43*, 56–69.

Kumar, P., Gupta, D. K., Mishra, V. N., & Prasad, R. (2015). Comparison of support vector machine, artificial neural network, and spectral angle mapper algorithms for crop classification using LISS IV data. *International Journal of Remote Sensing, 36*(6), 1604–1617.

Kurdthongmee, W. (2011). Utilization of a rational-based representation to improve the image quality of a hardware-based K-SOM quantizer. *Journal of Real-Time Image Processing, 6*(3), 199–211.

Liu, M., Wang, M., Wang, J., & Li, D. (2013). Comparison of random forest, support vector machine and back propagation neural network for electronic tongue data classification: Application to the recognition of orange beverage and Chinese vinegar. *Sensors and Actuators B: Chemical, 177*, 970–980.

Mboga, N., Georganos, S., Grippa, T., Lennert, M., Vanhuysse, S., & Wolff, E. (2019). Fully convolutional networks and geographic object-based image analysis for the classification of VHR imagery. *Remote Sensing, 11*(5), 597–613.

Petropoulos, G. P., Vadrevu, K. P., Xanthopoulos, G., Karantounias, G., & Scholze, M. (2010). A comparison of spectral angle mapper and artificial neural network classifiers combined with Landsat TM imagery analysis for obtaining burnt area mapping. *Sensors, 10*(3), 1967–1985.

Pham, B. T., Nguyen, M. D., Bui, K. T. T., Prakash, I., Chapi, K., & Bui, D. T. (2019). A novel artificial intelligence approach based on Multi-layer Perceptron Neural Network and Biogeography-based Optimization for predicting coefficient of consolidation of soil. *Catena, 173*, 302–311.

Srivastava, P. K., Han, D., Rico-Ramirez, M. A., Bray, M., & Islam, T. (2012). Selection of classification techniques for land use/land cover change investigation. *Advances in Space Research, 50*(9), 1250–1265.

Tan, K. C., San Lim, H., MatJafri, M. Z., & Abdullah, K. (2010). Landsat data to evaluate urban expansion and determine land use/land cover changes in Penang Island, Malaysia. *Environmental Earth Sciences, 60*(7), 1509–1521.

Thomson, A. G., Fuller, R. M., & Eastwoods, J. A. (1998). Supervised versus unsupervised methods for classification of coasts and river corridors from airborne remote sensing. *International Journal of Remote Sensing, 18*, 3423–3431.

Tuia, D., Volpi, M., Copa, L., Kanevski, M., & Munoz-Mari, J. (2011). A survey of active learning algorithms for supervised remote sensing image classification. *IEEE Journal of Selected Topics in Signal Processing, 5*(3), 606–617.

Vamsidhar, E., Varma, K. V. S. R. P., Rao, P. S., & Satapati, R. (2010). Prediction of rainfall using backpropagation neural network model. *International Journal on Computer Science and Engineering, 2*(4), 1119–1121.

Weller, A. F., Harris, A. J., & Ware, J. A. (2006). Artificial neural networks as potential classification tools for dinoflagellate cyst images: A case using the self-organizing map clustering algorithm. *Review of Palaeobotany and Palynology, 141,* 287–302.

Yang, C., Everitt, J. H., & Murden, D. (2011). Evaluating high resolution SPOT 5 satellite imagery for crop identification. *Computers and Electronics in Agriculture, 75*(2), 347–354.

Yuan, H., Van Der Wiele, C. F., & Khorram, S. (2009). An automated artificial neural network system for land use/land cover classification from Landsat TM imagery. *Remote Sensing, 1*(3), 243–265.

Zare, M., Pourghasemi, H. R., Vafakhah, M., & Pradhan, B. (2013). Landslide susceptibility mapping at Vaz Watershed (Iran) using an artificial neural network model: a comparison between multilayer perceptron (MLP) and radial basic function (RBF) algorithms. *Arabian Journal of Geosciences, 6*(8), 2873–2888.

Zhang, X., & Li, P. (2014). Lithological mapping from hyperspectral data by improved use of spectral angle mapper. *International Journal of Applied Earth Observation and Geoinformation, 31,* 95–109.

Zhang, Y., Huang, D., Ji, M., & Xie, F. (2011). Image segmentation using PSO and PCM with Mahalanobis distance. *Expert Systems with Applications, 38*(7), 9036–9040.

8 Viability of Information and Correspondence Innovation for the Improvement of Communication Abilities in the Healthcare Industry

Pinki Paul and Balgopal Singh
Banasthali Vidyapith

CONTENTS

8.1 Introduction 142
    8.1.1 Effective Ways to Improve Communication Skills 143
    8.1.2 Technologies of IT in Developing Communication Skills 144
    8.1.3 Need for Communication in Healthcare 145
8.2 Literature Review 146
8.3 Research Design 147
    8.3.1 Problem Statement 147
    8.3.2 Objectives 148
    8.3.3 Importance of This Study 148
    8.3.4 Research Questions 148
8.4 Research Methodology 148
    8.4.1 Survey Approach 148
    8.4.2 Populations and Samples 149
    8.4.3 Data Collection Methods 149
    8.4.4 Tools for Data Analysis 149
8.5 Results 149
8.6 Findings 150

DOI: 10.1201/9781003224068-8

8.7   Limitations.. 151
8.8   Conclusions.. 151
References.. 152

8.1   INTRODUCTION

In this digital time, individuals can, without much of a stretch, look for, access, learn from, and speak with others within a short range of time. This makes training available, accessible, and open to all. Well-being training makes mindfulness among people generally about transmittable infections, well-being status, anticipation measures, and different current indicative and helpful methods. This allows the individuals to pick the best medical clinics and specialists to approach for therapy and to have their life soundly. The United Nations Educational, Scientific and Cultural Organization (UNESCO, 2004) emphasizes that *information and communications technology* (ICT) has evolved from being a technology of communication and information alone, but it is a curriculum creation and delivery system for educators and learners. The last few decades showed a revolutionary change in information technology (IT) in every field of the modern world. This change comes from adapting new technologies by society in teaching, learning, and personality development. Many institutions utilize information and communications technologies in personality development, transforming traditional methods into technology-enabled service for humankind [1]. To encourage implementation and adaptation, the government should make new policies and strategies to raise awareness among ordinary people. Earlier systems are based on conventional human involvement with less technological interference in skill development and training, but today, we can achieve our target with less human resources intervention and greater technical involvement [2]. This will reduce the burden on training centers and increase the efficiency and impact on the targeted area.

As we know, communication skill is an essential aspect in family, society, school, university, healthcare centers, and place of work, i.e., organization everywhere. It is a way of conveying a message to the person in front to teach him effectively what we are trying to say. It is also a process of how well we understand the word said by others. It is an individualized characteristic, but it can be improved by expressing an individual message. Figure 8.1 depicts that several ways of communication are in use in communication and information exchange [3]. Among all these, the verbal means of communication is best, including its different forms such as speech, group meeting, debate, or group discussion under the oral communication category [4]. Other types of communication include electronic mail and online databases. In today's world, IT is one of the most significant technological interventions for humankind [5–7]. Knowledge and acquisition of new skills are doubling and tripling in a short period [8]. Software is now in use as a tool for communication processes [9]. A considerable amount of information is available on any topic over the Internet, making it very simple to learn a new language to be used in communication and practice. Being an entrepreneur,

FIGURE 8.1   Types of communication.

verbal and non-verbal communication should be excellent as the employees follow their upper-level order [10]. If one is lagging in any skills, it may affect the
organization's smooth running, productivity, task completion time, management,
and other aspects that may hamper the organization's name, fame, and reputation
directly or indirectly. Management-level and human resources communications
are crucial to a productive and enjoyable work environment in healthcare. The
art of communication is not a birth property, but it can be learned, improved, and
modernized with time and experience [11].

8.1.1   EFFECTIVE WAYS TO IMPROVE COMMUNICATION SKILLS

Communication does not only mean written and verbal communication, but
also lots of information from non-verbal communication and physical cues
such as facial expression and body language [12]. Figure 8.2 depicts that one
can learn from observation, and one has to pay attention to the hand–eye movement, body movement, facial expression during a presentation, and others to

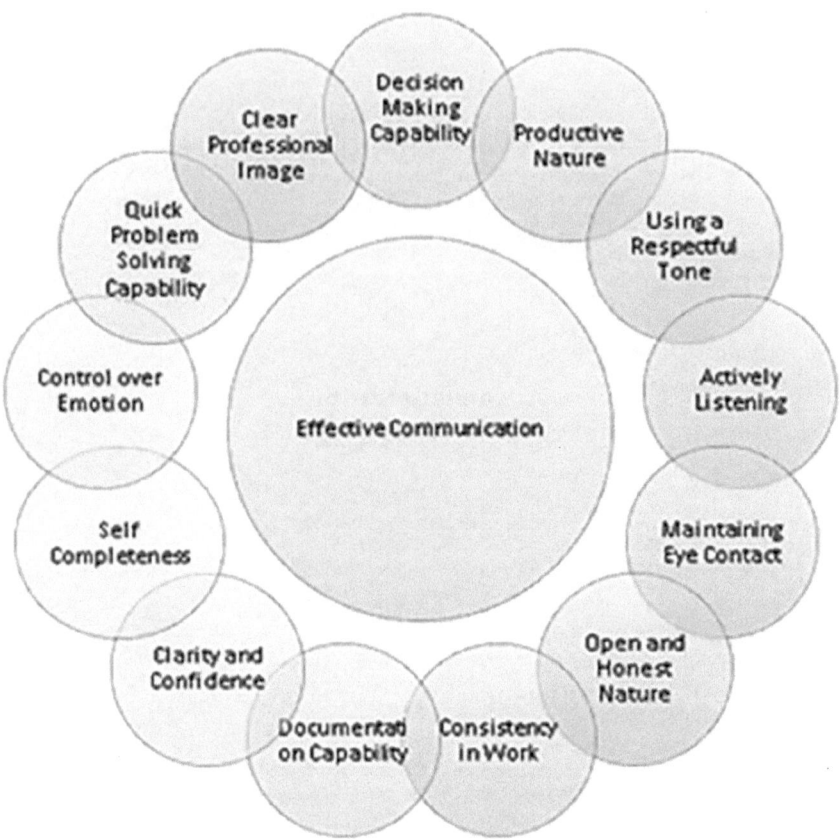

FIGURE 8.2   Factors affecting effective communication.

be an excellent communicator. Many things are to be avoided when attending a conference or giving a presentation in front of others. This is also important to make good eye contact while speaking. To create a good attention towards the listeners always the eye contact with full of confident and happiness. Make good eye contact while speaking, making open, conversant, and enjoyable to listeners.

8.1.2   Technologies of IT in Developing Communication Skills

Many soft skill development professionals suggested that family relationship courses help achieve excellent interpersonal skills. Figure 8.3 depicts the efforts to assist students in gaining excellent communication skills. Different methods and processes for soft skill acquisition are crucial in developing interpersonal communication and behavioral changes in communication skills [13].

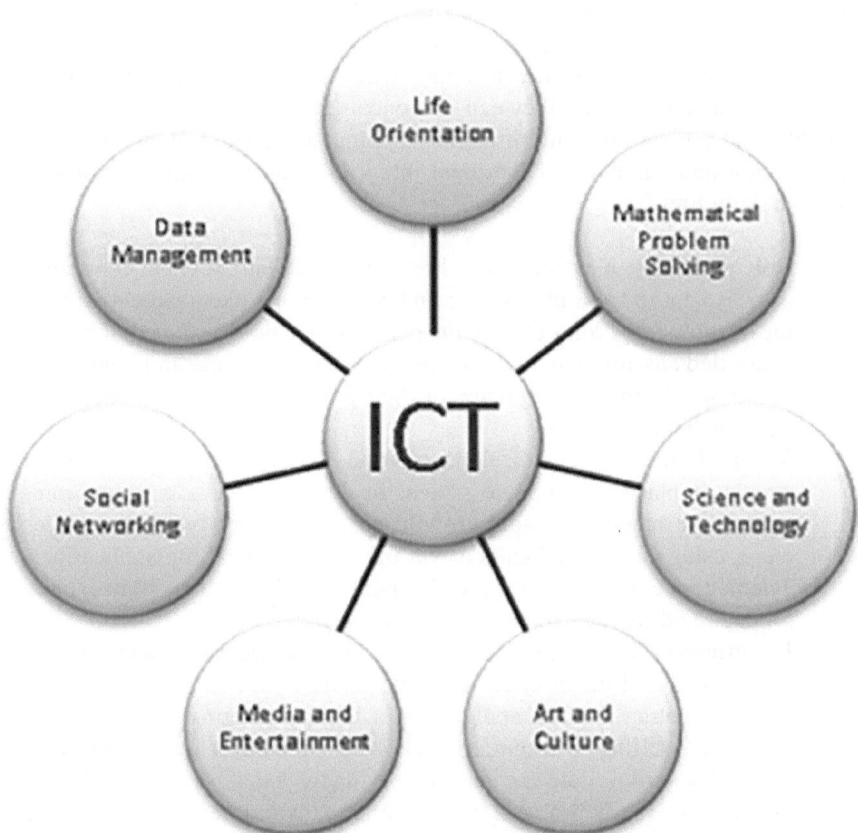

FIGURE 8.3   Information and communications technologies around our daily life.

8.1.3   NEED FOR COMMUNICATION IN HEALTHCARE

Figure 8.3 represents the information and communications technologies around our daily life. A good relation is an essential component for an industry like healthcare. It is helpful for developing better skills and interpersonal communication. This is an efficient driver for booming employees' performance and outcome. Establishing a significant and good conversation relationship among health professionals and patients makes it more comfortable for the clinician to diagnose and treat appropriately. The communication gap between doctor and patient has been seen as a typical one, specifically in semi-urban and rural areas, and adversely affects the treatment procedure [14]. Healthcare professionals learn how to communicate with patients from their senior fellows or under whom they are working as a trainee.

In some cases, it has been seen that some professionals are very good at patient preparation skills and professionalism, but they lack communication skills and interpersonal skills. It needs to be given more emphasis on improvement. New strategies should be implemented in this field to enhance and ensure excellent communication skills. Generally, competent professors and academicians are role models for the students in their academic careers to watch and learn from their respective teachers because they can influence students. Meanwhile, the learners should avoid eye-rolling, nail biting, toe-tapping, and lack of attentive listening. Observation and watching of the ideal interview may motivate to mimic such skills in daily life.

- **Role play**: Role play is a clinical practice where an excellent communicator is kept in front of a trainer and he has to act like a patient, and on the next time, the roles of both the persons have been reversed. It gives a controlled environment to the students to practice in safe and secure surroundings. It provides a review of their performance by the employees with his/her colleagues.
- **Telemedicine**: It is a medical information exchange between two distant people or places to improve patient health status via electronic communication modes. Its rapid growth includes various types of services either using audiovisual calls or via Internet-based calling applications. Telemedicine is also a part of the hospital setting to deliver clinical care, product, and services to patients and society.
- **E-commerce**: E-commerce or electronic commerce was first introduced in the middle of the 20th century, but it gets a pace at the beginning of the 21st century. It is an application of IT in marketing, or we can say buying and selling. The communication between buyer and seller is done through electronic data exchange removing the barriers of physical communication [15]. Nowadays, one can find many online stores where one will find a product of their choice without going to the shop. Moreover, one can get a customized creation of their choice.
- **Enabling communication skills among disabled**: Communication disability may arise due to inappropriate voice, language difficulty, cognition, memory and learning, reading, writing, understanding, and listening or social skills. These people suffer from many problems as society thinks they are stupid, which is a significant communication barrier. Every individual has the right to speak and express, so how can we stop them from having their birthright. We can communicate with this population by paying attention through the person who knows him very well and what he is trying to say. Other ways of communication through expression and gesture can be used, or we can use modern technologies to improve communication [16–18].

8.2 LITERATURE REVIEW

There is a rapid change in skill acquisition in the past few years by introducing new and advanced technologies in this field. In this shift, IT plays crucial roles, such as the World Wide Web (WWW) from where you can get lots of information. Many institutions and specialized education centers take these technologies in skill development, changing the traditional methods into electronic format [19]. To fasten and improve skill acquisition needs increased awareness with flexible and creative methodologies. Skill development and essential knowledge when applying the best fit strategy enhance the student's learning. Initially, new technology in any society faces reluctance from people, and after some time, it becomes part of the system.

It is seen that there is less use of WWW services in capacity development and skill training, mostly in undeveloped and developing countries. In general, we find that IT and WWW services are taken in other means, but not in skill development and training [20, 21, 22]. Current trends are changing day by day as the awareness of all these things is spreading. This is a good sign for the future as this will reduce the barriers of learning to some extent, such as traveling, availability of resources and time. The later outcome will give a better result in terms of skill acquisition and efficacy. This requires only one-time training and can be used by the learner several times till he/she learns or improves his/her performance [23].

Further research and development in creative strategies across initial to a final level must incorporate perceived barriers to making IT a tool for skill development and training [24]. Unwilling to learn new strategies and reluctance to shift from the conventional approach are the major problems in IT-based skill acquisition. The primary aim of this chapter is to enhance learning to improve communication skills in children and adults.

After an extensive literature review for this article, it was found that IT has been playing an essential role in every field of daily living in recent years. It uses start from the time we wake up and continues up to the time to sleep from mobile phone to turn on and turn off of different electronic devices like a light bulb, fan, air-condition, music system, etc. Its use and utility are increasing day by day as it has enormous potential to solve complex problems with ease and with minimal hardware requirements by reducing the human resources requirement.

Again, its application can be seen in banking, healthcare, defense, agriculture, education, communication, and other fields. Its extensive use in every area makes it the right candidate for skill development also. Its potential benefit enhances a person's quality of living and communication, too. Information and communications technologies can become an essential tool to improving interpersonal communication skills in an average person and a mildly disabled population [25].

8.3   RESEARCH DESIGN

8.3.1   PROBLEM STATEMENT

In today's scenario, communication is one of the most critical aspects to fulfill one's feelings. Communication can be used in different ways, such as verbal and communication through ICT. There are many challenges involved with communication, and hence, it is an essential part of this study to evaluate the most effective way of communication.

8.3.2 OBJECTIVES

- To study the different ways of communication and its significance.
- To evaluate the effectiveness of ICT in communication.
- To measure the most effective communication skills.

8.3.3 IMPORTANCE OF THIS STUDY

a. Information and correspondence developments can accept an essential piece of improving clinical consideration for individuals and organizations by giving new and more successful strategies for getting to, passing on, and taking care of information [26].

b. ICTs can help interface the information secludes that have ascended in the prosperity territory in making and new present-day countries—between prosperity specialists and the organizations they serve and between the creators of prosperity research and the experts who need it [27]. Through the headway of databases and various applications, ICTs also improve prosperity system efficiencies and hinder clinical bumbles.

c. A specialist in a faraway nation crisis facility is from the outset unsuitable to decide a patient to have an eccentric bunch of signs. In any case, using his MEDLINE search planning and the center's Internet affiliation, he can break down and adequately treat the patient for a tropical disease the patient got while traveling abroad.

d. A specialist, who sends CT channels and other clinical pictures by email to his association of individual contacts worldwide to help diagnose and treat inauspicious infants, assesses that teleconsultations have empowered him to save different lives an earlier year.

8.3.4 RESEARCH QUESTIONS

ICT has a significant impact on communication; also, ICT has certain limitations. This chapter responds to the following questions: What is the importance of ICT in healthcare? What are the essentials for determining the effectiveness of ICT in the healthcare industry?

8.4 RESEARCH METHODOLOGY

8.4.1 SURVEY APPROACH

An examination of the social exploration approaches proposes that a survey is a helpful instrument to gather essential information, as well as meetings with the respondents. In this study, survey method is used for collecting information from healthcare professionals.

8.4.2 Populations and Samples

The universe or population is the entire group of items in which the researcher is interested and wishes to plan to generalize. The population of interest in this research project consisted of all the 'students' of Political Science Department of the Gomal University, Khyber Pakhtunkhwa, Pakistan, while from this population, a sample of 50 students was selected by simple random technique.

A sample of 100 medical professionals was chosen by convenience sampling method from the Northeast Region of India.

8.4.3 Data Collection Methods

Primary data: They are collected for the first time for this particular research. They are defined as raw data. The data in this study are collected through observations, personal interviews, and telephonic interviews.

Secondary data: These data are collected through books, ICT journal articles, websites, healthcare reports, etc.

8.4.4 Tools for Data Analysis

The data presentation and analysis is done through pie charts, bar charts, frequency tables, and percentage.

8.5 RESULTS

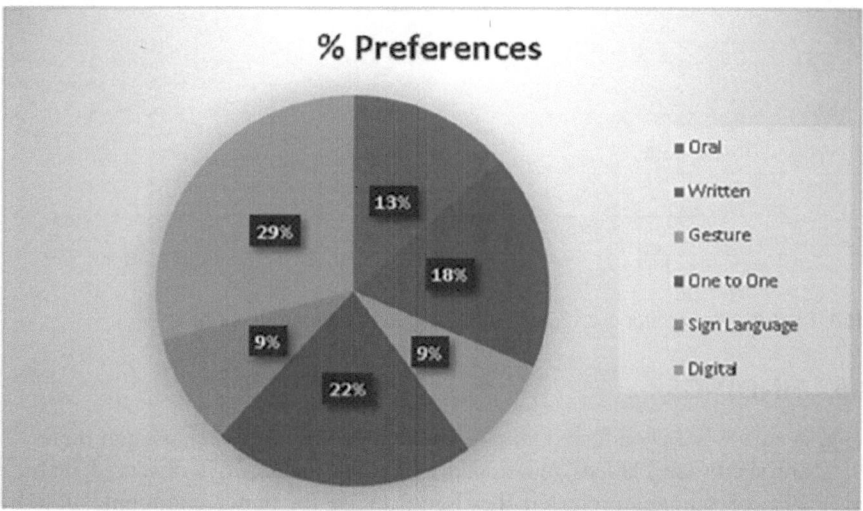

FIGURE 8.4   Preference for communication mode.

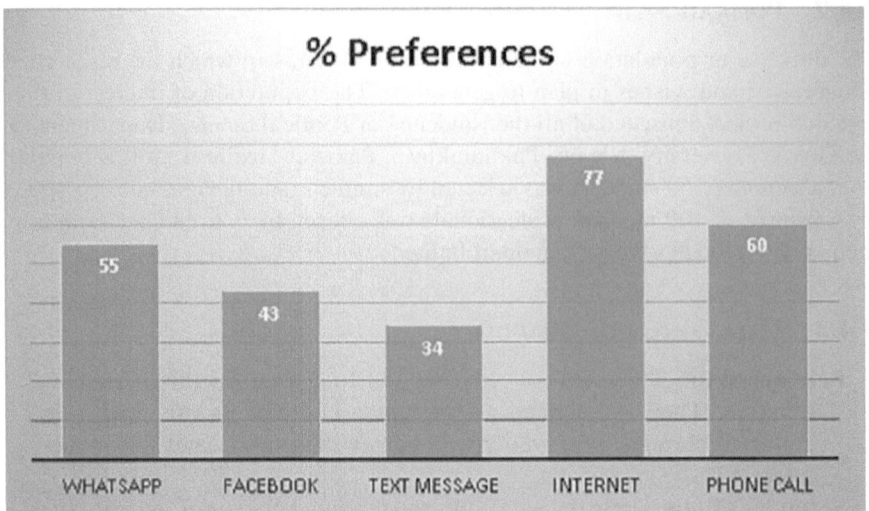

FIGURE 8.5 Preference for communication technology/tools.

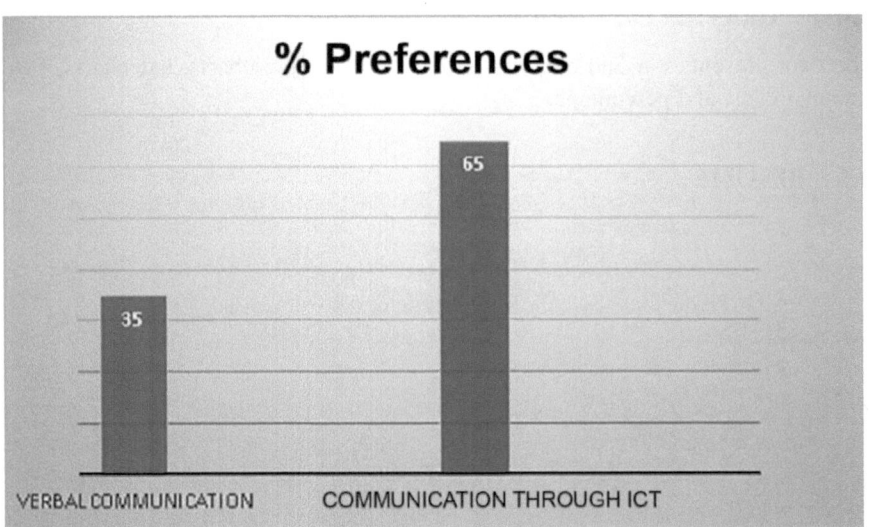

FIGURE 8.6 Preference for verbal vs. ICT mode of communication.

8.6 FINDINGS

a. A survey was conducted among 100 healthcare professionals, and there are six broad choices mentioned by professionals; it is observed that 29% of samples preferred the digital mode for better communication (Figure 8.4). Hence, one can say that the digital platform is one of the most effective ways of communication.

b. Figure 8.5 represents the most effective communication medium, and it is found that 77% of the total respondents in this study mentioned Internet as it is the commonly available communication technology compared to others.

c. Analyzing whether verbal communication/communication through ICT is useful for communication, further as per the discussion with professionals, it was (Figure 8.6) found that more than 65% prefer ICT than the verbal communication process.

8.7 LIMITATIONS

a. There are certain limitations for communication through ICT, such as network problem, which is the most crucial aspect of communication technology.

b. On the other hand, communication through ICT has some disadvantages such as lack of privacy, distraction from real life, potential for misunderstanding.

c. In the healthcare sector, patient data privacy is essential, so ICT-based communication may sometimes cause data loss, data storage, data misinterpretation, etc.

d. Telehealth is one kind of important healthcare technology, but it is not compatible with all the persons living in rural places and villages. Hence, verbal communication like face-to-face communication is easier for them in disease detection and diagnosis.

8.8 CONCLUSIONS

IT can become an excellent candidature to fulfill the present day's demand in every field. It is implemented in education, health, transportation, hospitality, public management system, aviation, and countries' growth and development. ICT for well-being (or e-wellbeing) programs are frequently viewed as costly, tedious, dangerous, and in any case, diverting from the essential concentration and plan of well-being segment programs. Sometimes these reactions might be substantial. Thus, ICT can assume an essential part to improve medical care for people just as networks. ICT can help connect the data holes that have surfaced in the well-being area in developing nations such as India by giving novel and proficient methods of getting to, conveying, and putting away data. The gap between well-being experts and the networks served by them can be tended to by actualizing ICT in medical care [28–30]. Further, with the advancement of information bases and comparable applications, ICT can improve well-being framework efficiencies and forestall clinical mistakes accordingly [31, 32].

Henceforth, after getting all the results and observations in this chapter, one can be sure that communication through ICT is better than verbal communications as a whole. It will increase the importance of communication skills and provide better consistency.

REFERENCES

1. Orlikowski, W. J., & Gash, D. C. (1994). Technological frames: Making sense of information technology in organizations. *ACM Transactions on Information Systems (TOIS), 12*(2), 174–207.
2. Powell, T. C., & Dent-Micallef, A. (1997). Information technology as competitive advantage: The role of human, business, and technology resources. *Strategic Management Journal, 18*(5), 375–405.
3. Oliner, S. D., & Sichel, D. E. (2000). The resurgence of growth in the late 1990s: Is information technology the story? *Journal of Economic Perspectives, 14*(4), 3–22.
4. Fichman, R. G. (1992). Information technology diffusion: A review of empirical research. *Paper Presented at the ICIS.*
5. Agarwal, R., & Prasad, J. (1997). The role of innovation characteristics and perceived voluntariness in the acceptance of information technologies. *Decision Sciences, 28*(3), 557–582.
6. Borda, O. F. (2006). Participatory (action) research in social theory: Origins and challenges. In Peter Reason and Hilary Bradbury (eds.), *The SAGE handbook of action research: Participative inquiry and practice*, 27–37. London: SAGE.
7. Brown, R. F., & Bylund, C. L. (2008). Communication skills training: Describing a new conceptual model. *Academic Medicine, 83*(1), 37–44.
8. Sanz, C. (2000). Bilingual education enhances third language acquisition: Evidence from Catalonia. *Applied Psycholinguistics, 21*(1), 23–44.
9. Sife, A., Lwoga, E., & Sanga, C. (2007). New technologies for teaching and learning: Challenges for higher learning institutions in developing countries. *International Journal of Education and Development Using ICT, 3*(2), 57–67.
10. Gabbott, M., & Hogg, G. (2000). An empirical investigation of the impact of nonverbal communication on service evaluation. *European Journal of Marketing, 34*(3/4), 384–398.
11. Krishna, V., & Morgan, J. (2004). The art of conversation: Eliciting information from experts through multi-stage communication. *Journal of Economic theory, 117*(2), 147–179.
12. Duffy, F. D., Gordon, G. H., Whelan, G., Cole-Kelly, K., & Frankel, R. (2004). Assessing competence in communication and interpersonal skills: The Kalamazoo II report. *Academic Medicine, 79*(6), 495–507.
13. Fryer-Edwards, K., Arnold, R. M., Baile, W., Tulsky, J. A., Petracca, F., & Back, A. (2006). Reflective teaching practices: An approach to teaching communication skills in a small-group setting. *Academic Medicine, 81*(7), 638–644.
14. Button, D., Harrington, A., & Belan, I. (2014). E-learning & information communication technology (ICT) in nursing education: A review of the literature. *Nurse Education Today, 34*(10), 1311–1323.
15. Sarkis, J., & Talluri, S. (2004). Evaluating and selecting e-commerce software and communication systems for a supply chain. *European Journal of Operational Research, 159*(2), 318–329.
16. Roberts, T., & Billings, L. (2009). Speak up and listen. *Phi Delta Kappan, 91*(2), 81–85.
17. Robin, B. R. (2008). Digital storytelling: A powerful technology tool for the 21st century classroom. *Theory into Practice, 47*(3), 220–228.
18. Russo, A., Watkins, J., Kelly, L., & Chan, S. (2008). Participatory communication with social media. *Curator: The Museum Journal, 51*(1), 21–31.
19. Brown, S., Hardaker, C., & Higgett, N. (2000). Designs on the Web: A case study of online learning for design students. *ALT-J, 8*(1), 30–40.

20. Hakkarainen, K., Muukonen, H., Lipponen, L., Ilomäki, L., Rahikainen, M., & Lehtinen, E. (2001). Teachers' information and communication technology (ICT) skills and practices of using ICT. *Journal of Technology and Teacher Education, 9*(2), 181–197.
21. Hamel, G., Doz, Y. L., & Prahalad, C. K. (1989). Collaborate with your competitors and win. *Harvard Business Review, 67*(1), 133–139.
22. Kaiser, A. P., & Hancock, T. B. (2003). Teaching parents new skills to support their young children's development. *Infants & Young Children, 16*(1), 9–21.
23. Khubchandani, L. M. (2003). Defining mother tongue education in plurilingual contexts. *Language Policy, 2*(3), 239–254.
24. FernáNdez-LóPez, Á., RodríGuez-FóRtiz, M. J., RodríGuez-Almendros, M. L., & MartíNez-Segura, M. J. (2013). Mobile learning technology based on iOS devices to support students with special education needs. *Computers & Education, 61,* 77–90.
25. Afshari, M., Bakar, K. A., Luan, W. S., Samah, B. A., & Fooi, F. S. (2009). Factors affecting teachers' use of information and communication technology. *Online Submission, 2*(1), 77–104.
26. Fischer, B. B., & Fischer, L. (1979). Styles in teaching and learning. *Educational Leadership, 36*(4), 245–254.
27. Owston, R. D. (1997). Research news and comment: The world wide web: A technology to enhance teaching and learning? *Educational Researcher, 26*(2), 27–33.
28. Lawson, T., Comber, C., Gage, J., & Cullum-Hanshaw, A. (2010). Images of the future for education? Videoconferencing: A literature review. *Technology, Pedagogy and Education, 19*(3), 295–314.
29. Madon, S. (2000). The internet and socio-economic development: Exploring the interaction. *Information Technology & People, 13*(2), 85–101.
30. Noe, R. A., Hollenbeck, J. R., Gerhart, B., & Wright, P. M. (2017). *Human Resource Management: Gaining a Competitive Advantage.* McGraw-Hill Education, New York.
31. Clark, H. H., & Krych, M. A. (2004). Speaking while monitoring addressees for understanding. *Journal of Memory and Language, 50*(1), 62–81.
32. Dela Cal-Fasoni, L. (2001). *A Technology to Enhance Teaching and Learning. Front Row Phonics: Acal Filed Test, Mal.* California State University.

9 Application of 5G/ 6G Smart Systems to Overcome Pandemic and Disaster Situations

Jayanta Kumar Ray
Sikkim Manipal University

Sanjib Sil
University of Calcutta

Rabindranath Bera
Sikkim Manipal University

Quazi Mohmmad Alfred
Aliah University

CONTENTS

9.1 Introduction 155
9.2 4G, 5G and 6G... 158
    9.2.1 4G 158
    9.2.2 5G... 159
        9.2.2.1 Why 5G? 161
        9.2.2.2 5G Is Far Superior to 4G... 162
    9.2.3 6G 164
        9.2.3.1 Why 6G? 165
9.3 Smart Environment.. 166
9.4 Summary and Conclusions 173
References..173

9.1 INTRODUCTION

The mobile communication technology is being developed day by day, and in future, it has no end. From 1G to 4G, various developments have taken place in a reasonable procedure. Due to the development of mobile communication technology,

various facilities and applications for user equipment will increase. Fourth generation is basically the update from 3G technology, and a technology called mobile
broadband Internet access [1] was introduced. The data rates are from 200 Mbps to
1 Gbps. Here, Long-Term Evolution (LTE) is introduced [2]. The main network is
Internet. In 2020, due to COVID-19, many countries declared lockdown for many
days to restrict the spread of the virus COVID-19 [3]. Due to lockdown shown in
Figure 9.1, many offices and transportations are closed, physical attendance of the
workers became obsolete, and the duty of the workers was done from home in
virtual mode.

After few days, many schools and colleges were starting their education in
online mode, which is called online education system. But in 4G, there remains a
limitation; factories and vehicles were not executed in online mode. On the other
hand, a lack of communication in disaster condition (Figure 9.2) occurred due to
cyclone, i.e., Amphan, Yash in West Bengal, Fani in Odisha.

FIGURE 9.1   Lockdown during COVID-19. (Courtesy [3].)

FIGURE 9.2   Disaster condition during Yash cyclone in 2021. (Courtesy [4].)

Due to lack of communication, a large number of problems faced the people in remote areas. Many mobile towers had fallen due to disaster conditions. At present, the mobile communication totally depends on mobile towers. In 5G technology, device-to-device (D2D) [5] communication will be introduced. As a result, communication will not be disturbed during disaster conditions. Basically in D2D [6], there will be no concept of uplink and downlink. In 4G, only eMBB is present. But for the execution of factory and vehicles in online mode, in 5G, another two terms exist in the real world, i.e., mMTC and URLLC (Figure 9.3) [7]. In 5G, LTE will be upgraded to New Radio (NR).

The aim of 5G is to convert real world into smart world. In the smart world, smart city, smart vehicle, smart factory, smart home, etc., are included. The smart world contains various intelligence technologies. The smart world is based on the execution of smart data. Smart data are massive data which are present in the cloud. Another important concept that comes into existence in 5G is Internet of things (IoT). In IoT, any kind of device can be executed with the help of Internet. Basically up to 4G, communication is from human to human. In 5G, the communication will be from human to machine and human to human. From IoT, another

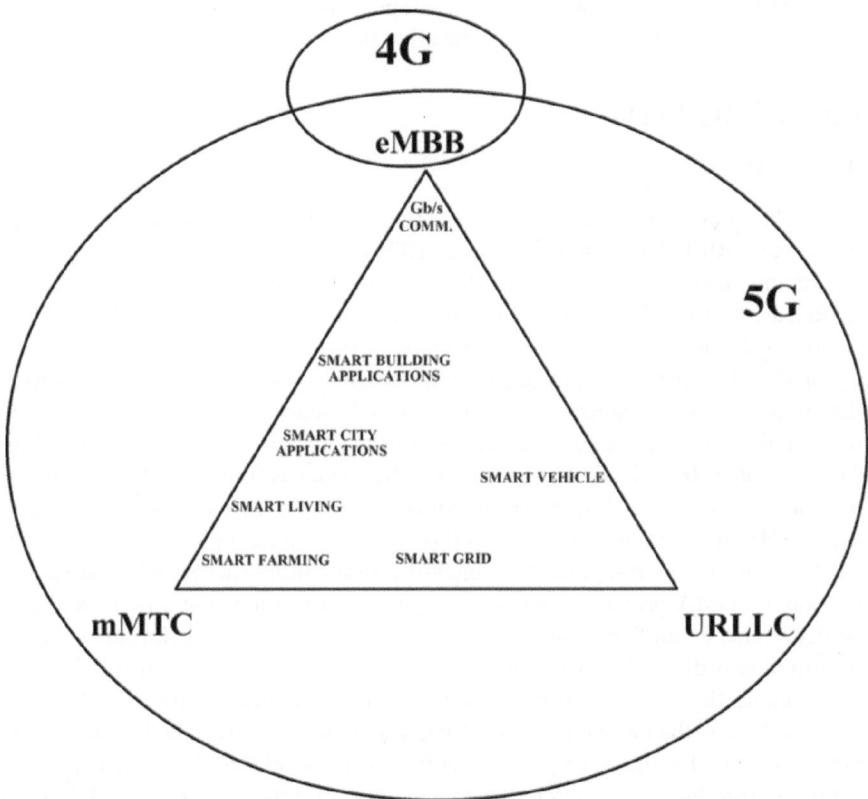

FIGURE 9.3   4G and 5G use cases [18].

new concept arises, which is called Internet of vehicles (IoV). In IoV, the concept of Doppler is included. In the smart world, by applying IoT, many smart environments such as smart factories, smart homes and smart hospitals can be developed; side by side by applying IoV, smart vehicles can be developed. Thus, by applying online mode in factories and vehicular applications, the problems faced during the lockdown in the real world due to COVID-19 will be fulfilled to a large extent in comparison with 4G [9]. In smart factories, mMTC and eMBB are required, and in smart vehicles, eMBB and URLLC are required. So it is expected that 5G [9] can handle not only COVID-19, but any kind of disaster situations such as Amphan and Fani in West Bengal and Odisha. It means that there will be less problems in disaster conditions if 5G technology is applied. On the other hand, the problem will be totally solved if 6G technology is introduced. Nowadays, in many countries, the 5G technology is already introduced. In the future when 6G will be introduced, the world will become smarter than the world of 5G. In 5G smart factories, only the machines can be executed in online mode, but in 6G smart factories, more facilities can be provided than the 5G. On the other hand, 6G will follow the principle of specificity. For example, the operations done by the doctor will become more successful in online mode. In 5G, smart vehicles can be stopped or moved in online mode, but in 6G, smart vehicles will be smarter than the previous stage. As a result, the accident cases will be reduced to nearly zero.

9.2   4G, 5G AND 6G

9.2.1   4G

In 4G, the Internet is the main network. Microwave signal is used. It was launched in the year 2010. The state of the art is LTE (Long-Term Evolution). It provides broadband data access, which facilitates millions of users around the world. In Release 8, the LTE standard was first introduced. Here, LTE Advanced was standardized by Third Generation Partnership Project (3GPP) [9]. When 3G was updated to 4G, there was a huge rise in the facilities such as access methodology, data transfer rate, transmission terminology and security [10]. Due to these facilities, multimedia data can be accessed by the users at their own place and own time. The data rate is from 2 Mbps to 1 Gbps. The frequency is from 2 to 8 GHz. The core network belongs to all Internet Protocol (IP) network. By using high-speed data access (HSPA), the Internet is accessed by the user. Here, high spectral efficiency can be achieved by orthogonal frequency-division multiple access (OFDMA). Side by side, OFDMA provides good support for channel allocation scheme and reliability of the signal from source to destination. During OFDM transmission, the information is divided into a number of parallel sub-streams. As a result, interference among the signals will be reduced and the reliability of the signal will be increased. Now the mobile phones having 4G technology are called smartphones. Hence, we say that the concept of smart had been introduced. Here the communication is from human to human. Due to the opportunities provided by 4G, many applications such as banking, education, administration and shopping are facilitated by online mode during pandemic situation. Hence, work from home for many organizations is carried out successfully. But in many applications such as factory, robots and vehicle, 4G wireless communication cannot be applied.

9.2.2 5G

While shifting from 4G to 5G, the smartness will be increased. From smartphones, the smartness of the system will be applied to smart factories, smart vehicles, smart robots, etc. It is possible with the help of smart data [6]. A device can be converted to a smart device by connecting it to Internet. Hence, a new term arises called IoT. It means that the smart devices can be executed with the help of Internet.

In a simple way, the devices can be operated in online mode. The main difference between 5G and 4G is that 5G concentrates on machine type communication and the IoT. Hence, new services are included in 5G, which are mMTC and URLLC having the connectivity of IoT in a massive manner. The availability of massive bandwidth, low latency and high reliability facilitates the essentials for manufacturing and critical applications. As a result, the 5G technology will become more flexible and require low cost and less time for deployment. The aim of 5G is to convert the environment into a smart environment which includes smart vehicles, smart education, smart life, smart health, smart agriculture, smart factories, smart irrigation and smart cities (Figure 9.4). The smart factory includes automation of factory, process, logistics and warehousing, monitoring and maintenance. 5G technologies are integrated in such a way that smart factory will become more efficient, productive and automative. In a smart factory, the segmented technologies and standards are unified. The smart vehicle refers to autonomous vehicles. In 5G mobile communication, the smart vehicle will be able to support both transportation infrastructure and personal vehicles. Using V2V and V2I, the efficiency of the automobile will increase and the road safety and fuel economy will be improved. Here sensors are embedded into the tracks, railways

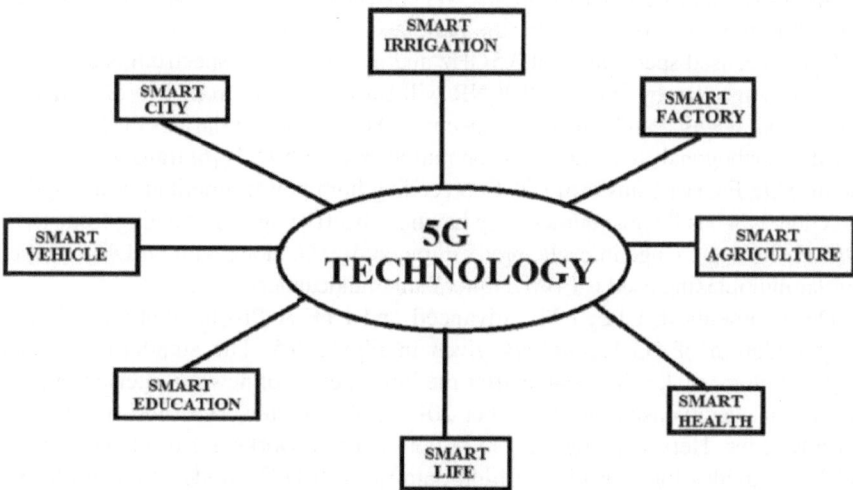

FIGURE 9.4 5G technologies with their impact.

and aerodromes so that the smart vehicles are able to communicate efficiently. The smart irrigation includes irrigation automation and sensor monitoring. The smart city includes road traffic, electric, water system, waste management and public safety. In the smart city, the researchers are focused on cloud and edge computing, IoT, D2D communication, machine-to-machine (M2M) communication, cooperative multi-cell solution, network slicing, security, etc. The smart health include health care and hospitality. In smart health, various items such as wearable health monitoring devices, secure remote consultation and robotic applications are utilized. The smart life includes self-automation and self-monitoring. The smart grid includes primary frequency control, secondary frequency control and distributed voltage control. In the smart grid, the researchers have dedicated themselves to network slicing, IoT, edge computing, aerial maintenance of critical infrastructure, artificial intelligence (AI), machine learning, vehicle to grid (V2G), communication test bed, etc. The communications are from human to human and human to machines. The main requirement for Industry 4.0 is the connectivity in which the interaction between machine, people and objects is included. The 5G wireless communication has already been launched in the year 2020. There are three main benefits of 5G wireless technologies, i.e., faster speed, shorter delays and increased connectivity [10]. The requirements for 5G technology include better revenue for global operators, improved data coding, wireless access and backhaul, lower battery consumption, lower outage probability, better coverage and high data rates, and smart beam antenna system. During the implementation of 5G, NR is one of the crucial supporting technologies. NR is the new interface and access method that provides facilities according to the requirements in the future. With the help of NR, new applications having the requirements of enhanced data rates, latency, coverage, capacity and reliability can be enabled. As a result, improved network energy performance and the use of spectrum in very high-frequency bands can be obtained. In the 3GPP, NR is specified as fifth generation (5G) radio interface [11]. This fulfills the updation from mobile broadband (MBB) service to enhanced mobile broadband (eMBB) for higher system capacity, higher data rates, etc. By expanding the range of spectrum, the radio access technology can be deployed. In LTE, the licensed spectrum is at 3.5 GHz and the unlicensed spectrum is at 5 GHz. The decision taken by 3GPP is that NR will start operating from 1 to 52.6 GHz in case of both licensed and unlicensed spectrum. Having the similarities in LTE, NR is based on orthogonal frequency-division multiplexing (OFDM) [6] transmission with the discrete Fourier transform (DFT) preceding high-power amplifier in the uplink direction. NR has flexible numerology having subcarrier spacing ranging from 15 to 240 kHz having change in cyclic prefix duration. In 5G, along with OFDM, another similar multiplexing used is FBMC (filter bank multicarrier).

Developments in LTE, LTE Advanced and LTE-A Pro technologies [9] are the foundation of 5G technology given in Figure 9.5. The standardization of 5G technology makes it possible after the introduction of new air interface called 5G NR. It was released on December 2017 applicable for NSA (non-standalone) 5G NR cases. Here existing LTE radio and core networks are used. The use of 5G NR provides the facilities of high data rates in NSA mode. As a result, 5G technology can be implemented using the existing 4G setup.

5G RADIO ACCESS

FIGURE 9.5   5G radio access.

TABLE 9.1
5G Numerology

μ	Number of Slots	Subcarrier Frequency (KHz)	Time
0	1	15	1 ms
1	2	30	0.5 ms = 500 μs
2	4	60	0.25 ms = 250 μs
3	8	120	0.125 ms = 125 μs
4	16	240	0.0625 ms = 62.5 μs

9.2.2.1   Why 5G?

While shifting from 4G to 5G, there is a great difference in key features such as peak data rate, mobility, latency, connection density and area traffic capacity. In 4G [12], the peak data rate is 1 Gbps, but in 5G [12], it will be 20 Gbps. The user experienced data rate is 10 Mbps in 4G and 100 Mbps in 5G [11]. On the other hand, the mobility is 350 kph in 4G and 500 kph in 5G and latency is 10 ms in 4G and less than 1 ms in 5G [11]. In case of connection density, the number of devices is 10^5 devices/km² in 4G and 10^6 devices/km² in 5G [13]. The area traffic capacity for 4G is 0.1 Mbps/m², and for 5G, it is 10 Mbps/m². In 4G, the bandwidth ranges from 1.4 to 20 MHz. In 5G, the bandwidth is extended up to 60 GHz. In 4G, for forward error correction, turbo coding [14] is used, but in 5G, LDPC (low-density parity-check) [8] coding is used. In 4G, there is a presence of moderate code rate, but in 5G, high code rates are present. For moderate code rate, the turbo codes [15] have better performance than LDPC [16]. In 5G, LDPC shows better performance. The checking procedure of decoding is easier in LDPC than in turbo codes. In 4G, the subcarrier spacing is 15 KHz, and in 5G, the subcarrier ranges from 15 to 240 KHz as given in Table 9.1 (Table 9.2).

In 4G, the frequency band is only sub-6 GHz, and in 5G, it is sub-6 GHz and 28 GHz. In 4G, the multiple access techniques are CDMA (code-division multiple access) and OFDMA, but in 5G, the multiple access techniques are CDMA and BDMA (beam-division multiple access). In 4G, the smartness belongs to mobile phones. But in 5G, the smartness will be extended to machines and vehicles.

TABLE 9.2

Comparison of 4G, 5G and 6G

Issue	4G	5G	6G
Data rate	1 Gbps	10 Gbps	1 Tbps
Latency	100 ms	10 ms	1 ms
Spectral efficiency	15 bps/Hz	30 bps/Hz	100 bps/Hz
Application types	eMBB	eMBB, URLLC, mMTC	MBRLLC, mURLLC, HCS, MPS
Device types	Smartphones	Smartphones, sensors, drones	Sensors and DLT devices, CRAS, smart implants
Frequency bands	Sub-6 GHz	Sub-6 GHz, millimeter wave	Sub-6 GHz, millimeter wave, sub-millimeter wave
Mobility support	350 km/h	500 km/h	1000 km/h
Satellite integration	Not applicable	Not applicable	Applicable
Artificial intelligence	Not applicable	Partly applicable	Fully applicable
Smart vehicle	Not applicable	Partly applicable	Fully applicable
Extended reality	Not applicable	Partly applicable	Fully applicable
Haptic communication	Not applicable	Partly applicable	Fully applicable
Terahertz communication	Not applicable	Partly applicable	Fully applicable
Service level	Video	Virtual reality, augmented reality	Tactile
Architecture	MIMO	Massive MIMO	Intelligent surface
Maximum frequency	6 GHz	90 GHz	10 THz
Reliability	99.99%	99.9999%	99.999999%
Communication	Human to human	Human to human, human to machine	Human to human, human to machine, machine to machine
Year of deployment	2010	2020	2030
Technology	LTE Advanced	LTE Advanced + New Radio	New Radio access technologies
Handoff	Horizontal and vertical	Horizontal and vertical	Horizontal and vertical

As a result, the smartness in the society will increase and 5G will provide good support during situations such as pandemic.

9.2.2.2 5G Is Far Superior to 4G

In SMIT, East Sikkim, the authors had done two experiments in hardware mode using both millimeter wave antennas and microwave antennas. The type of software utilized for these experiments was SystemVue. The signals are downloaded into the computer with the help of signal downloader present in SystemVue software and then transferred to the vector signal generator (VSG). From VSG, the signals are sent to the transmitting antenna. On the other hand, the signals are received by the receiving antenna. Then the signals are transferred to the vector signal analyzer. From the

FIGURE 9.6   Microwave horn antenna at 3 GHz.

FIGURE 9.7   Millimeter wave dish antenna at 28 GHz.

vector signal analyzer, the signals are uploaded into the SystemVue software. As a result, the received signal will be displayed on the data sink present in the SystemVue.

In case of two microwave antennas, the applied frequency is 3 GHz and the type of antenna is horn antenna (Figure 9.6).

Case 1: Distance between the transmitter and the receiver is 100 ft
> Received signal strength indicator (RSSI): −97.6 dBm.
> Signal-to-noise ratio (SNR): 30.6 dB.

Case 2: Distance between the transmitter and the receiver is 99.6 ft
> RSSI: −97.2 dBm.
> SNR: 31.3 dB.
> In case of two millimeter wave antennas, the applied frequency is 28 GHz and the type of antenna is dish antenna (Figure 9.7).

Case 3: Distance between the transmitter and the receiver is 100 ft
> RSSI: −44 dBm.
> SNR: 50.7 dB.

Case 4: Distance between the transmitter and the receiver is 99.6 ft
> RSSI: −43.8 dBm.
> SNR: 49.4 dB.

From the above cases, by comparing case 1 and case 3 (distance between the transmitter and the receiver is 100 ft), it is seen that the RSSI of the millimeter wave antenna is 53.6 dBm more than that of the microwave antenna.

The SNR of the millimeter wave antenna is 20.1 dB more than that of the microwave antenna.

By comparing case 2 and case 4 (distance between the transmitter and the receiver is 99.6 ft), it is seen that the RSSI of the millimeter wave antenna is 53.4 dBm more than that of the microwave antenna.

The SNR of the millimeter wave antenna is 18.1 dB more than that of the microwave antenna.

9.2.3   6G

Sixth-generation (6G) [17] system is the new updated technology of wireless communication. Here the smartness of the systems will be increased. By increasing the smartness, the super-smart society will be formed. It provides full support of AI and is expected around 2030. The aim of 6G [18] is to fulfill the higher system capacity, higher data rate and lower latency and to provide higher quality of service (QoS) than that of the previous stage. Here the IoT will be converted to Internet of everything (IoE). From 1G to 2G, there is a transition from analog system to digital system. From 3G onward, the digitalization is increased. It is expected that the digitalization will be completely fulfilled in 6G. During the shift from 5G to 6G, there will be lot of changes in updation and development. In 5G, the important terminologies are URLLC, mMTC and eMBB. In 6G, URLLC will be upgraded to uHSLLC (ultrahigh-speed and low-latency communication). In uHSLLC, the end-to-end delay will be less than 1ms and the reliability will be 99.99999 [19]. eMBB will be upgraded to uMUB (ubiquitous mobile ultra-broadband) (Figure 9.8). In mMTC, the number of devices will be up to 10 million/km^2. Another new term arises in 6G is ultrahigh data density (uHDD) [20].

Here the conversion from "connected things" to "connected intelligence" takes place. On the other hand, IoT is converted to IoE. In 6G, there are few additional requirements such as AI-integrated communication, tactile Internet, high energy efficiency, enhanced data security and low backhaul and access network congestion. 6G technology is applicable to not only smart vehicles, but also unmanned aerial vehicles (UAVs) [21]. In 6G, battery lifetime will be longer and the procedure of charging the battery will be in wireless mode. Hence, we hope that the process of electricity will be in wireless mode and the environment will be totally wireless (Figure 9.9). In 6G mobile communication, the industrial revolution will be upgraded, which means the control techniques with automation function should be present in Industry 4.0. For the control techniques, the latency

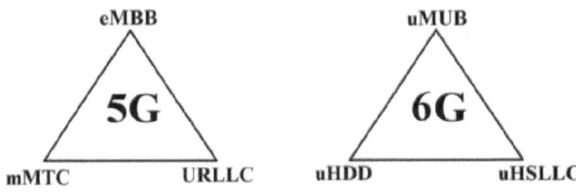

FIGURE 9.8   5G and 6G key requirements.

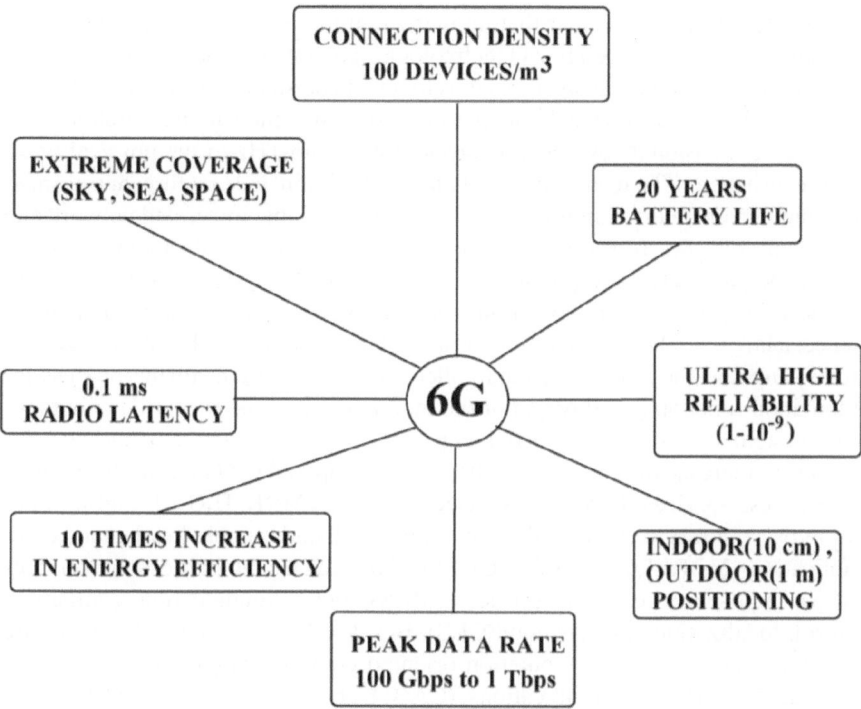

FIGURE 9.9 6G technologies.

should be 0.1 ms and the reliability should be 1–10−9. For automation function in the industrial event and for the maintenance of system stability, the delay jitter is the new item proposed in 6G. In 6G mobile communication, there are various visions such as wireless brain–computer interaction, fully autonomous vehicles, connected robotics, integrated smart city and different realities (augmented reality, extended reality, machine reality and virtual reality). The functionalities of IoT are increased, which include sensing, data collection, analysis and storage. Multiple mobile system technologies form the 6G technology.

9.2.3.1 Why 6G?

In 5G, the data rate is 10 Gbps, but in 6G, the data rate will be 1 Tbps. In 5G, the maximum spectral efficiency is 30 bps/Hz, but in 6G, it will be 70 bps/Hz or more. In 6G, the end-to-end latency will be 1 ms, which is 9 ms less than the previous generation, i.e., 5G. In smart vehicles, the mobility support in 5G is up to 500 km/h, while in 6G [22], it will be 1000 km/h. Hence, we hope that not only smart vehicles, but also smart aircraft will be invented. Few technologies that will be partly utilized in 5G can be fully utilized in 6G. These technologies are AI, autonomous vehicles, extended reality and haptic communication. In 5G, millimeter wave is used, but in 6G, the sub-millimeter wave [12] is used. In 5G, the maximum frequency is 90 GHz, but in 6G [15], the frequency ranges from 0.1 to 10 THz. It is clear that terahertz communication [10] is widely used. The new facilities provided by 6G technologies are holographic communications, high-precision manufacturing, rapid development and smart environment. To activate 6G technologies, few technologies are required, which are new architecture, AI, three-dimensional coverage, incorporation of sub-THz to the physical layer, high security, etc. The techniques in 6G technologies include intelligent transmission, multi-node multi-domain joint transmission, reliable integrated network and multi-band ultrafast transmission [17]. In 6G technology, control and optimization of the parameters are possible. The visions for 6G will be intelligence and openness [15]. For robotic operation, accurate precise positioning is mandatory. Hence, a large number of sensors are required for applying in 3D environment for monitoring and the amount of data collected during manufacturing process for automation function in industry. For the maintenance of the execution of autonomous system, system security and data privacy are main function items. The aim of 6G is to increase the capacity 10–100 times compared to 5G due to the application of these services. In 5G, the service types are eMBB, URLLC and mMTC. But in 6G, the service types will be mobile broadband reliable low-latency communication (MBRLLC), massive URLLC (mURLLC), human-centric services (HCS), multi-purpose energy services (MES), and communication, computing, control, localization and sensing (3CLS). Besides these service types, there are additional services, i.e., computation-oriented communications (COC), contextually agile eMBB communications (CAeC) and event-defined URLLC (ED URLLC). In comparison with 5G, the features of 6G include very high data rates (1 Tbps), very high energy efficiency, massive low latency control, very broad frequency bands and connected intelligence having machine learning capability. The principle of "new services" denotes the utilization of spectrum in infrastructure. In IoE, the IoT is the main pillar. In IoT, the things refer to devices, sensors and machines, and these devices are connected with the Internet. The IoE involves machine-to-machine, human-to-machine and human-to-human communications (Figure 9.11). Due to M2M communication in 6G, the conversion from IoT to IoE comes into existence. The 6G connectivity vision can be specified with the help of four terms [23,24], viz. deep, intelligent, holographic and ubiquitous (Figure 9.10).

9.3 SMART ENVIRONMENT

The importance of smart basically means the execution by the devices in place of human. As a result, the physical work for the human in the society will be reduced. For example, in the past, people had to submit electricity bills in particular offices. But at present, people can submit electricity bills via online sitting at home or anywhere in the world and get the proof of payment in digitized format. This procedure is called online submission. In the past, people had to go to shopping malls for purchasing various items. But nowadays, people can choose and purchase any items online sitting at home. From these two examples, we say that during situations like pandemic necessary requirements can be fulfilled online

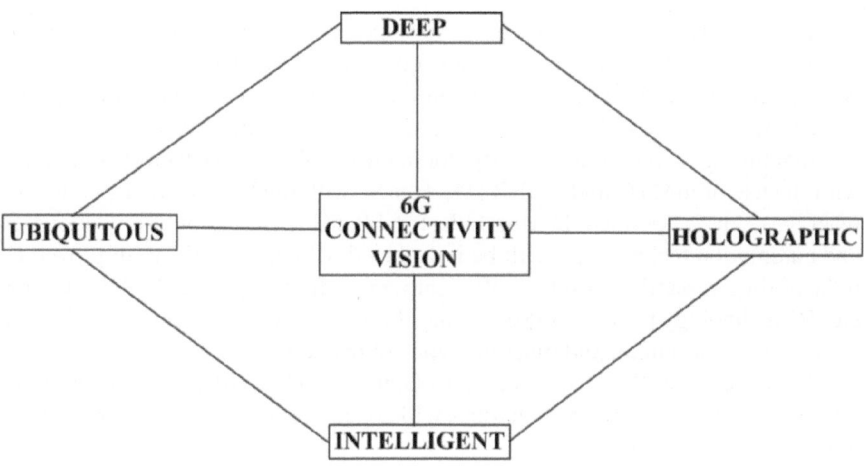

FIGURE 9.10   6G connectivity vision.

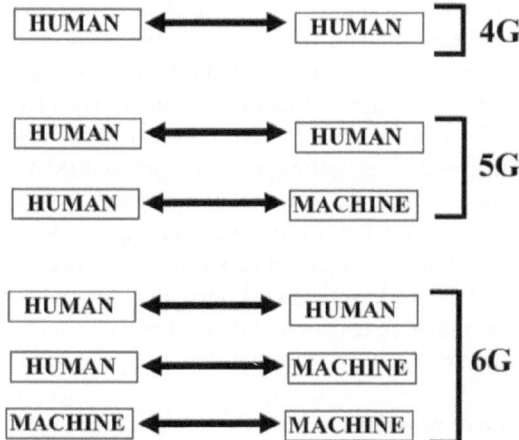

FIGURE 9.11   4G, 5G and 6G communications.

sitting at home. Because in pandemic situations, leaving residence is totally pro-
hibited. Online systems can make our life easy in critical situations. This facility
is possible due to the presence of 4G smartphones. But 4G technology is applicable to smartphones, but not more than that. When there will be transition from
4G to 5G, world is thinking to increase the smartness in the society and aim to
create a smart world. The smart world includes smart cities, smart homes, smart
factories, smart health and smart vehicles. During disaster situations such as tor-
nadoes and cyclones the problem will be solved using D2D communication.

Smart factory: The smart factory [25] means the factory can be operated
online. It includes the automation of factory, process, production, monitoring and

maintenance. Here 4G technology is not applicable. From 5G technology, the concept of smart factory [13] is introduced. In the smart factory, every machine is operated online. As a result, smart production and smart manufacturing takes place [26]. As a result, during pandemic situations, the automatic production and manufacturing in the smart factory commences. The smart factory is possible with the use of mMTC and eMBB [14]. Hence, with the help of smartphones, the machines can be operated. During Industry 4.0, 5G wireless communications, AI and automation technologies will be introduced, which will lead to improvement in flexibility, versatility, resource efficiency, cost efficiency, industrial production, etc. 5G technology provides lower latency, high reliability, seamless and high-end connectivity for human and machine type connections.

The world hopes that in 6G, the smart factory will be further developed. In 6G, the concept of sensing by the machines will appear; as a result, the 6G will be able to facilitate more than the 5G.

Tactile Internet: Tactile Internet is the applicable Internet that is used for perception, accession, manipulation and control of objects or processes in perfect time. For autonomous industry, it is the reliable Internet [27]. In the tactile Internet, human-to-machine interaction is enabled. It provides the transmitting facility of touch and actuation in addition to audiovisual information. The new features include ultrafast action, high availability, high security, ultra-reliability and autonomous control of haptic or tactile machines. The enabling technologies in the framework for tactile Internet have physical/MAC layer techniques and network layer with cloud level techniques. The physical/MAC layer techniques are dynamic resource allocation, advanced signal processing, link adaptation, multiple access scheduling, latency reduction techniques and reliability enhancement techniques [28]. The techniques of network layer with cloud level techniques are software-defined networking, network virtualization, edge cloud processing, caching techniques, machine learning and novel network architectures. The tactile Internet has three domains, i.e., master domain, network domain and control domain.

Interconnection of massive devices: Factories are arranged with a large number of devices having ubiquitous connectivity. These devices are connected to a single network, and the industrial IoT network [13] is established. Heterogeneous networks and heterogeneous devices are interconnected to each other. The situation is critical, and it is solved by the installation of massive portable antennas having antenna arrays.

Robotics: Here the robots perform their specific task according to the instructions given by the user [13]. The performance of the robots will be improved if 5G technology provides huge data rate, low latency and ultra-reliable connection. As a result, the productivity of the factory will significantly be increased. The process of synchronization and management of robots is supported by 5G technology.

Edge computing: It is the intelligent computing that produces intelligent services for the purpose of intelligent manufacturing. It enables dynamic monitoring and controls the manufacturing processes. The edge computing [13] requires IT ecosystem, which is used for transfer of data in a dynamic manner. It is basically the distributed autonomy in industry. Here the combination of disperse treatment with centralized upload takes place. As a result, there will be an increase in the usage of the network bandwidth, which will increase the autonomy, security and robustness of the manufacturing system.

Virtual reality/augmented reality (VR/AR): Due to the presence of AR/VR [29], the production efficiency, product quality and design will be improved. Here the production cost will be less than that before. VR is basically software which guides staff for execution using 3D environment. Safety and profit of the organization will increase. It requires low latency and ultrahigh data rate. It helps the employees to work in an easy way.

Big data: The big data [29] basically means smart data. The type of big data is sensor data, machine log data and manufacturing process data. It is used in intelligent manufacturing, industrial supply chain analysis, product quality control, active maintenance and optimization. Due to the active maintenance based on big data, production continuity is confirmed. The production continuity depends on the utilization of the equipment and its productivity. The product design optimization based on manufacturing big data includes the optimization and analysis of big data. It is helpful for the collection of product data.

Artificial intelligence (AI): It helps the network to collect data in specified time from various devices such as robots and sensors. During the manufacturing process, if any problem arises, it delivers a warning to help the industrial worker. In support of URLLC, AI [29] can execute the decision in autonomous mode.

Time-sensitive networking (TSN): A group of IEEE standards that allow Ethernet to be applied for time-sensitive applications having the need of both latency and bandwidth. As a result, the critical aspects are fulfilled. The usage of TSN [28] is in various applications such as audio/video distribution, automotive and industrial. The time synchronization is enabled by IEEE 802.1AS. Hence, the behavior of TSN is determined. IEEE 802.1Qbv is the time-aware scheduling that produces the QoS facility. As a result, the coexistence of high-priority traffic is enabled. TSN can provide low latency. TSN can be used for fronthaul as well as backhaul. A TSN standard IEEE 802.1CM has the specificity toward fronthaul. There are many protocols that can run over the Ethernet without Internet Protocol, such as PROFINET and EtherCAT. 5G is integrated with TSN as TSN is interrelated to URLLC in terms of four components, i.e., latency, resource management, synchronization and reliability.

CoMP (coordinated multipoint): With the help of CoMP [28], spatial diversity is produced, which confirms the high reliability and low latency. The utilization of CoMP technique is according to various requirements such as network architecture and fronthaul requirement. The usage of Release 15 NR gives the opportunity for the implementation of CoMP techniques to the user equipment. Separate reference signals are used to obtain the channel information at the transmitter. The instructions to the user are delivered by transmission and receiving points (TRPs) having SRS (sounding reference signal). Different CoMP options are coherent joint transmission with interference nulling (CJT-MU) for multiple users, coherent joint transmission without interference nulling (CJT-SU) and non-coherent joint transmission (NCJT) for single user. Due to interference nulling in CJT-MU, the scheduling is on the same frequency resources. As there is no interference nulling in CJT-SU and NCJT, the scheduling is on a given set of time–frequency resources.

Device-to-device communication: This type of communication is basically a peer-to-peer network. In this communication, no mobile towers are required. Disaster situations occur in any place in the world at any time. For example, Fani in Odisha (India) and Amphan in West Bengal (India) are two events that occurred in 2019 and 2020. Due to these events, lot of mobile towers were broken and lack of communication took place. In 4G, the mobile phones are dependent upon mobile towers which are the base station networks. Here the concept of uplink and downlink is included. But in 5G, the concept of device to device is included, which means the communication from mobile to mobile in a direct manner. Hence, no mobile tower is required. Basically, it will use Sidelink communication. It will be of great help during disaster conditions. D2D [21] communication has the following features:

One-hop communication: The occurrence of communication will be in a single hop. The resource requirements are less. The efficient utilization of the spectrum will be produced. Latency will be reduced.

Usage of same spectrum: D2D users and cellular users share the same spectrum in the communication. As a result, the spectrum reuse ratio will be improved.

Reduction in power: D2D communication happens over a small distance. Hence, the transmission power required is less and the battery life will increase (Figure 9.12a).

Long-distance communication: There will be the possibility of long-distance communication with the help of relays. As a result, the range of communication will increase as shown in Figure 9.12b.

D2D communication was standardized in 3GPP Release 12. The mission-critical push-to-talk (MCPTT) [17] was standardized in Release 13 in the year 2016. In 2017, some mission-critical services and their enhancement are updated. In Release 15, the mission-critical services are further developed, which include the interconnection between mission-critical systems and its service requirements from various industries. The data rate ranges from 5 to 10 Gbps. The maximum transmission power is 24 dBm. LDPC codes are used.

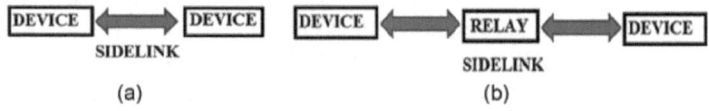

FIGURE 9.12   (a) Device-to-device communication (short range); (b) device-to-device communication (long range).

LDPC: In LDPC [15], the sparse parity check matrix is represented. Two types of nodes, i.e., check node and variable node, are present. "1" in the matrix is represented as connection, and "0" is represented as disconnection.

As a result, the complexity of encoding and decoding will be low.

Encoding: The output codeword is represented by

$$C = uG \tag{9.1}$$

where u is the input and G is the generator matrix.

The generator matrix is not a design parameter. It is executed by arranging the parity check matrix in a systematic form.

Decoding: When the message is passed between check node and variable node, the decoding of LDPC will be performed. First, the channel log likelihood ratio (LLR) L_a is sent by the variable node to the bth part of the check node L_b. After processing by the check node, messages will be sent to the variable node and it is expressed by

$$L_{b \text{ to } a} = 2\tanh^{-1}\left[\Pi_{b \,\in\, N(b)-\{a\}} \tanh\left(L_{a \text{ to } b}/2\right)\right] \tag{9.2}$$

where $L_{a \text{ to } b}$ is the message from variable node to check node. $N(b)$ is the set of variable nodes connected to the bth part of the check node. The message received by the variable node is processed and again sent as new messages to the check node. And here the sending and receiving by these two nodes will be in a serial mode and the performance will be created. Message updation happens in parallel between all check nodes and all variable nodes, and it is called flood schedule. The achievement of improved performance is possible when the performance of serial scheduling is improved.

A device having a 28 GHz antenna achieves more coverage area than a device having a sub-6 GHz antenna, as given in Figure 9.13.

Smart vehicles: Smart vehicles are basically unmanned vehicles. A smart vehicle is formed by the application of the IoV. Here the concept of Doppler is included. It can be done by the application of radar, URLLC and eMBB [14]. It is basically a critical communication in 5G. In the real world, besides pandemic and disaster situations, accidents are the main problem. In smart vehicles, two things must be applied, i.e., LOS (line of

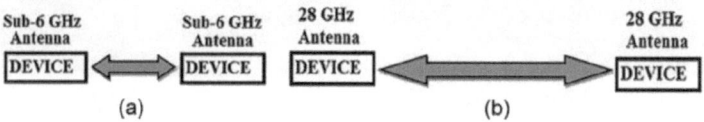

FIGURE 9.13 (a) Device-to-device communication (sub-6 GHz antenna); (b) device-to-device communication (28 GHz antenna).

sight) and NLOS (non-line of sight). In LOS, radar is used, but in NLOS, URLLC is used.

To build a smart vehicle, lot of sensors, actuators and automotive Ethernet (TSN-enabled Ethernet) connections are required. The smart vehicle works in three steps, i.e., sense, understand and act. In the sense step, the raw data obtained from various devices such as cameras, radars, lidars and sensors are sent for sensor processing. In the understand step, the object parameters such as time stamp, dimensions and position/velocity are obtained. After sensor processing, a part of data are sent to the sensor fusion and the remaining compressed data are sent for visualization. The sensor fusion is interconnected with vehicular communication and maps. On the other hand, the act part performs various actions, i.e., do nothing, warn, complement and control. The data from sensor fusion are sent to action engine. The action engine is connected with the driver state. Following the action engine, vehicles can be controlled, which includes brake, accurate steering. Between action engine and vehicle control, visualization is present.

On the other hand, in the smart vehicle, vehicle can be driven by the driver sitting at home. It had been achieved by IIT Delhi in IMC 2018 [30], Jio Showcase exciting 5G use cases, IOT solution and wireless services (Figure 9.14).

Figure 9.14a shows a driver in a specific place. The driver drives the car by moving and steering. The driver is able to see the objects present in front of the car with the help of a digital TV. Figure 9.14b shows the driverless car moving on the road. Above the car is a box in which 5G NR is present. This type of car will be helpful in pandemic situations. A car is an important thing which will be helpful during emergency periods. Suppose a patient had fallen ill during a pandemic. Then he or she will have to go to hospital for treatment. But the driver will not want to go to hospital due to scarcity. In this situation, a smart car will give a

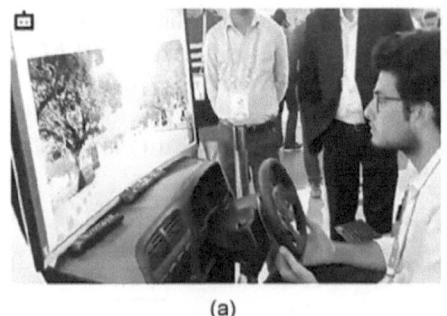

(a)   (b)

FIGURE 9.14   (a) Driver in a specific place (courtesy [30]); (b) driverless car (courtesy [30]).

good relief to the patient. With the help of the smart car, the driver can drive the car sitting at home without having any scarcity and the patient will be able to go to hospital.

9.4 SUMMARY AND CONCLUSIONS

In 2020, the real world faced a pandemic situation. 4G technology facilitates few applications such as office, information technology, meeting and education. 5G technology removes the limitations of 4G and can be used in areas such as smart factories, smart vehicles and smart health during pandemic situations. It also enables D2D communication during disaster situations. When 5G will shift to 6G, the facilities will increase far more.

REFERENCES

1. E. Ezhilarasan and M. Dinakaran "A review on mobile technologies: 3G, 4G and 5G," *2017 Second International Conference on Recent Trends and Challenges in Computational Models (ICRTCCM),* Tindivanam, 2017, pp. 369–373, Doi: 10.1109/ICRTCCM.2017.90.
2. K. Gopal "A comparative study on 4G and 5G technology for wireless applications," *IOSR Journal of Electronics and Communication Engineering (IOSR-JECE),* e-ISSN: 2278-2834, p- ISSN: 2278-8735.vol. 10, no. 6, Ver. III (Nov–Dec.2015), pp. 67–72, www.iosrjournals.org.
3. https://www.thehindubusinessline.com/news/national/corona-threat-west-bengal-announces-lockdown-measures/article31133982.ece.
4. https://www.thetimesofbengal.com/2021/05/21/3-lakh-people-are-being-relocated-amfans-education-is-a-big-hope-to-stop-yash/.
5. M. Usman, F. Granelli and M.R. Asghar "5G and D2D communications at the service of smart cities", *Transportation and Power Grid in Smart Cities: Communication Networks and Services*, chapter 5, First Edition., Edited by H.T. Mouftah, M. Erol-Kantarci and M.H. Rehmani, Published 2019 by John Wiley & Sons Ltd, Hoboken, NJ.
6. A. Gupta and R. K. Jha, "A survey of 5G network: architecture and emerging technologies," *IEEE Access*, vol. 3, pp. 1206–1232, 2015, Doi: 10.1109/ACCESS.2015.2461602.
7. M. Noohani "A review of 5G technology: architecture, security and wide applications," *International Research Journal of Engineering and Technology (IRJET)*, e-ISSN: 2395-0056, vol. 07, no. 05, May 2020, https://www.researchgate.net/publication/341541673.
8. G. Prasad, H. A. Latchman, Y. Lee and W. A. Finamore, "A comparative performance study of LDPC and Turbo codes for realistic PLC channels," *18th IEEE International Symposium on Power Line Communications and Its Applications*, Glasgow, UK, 2014, pp. 202–207, Doi: 10.1109/ISPLC.2014.6812365.
9. A. Jain, E. Lopez-Aguilera and I. Demirkol, "Evolutionary 4G/5G network architecture assisted efficient handover signaling," *IEEE Access*, vol. 7, pp. 256–283, 2019, Doi: 10.1109/ACCESS.2018.2885344.
10. G. Barb and M. Otesteanu, "4G/5G: a comparative study and overview on what to expect from 5G," *2020 43rd International Conference on Telecommunications and Signal Processing (TSP)*, Milan, Italy, 2020, pp. 37–40, Doi: 10.1109/TSP49548.2020.9163402.

11. A. Gohil, H. Modi and S. K. Patel, "5G technology of mobile communication: a survey," *2013 International Conference on Intelligent Systems and Signal Processing (ISSP)*, Vallabh Vidyanagar, India, 2013, pp. 288–292, Doi: 10.1109/ISSP.2013.6526920.

12. A. J. Seeds, H. Shams, M. J. Fice and C. C. Renaud, "TeraHertz photonics for wireless communications," *Journal of Lightwave Technology*, vol. 33, no. 3, pp. 579–587, 1 Feb.1, 2015, Doi: 10.1109/JLT.2014.2355137.

13. J. Cheng, W. Chen, F. Tao and C.-L. Lin, "Industrial IoT in 5G environment towards smart manufacturing," *Journal of Industrial Information Integration*, vol. 10, pp. 10–19, 2018, ISSN 2452-414X, Doi: 10.1016/j.jii.2018.04.001 (www.sciencedirect.com/science/article/pii/S2452414X18300049).

14. P. Popovski, K. F. Trillingsgaard, O. Simeone and G. Durisi, "5G wireless network slicing for eMBB, URLLC, and mMTC: a communication-theoretic view," *IEEE Access*, vol. 6, pp. 55765–55779, 2018, Doi: 10.1109/ACCESS.2018.2872781.

15. B. Tahir, S. Schwarz and M. Rupp, "BER comparison between convolutional, turbo, LDPC, and polar codes," *2017 24th International Conference on Telecommunications (ICT)*, Limassol, Cyprus, 2017, pp. 1–7, Doi: 10.1109/ICT.2017.7998249.

16. M. H. Alwan, M. Singh and H. F. Mahdi, "Performance comparison of turbo codes with LDPC codes and with BCH codes for forward error correcting codes," *2015 IEEE Student Conference on Research and Development (SCOReD)*, Kuala Lumpur, Malaysia, 2015, pp. 556–560, Doi: 10.1109/SCORED.2015.7449398.

17. https://www.3gpp.org/NEWS-EVENTS/3GPP-NEWS/1875-MC_SERVICES.

18. E. Calvanese Strinati, S. Barbarossa, J. L. Gonzalez-Jimenez, D. Kténas, N. Cassiau and C. Dehos, "6G: the next frontier," *arXiv:1901.03239v2 [cs.NI]* 16 May 2019.

19. Y. Yuan, Y. Zhao, B. Zong, et al. "Potential key technologies for 6G mobile communications," *Science China Information Sciences* vol. 63, no. 183301, 2020. Doi: 10.1007/s11432-019-2789-y.

20. S. Elmeadawy and R. M. Shubair, "6G wireless communications: future technologies and research challenges," *2019 International Conference on Electrical and Computing Technologies and Applications (ICECTA)*, Ras Al Khaimah, United Arab Emirates, 2019, pp. 1–5, Doi: 10.1109/ICECTA48151.2019.8959607.

21. A. Asadi, Q. Wang and V. Mancuso, "A survey on device-to-device communication in cellular networks," *IEEE Communications Surveys & Tutorials*, vol. 16, no. 4, pp. 1801–1819, Fourthquarter 2014, Doi: 10.1109/COMST.2014.2319555.

22. K. B. Letaief, W. Chen, Y. Shi, J. Zhang and Y. A. Zhang, "The roadmap to 6G: AI empowered wireless networks," *IEEE Communications Magazine*, vol. 57, no. 8, pp. 84–90, August 2019, Doi: 10.1109/MCOM.2019.1900271.

23. W. Saad, M. Bennis and M. Chen, "A vision of 6G wireless systems: applications, trends, technologies, and open research problems," *IEEE Network*, vol. 34, no. 3, pp. 134–142, May/June 2020, Doi: 10.1109/MNET.001.1900287.

24. P. Yang, Y. Xiao, M. Xiao and S. Li, "6G wireless communications: vision and potential techniques," *IEEE Network*, vol. 33, no. 4, pp. 70–75, July/August 2019, Doi: 10.1109/MNET.2019.1800418.

25. C. Bockelmann et al., "Towards massive connectivity support for scalable mMTC communications in 5G networks," *IEEE Access*, vol. 6, pp. 28969–28992, 2018, Doi: 10.1109/ACCESS.2018.2837382.

26. B. Chen, J. Wan, L. Shu, P. Li, M. Mukherjee and B. Yin, "Smart factory of industry 4.0: key technologies, application case, and challenges," *IEEE Access*, vol. 6, pp. 6505–6519, 2018, Doi: 10.1109/ACCESS.2017.2783682.

27. S. K. Sharma, I. Woungang, A. Anpalagan and S. Chatzinotas, "Toward tactile internet in beyond 5G era: recent advances, current issues, and future directions," *IEEE Access*, vol. 8, pp. 56948–56991, 2020, Doi: 10.1109/ACCESS.2020.2980369.

28. M. Khoshnevisan, V. Joseph, P. Gupta, F. Meshkati, R. Prakash and P. Tinnakornsrisuphap, "5G industrial networks with CoMP for URLLC and time sensitive network architecture," *IEEE Journal on Selected Areas in Communications*, vol. 37, no. 4, pp. 947–959, April 2019, Doi: 10.1109/JSAC.2019.2898744.

29. A. O. Salau, N. Marriwala and M. Athaee, "Data security in wireless sensor networks: attacks and countermeasures," *Lecture Notes in Networks and Systems*, vol. 140, pp. 173–186, 2021. Doi: 10.1007/978-981-15-7130-5_13.

30. https://www.youtube.com/watch?v=i28L70go9PM.

10 Risk Perception, Risk Management, and Safety Assessments
A Review of an Explosion in the Fireworks Industry

N. Indumathi, R. Ramalakshmi,
N. Selvapalam, and V. Ajith
Kalasalingam Academy of Research and Education.

CONTENTS

10.1 Introduction 178
10.2 Composition..178
10.3 Manufacturing Process... 179
10.4 Field Study 181
10.5 Hazards in Fireworks Industries...181
    10.5.1 Fire Accidents.. 182
    10.5.2 Chemical Risk Factors... 183
    10.5.3 Study of the Workers 184
    10.5.4 Analysis of Safety.. 184
    10.5.5 Workers Safety Using Regression Analysis........................... 185
    10.5.6 Safety Environment Prediction Using Chi-Square Analysis... 185
    10.5.7 Job Safety Analysis.. 185
10.6 Findings 186
    10.6.1 Lack of Training 186
    10.6.2 Usage of Personal Protective Equipment............................... 187
    10.6.3 Health Issues in Fireworks Industries.................................... 187
    10.6.4 Causes for Accidents.. 188
    10.6.5 The Age Group Dispersion in Fireworks Industries............... 190
10.7 Conclusions...191
References...191
Book 193
Conference 193
Website... 193

DOI: 10.1201/9781003224068-10

10.1   INTRODUCTION

India is the second largest country that has manufacturing units of fireworks, and Sivakasi is the town that acts as a hub for the manufacturing of fireworks, printing, and match works. Thus, this city was named as "Kutty Japan" (which means "Little Japan" in Tamil) by the late Prime Minister Jawaharlal Nehru. All these industries of Sivakasi provide employment for over 250,000 people with an estimated turnover of 20 billion (US\$290 million) (Rajathilagam and Azhagurajan, 2012). The city Sivakasi is hotter than the nearby cities, and thus, it could be considered the god's gift for the production of fireworks, which requires extensive drying process. Since it has a suitable climate of having low rainfall and a dry climate, it contributes to the large production of fireworks. The fireworks have been utilized by the people as a mark of joy and happiness and also on the occasions of marriage, political victory, and many other events. The fireworks industry has slowly grown up and occupies a large share in the Indian economy. The market potential for fireworks is in all likelihood to grow at a rate of 10% per annum (BBC, 23 Sep 2012). Figure 10.1 represents the study area of the present study—Sivakasi and nearby premises, which is covered by the dry land.

Explosive chemicals and flammable materials are used in the fireworks to produce light, sound, and smoke. Thus, accidents in fireworks industries are inevitable through expected and unexpected zones of the industries, even though considerable precautionary measures have been incorporated by these industries. In terms of accidents, some of the hurdles faced by these industries are by the illiterate workers who come to these industries from the nearby villages. Thus, these illiterate workers are not aware of the chemicals' explosive properties, when they perform the process of manufacturing them, which often creates problems for these industries especially from a safety point of view (Rajanna, 2015). The Factories Act, 1948, and the Explosives Act are enforced in the safety of fireworks industries to avoid an explosion. The effective implementation of the acts and periodic inspections of the factories may reduce accidents; however, human error cannot be ignored in most of the accidental occasions.

10.2   COMPOSITION

Firecrackers produce light, sound, color, and smoke based on the composition of pyrotechnic chemicals, which takes a major part in producing fireworks. It involves the fundamental concepts of physics and chemistry.

Figure 10.2 depicts a standard composition of a firework. Charcoal and the oxidizing agents stimulate the burning property. Reducing agents such as sulfur is required to burn the oxygen formed by the oxidizing agent to produce hot gases, while regulators speed up the method of burning. At high temperature, metals such as copper, strontium, and barium offer excellent colors of attraction to the

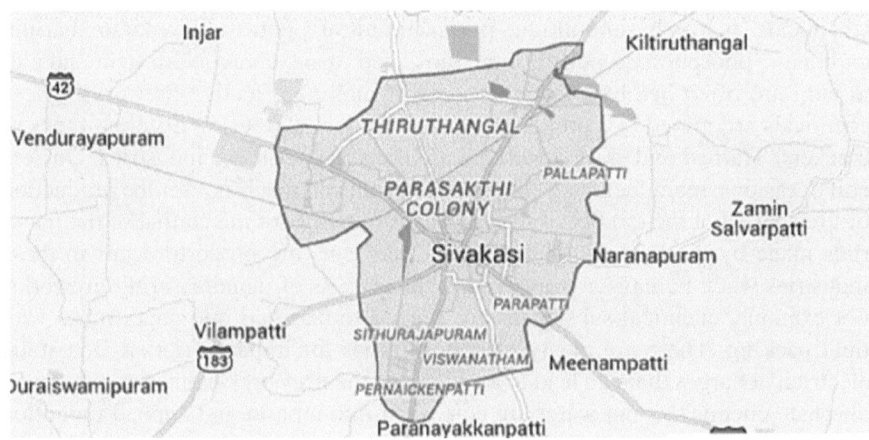

FIGURE 10.1   Study area.

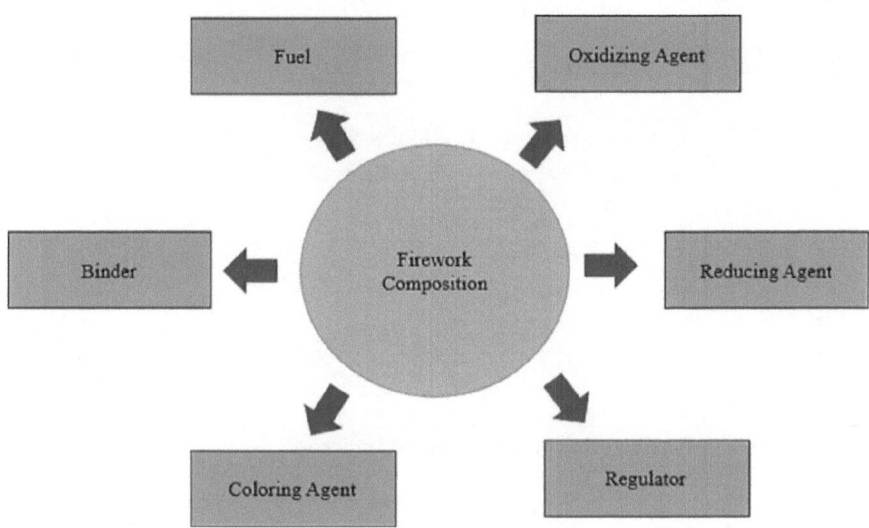

FIGURE 10.2   Firework chemical composition.

fireworks. Binders often have no role, but they will keep the combination together (Ghosh and Khatsuria, 1987).

10.3   MANUFACTURING PROCESS

Manufacturing fireworks requires certain skills, production methods, and often many undefined parameters, which are often trade secrets of these industries.

Chemicals such as barium nitrate, potassium nitrate, potassium chlorate, barium carbonate, phosphorous, sulfur, iron chips, lead oxide, charcoal, dextrin and aluminum are often involved in the process of making effective fireworks. These chemicals are mixed in a proper proportion for manufacturing different types of crackers. Trained and skilled workers are the assets of these industries. On several occasions, manufacturing companies introduced machines for the production of fireworks, but the quality of the crackers often did not match that of the materials made by the humans. Nevertheless, accidents are prone to occur in these industries when humans are involved in the process of manufacturing fireworks. For example, chemical substances are needed to be filled into paper tubes with tight packing. There are plenty of opportunities for impact, friction, and static electricity charges that can lead to accidents in the fireworks industries. Charcoal, alternate chemicals, and water are combined into a paste and applied on cotton wicks. After drying, the wicks are cut to the required length and attached to the crackers and specific fireworks. The fuses are inserted and allowed to dry. For

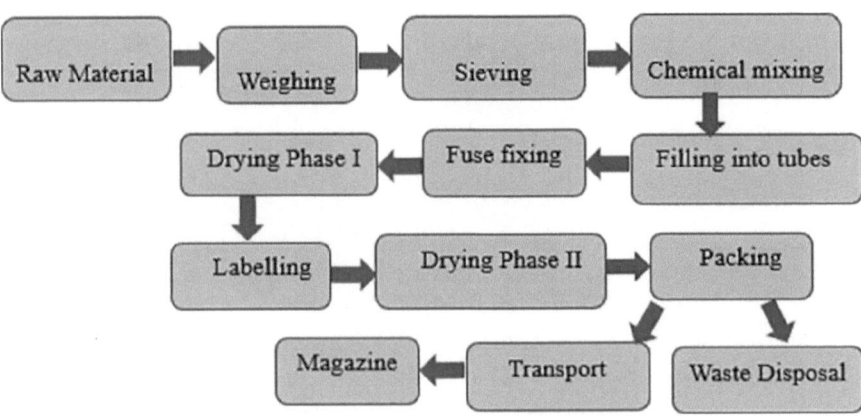

FIGURE 10.3 Manufacturing process of ground crackers.

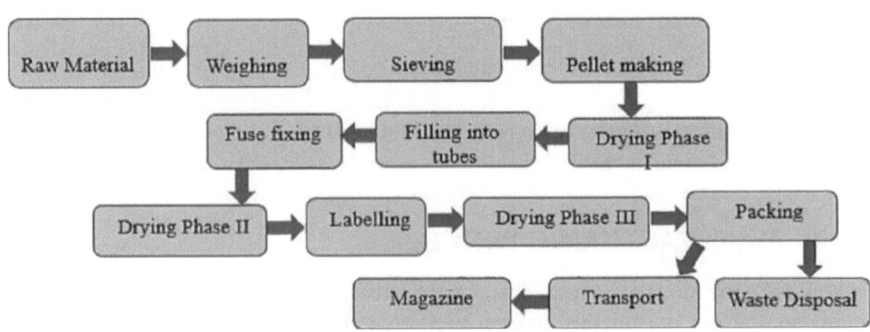

FIGURE 10.4 Manufacturing process of fancy crackers.

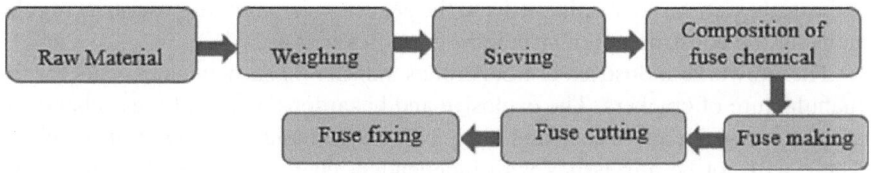

FIGURE 10.5   Manufacturing of fuses.

drying the goods, specially prepared systems are used. Typically, the common strategy is to dry the products twice inside the production unit, once during the process of fixing the fuse and again after the products are finished. Therefore, there is a risk of mud accumulation and overheating that can result in accidents (Ravi and Gandhinathan, 2013).

Fireworks are routinely dried in the bright sunlight and packed manually. These packed materials are held within the manufacturing unit, which is known as "magazine." The goods are shifted from the magazine to warehouses by means of trucks and other modes in closed/open containers. Appropriate precautions ought to be taken to avoid accidents during the loading of finished goods. Careless handling, impact on overloading, and dragging of materials might cause accidents. Every procedure involves manual handling of finished products, which is ready to explode by heat, impact, vibration, and collision, and thus, workers are at risk in every movement of the finished products. The withering powders from the finished products also cause health issues to the workers (Ajith et al., 2019). The fireworks are classified into four categories such as ground crackers, aerial crackers, sparklers, and amorces. The firework manufacturing process of ground and aerial crackers is shown in Figures 10.3 and 10.4, and the fuse making process is shown in Figure 10.5.

10.4   FIELD STUDY

Data collection was performed in fireworks industries, to quantify the sources of risk to the workers, environment, and manufacturers. The data for the study were collected from 20 firework manufacturing industries through the survey questionnaire. The survey was designed to gather knowledge concerning the knowledge of workers, infrastructure amenities, safety aspects, chemical handling, awareness, training, processing risk, and health issues. Finally, the risk factors that manufacturers must consider in order to comprehend the related issues of the fireworks industries have been discussed.

10.5   HAZARDS IN FIREWORKS INDUSTRIES

Fireworks factory construction layout is more important for fireworks manufacturing. Every fireworks industry must be constructed as per the norms of Factories Act, 1948. The manufacturing of firework products near residential areas is not allowed due to the explosive nature of the industry. For this reason, an isolated industrial estate has to be planned and the government has to impose strict regulations for the construction of such industries (Sekar et al., 2010).

The fireworks industries use hazardous and non-hazardous chemicals for the manufacture of crackers. The explosive and hazardous nature of these chemicals should be considered in each and every process of manufacturing of the crackers. There will not be any issues with independent chemicals until they have been mixed for the production of crackers.

During the mixing processes of chemicals and the drying of these mixed powders, there is a tendency for explosion, and several unknown factors need to be considered as they are the major cause of the firework explosion and also produce health issues to the workers (Poulton and Kosanke, 1995). The possible causes of accidents during mixing processes could be impact, friction, buildup of static electricity, and human factors leads to human errors. The products are dried by placing them on a dedicated place wherein an accumulation of dust and overheating might also induce health injuries. After the production of fireworks, they are stored locally by various modes of transportation, either within the factory or in nearby storage locations of the factory. During this period, careless handling, impact loading, overloading, and dragging of materials may lead to accidents (Katoria et al., 2013). The risk assessment takes place during the manufacturing processing, while the process from storage of chemicals to transportation have already been reported in Indian fireworks manufacturing. According to the reports, 36% of the accidents were due to ignition stimulus caused due to friction sensitiveness and 25% were due to impact sensitiveness. The third contributing factor could be the decomposition of chemicals, which occurs due to the presence of moisture or impurities during storing and processing, and the other sources of risk are lightning and static electricity (Bauer, 2013).

10.5.1 FIRE ACCIDENTS

Optimum temperature and humidity are needed for fireworks industries. Either high temperature or low humidity will cause the fireworks chemicals to self-ignite, which generally should be avoided. The environmental changes are unpredictable, and thus, environmental changes are of major concern to the firework manufacturers. The environmental conditions (weather conditions) of the calendar year will be given more importance by the manufacturers, since they play a major role in promoting the sources of risk to fireworks industries. Besides, selection of time for the chemical mixing, drying duration of the crackers, disposing time of the waste will be decided based on the environmental conditions such as temperature of the day and humidity. Excess heat and low humidity will often induce accidents as mentioned before, and thus precautions need to be taken when carrying out all these processes (Muthurajam and Sathiabama, 2015). These conditions majorly affect the places such as fancy crackers production unit and also the production spot of main raw materials such as gun powder. The gunpowder is a combination of oxidizers, sulfur, and charcoal. Very often, the accidents do not occur at the initial stage, but they are prone to happen during the drying process. On some occasions, some of the chemicals such as oxidizer could form a slag, which tends to self-ignite while sieving or transportation. So optimum drying duration of the day and extensive ventilation are essential to avoid heat accumulation. Even small negligence will lead to accidents. The mortality rate of gunpowder explosion burns from fireworks accidents is excessive (Chen et al., 2002). Most of the deadly accidents in the fireworks industry happened in filling and mixing sections. Mechanical effects, weather, flash composition, and chemical reactivity are the predominant contributors to accidents (Surianarayanan et al., 2008). There are potential reasons behind these accidents, which are most likely the physical (mechanical, thermal, electrical), natural, chemical, and organizational elements. Accidents that occur at fireworks storage locations and storage points would be the worst because it would have devastating effects on workers, facilities, people, and the environment (Pirone et al., 2016).

10.5.2 Chemical Risk Factors

One of the main responsibilities of the employer is the implementation of Occupational Safety and Health Act 1994 to protect the public and employees from the detrimental effects of chemical substances spread out inside the premises of the industry. A chemical fitness hazard evaluation (CHRA) on chemical use is performed periodically to evaluate the workers' physical fitness (Husin et al., 2012). The use of chemicals in the fireworks industry is inevitable and leads to many hazards such as fire and explosion, and occupational hazards and injuries such as chemical skin burns , respiratory diseases, reproductive problems, and even death (Freivalds and Johnson, 1990). Burning of waste products of the fireworks industries may produce carbon monoxide. Although there is less possibility of exposure to these smokes for the workers, if they get exposed to carbon monoxide, it may find its way to the bloodstream through inhalation and would lead to carboxyhemoglobin formation in the bloodstream. Continuous production of carboxyhemoglobin would lead to headache, damage to the central nervous system, and eventually death (Poulton and Kosanke, 1995). Similarly, at the explosion of fireworks, particles of metals such as iron and copper may have the possibility to enter the inner sections of the cornea, which will result in retinal trauma, cataract formation, and glaucoma (Al-Tamimi, 2014). The atmospheric contaminations in the form of particles are any form of risk quotients, risk indices, and cancer risks are by the particulate matter of PM_{10}. Although there are many other air pollutants that can cause breathing problems, cardiovascular problems, bronchitis, lung infection, lung fibrosis, deep vein thrombosis, and lung cancer. PM and other pollutants, such as O3 and NO2, are more likely to cause these unfavourable outcomes (Chalvatzaki et al., 2019). Fireworks industry uses flash powder. The flash powder mixture is highly sensitive and produces electrostatic discharge and friction. Flash powder mixture is highly likely to lead accidents and also illness (Azhagurajan, and Selvakumar, 2014). Dust explosions are a frequent hazard in pyrotechnics industries and other industrial environments. Dust explosions are one of the causes of mortality, contamination, and adverse environmental effects.

A flammable dust is any quality material that has the potential to seize hearth and explode when mixed with air. The waste disposal of chemical mixing produces lots of dirt or pollution to the surroundings (Taveau, 2014). Exposure to these chemicals often results in various short- and long-term health problems such as poisoning, pores and skin rashes, and disorders of the lung, kidney, and liver (Gouder and Montefort, 2014).

10.5.3 STUDY OF THE WORKERS

In the fireworks industry, employees are mostly illiterate and also the age group is around 20–58. The workers have chosen this particular work due to lack of education, interest, or economic constraints. Employee's concentration is more important for this work. Carelessness is the primary stimulus of an accident (Vijayaragavan, 2014). The industry wishes to regulate the employee reward system, and promotions should be granted solely on the basis of work involvement and experience. If those factors are given a little more attention, the company will be able to retain desirable employees who have a high level of satisfaction, organizational dedication, and involvement (Ananthi, 2017). The issues of excellence in work–life balance of the labors have been analyzed for providing better working conditions and also for providing a more efficient environment for working with flexibility and proper balance between their work and personal. The stress or lack of concentration and human error lead to accidents (Selvakumar and Jegatheesan, 2017). A majority of the workers inside the fireworks industry are women. The standard of living of the women workers' families depends on the earning of the women in the family. Simultaneously, women are facing social, economic, health, psychological, and organizational problems (Saravana Kumar and Karunanidhi, 2016).

10.5.4 ANALYSIS OF SAFETY

Over 90% of the workforce for the fireworks industries is generated from the nearby villages, and most of them have not completed their secondary education and thus they lack knowledge of the chemicals that they handle for making fireworks. A process safety analysis is carried out on the manufacturing steps and the system security. This safety analysis produces safe environment for manufacturing and also reduces hazards in the workplace (Moshashaei and Alizadeh, 2017). However, unsafe operations of chemicals always lead to accidents due to chemical spills, friction, electricity, dragging, transport moisture, waste disposal, etc. Industries and workers should follow the safety precautions to prevent accidents (Rajathilagam & Azhagurajan, 2012). To create a safe environment, it is crucial that the management ought to have the dedication to set up a secure environment. The management has to provide top priority to the safety of the workers. On the other hand, workers should be alert while handling these explosive chemicals while at work. They have to observe protection precautions and regulations of the factory with proper protection training, and thus, the employees should be offered an effective protection education time to time to prevent any kind of unexpected accidents (Sales et al., 2007).

10.5.5 Workers Safety Using Regression Analysis

Firework employees have to carefully handle the dangerous materials that lead to skin irritation, ulcers, asthma, and many other complications. The regression analysis has been used to find the chemical hazards of the composition in fireworks manufacturing. The regression model may be used to analyze the processes where accidents are prone to occur, and based on that, automated machineries can be substituted to handle those processes to prevent accidents. Proper training can be encouraged on the hazards of chemical substances they handle in the work environment. Awareness about personal hygiene and right protection devices such as gloves and apron may be developed, and training should be given for using the modern protective equipment (Porchelvi and Devi, 2015).

10.5.6 Safety Environment Prediction Using Chi-Square Analysis

Rajathilagam conducted a chi-square test to analyze the safety of the environment in the fireworks industries. The exclusive forms of speculation have been framed and distributed to the company and the contract employees. The chi-square test showed that it would not be appropriate to examine the company/contract employees with the action taken toward the employees.

All other assumptions, such as employee opinions, security budget, danger level, respondent knowledge, and firehouse assessments, were found to be highly accurate in terms of matching expectations, seducation, and firefighting checks, matched the expectations very well (Rajathilagam, 2016).

10.5.7 Job Safety Analysis

Job safety analysis (JSA) is a technique of accident prevention method that can identify the potential hazards associated with the job and give control measures to minimize the hazards. In this job safety study that was carried out, each potential possibility of the industrial hazards associated with the manufacturing work was defined under different sections and effective control measures were recommended by preparing JSA worksheet. This study clearly shows that the incidents were mostly due to human errors and the accidents can be considerably avoided with efficient safety management with adequate training and education for the employees (Singh and Banodha, 2019).

JSA is a systematic process that involves breaking the job/task into many phases and then evaluating each step of the process and finally predicting the steps that involve potential hazards. Consider the following example: the phases involved in a specific study can be characterized as follows: the work process, observation, dividing the work process, identification of hazards, and the control methods for each of these processes. This safety analysis mainly focused on the processes and the corresponding hazards involved in them. The result of analysis ensures the safety measures recommendation of each process to reduce the risk and provide the safe working condition (Pandey et al., 2019). Table 10.1 represents the identified processing hazards and also suggests recommendations for that work process.

10.6   FINDINGS

To analyze the risk factors in the fireworks industries, the authors have framed a questionnaire with definite key parameters that include age, process, training, chemical management, awareness, protective equipment, and health issues. The workers, foremen, and owners and other stakeholders were requested to examine the questionnaire, and some parameters were obtained from discussions.

10.6.1   LACK OF TRAINING

The Factories Act, 1948, states that proper training be given to the workers in the fireworks industries; however, the results of the questionnaire showed that 76% of employees were not given any form of training for the manufacturing process and on the safety issues, while 24% of respondents mentioned that good training

TABLE 10.1
Hazard Identification

Process	Identification of Hazards	Recommendations
Taking chemicals from storeroom	Slips and falls	Wear anti-slip boots, apron, and shoes when working
Mixing of chemicals	Irritation of the skin of the body, headache, and kidney problems; prolonged exposure to chemicals	Nose mask, protective gloves, and eye covers
Weighing for optimal product mixing	Lung problems due to dust accumulation	Wear anti-slip gloves and shoes while at work
Switch to adding fuse	Eye and nose irritation, back pain, kidney infection, and muscle weakness	Use fully closed goggles, wrist gloves, and apron
Blending in trays	Friction sparks and fire	Use wooden plates
Chemical tube filling	Static charges causing fire	Use metal plates outside the wall of the room
The method of pallet making	Fire hazard, stored gun powder paste causes thermal decomposition that can result in fire, drying in open sunlight for a long time can result in fire, dragging the bag can result in friction causing fire	Correctly use PPE, such as fire-retardant apron and neoprene gloves; wash hands properly with soap water every time after handling chemicals; train the staff on hazards and working procedures

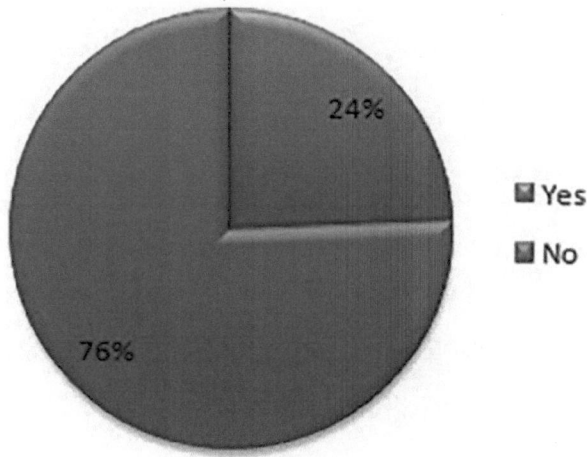

FIGURE 10.6   Lack of training.

has been offered to them on the manufacturing process and the safety-related precautions represented in Figure 10.6. In our opinion, those who are at the foremen level were given proper training, while workers at lower grades who do a very much monotonous job were not given any training because it does not require any skill for the performance of that job.

Safety-related training for those 76% people were felt not required because they rarely faced any accidents while performing the routine jobs that have been given to them. Nevertheless, a good training program to those employees may improve the overall safety of the industries, which they should consider in the long-term perspectives.

10.6.2   Usage of Personal Protective Equipment

Fireworks industries suggested the workers to use personal protective equipment (PPE), for the avoidance of health problems caused by the chemicals. However, Figure 10.7 illustrates that the workers are not interested in wearing PPE due to the uneasiness. From the data analysis, it was found that 82% of employees are not interested in wearing PPE during working hours despite the presence of potential health hazards due to airborne particles of the industry and 18% of population of the industries wear PPE or other personal protective equipment for their safety. A new type of comfortable protective equipment would solve this issue.

10.6.3   Health Issues in Fireworks Industries

The chemicals of the fireworks industries are not harmful as independent chemicals. For example, sulfur helps remove mercury-like toxic metals and charcoal has routinely been used by the people for cooking. The various chemical oxidants used in these industries are water-soluble products, but they are ignitable

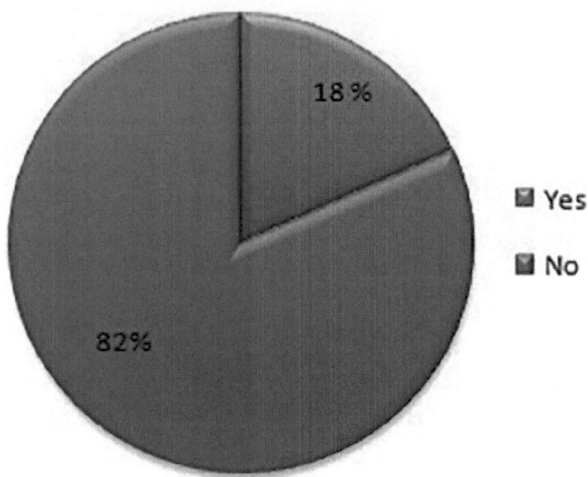

FIGURE 10.7 Use of personal protective equipment.

and act as explosive only in combination with the other two chemicals. However, long-term exposure to these chemicals will have an impact on the health of the workers for that the workers are expected to use PPE and masks. According to our survey, the workers get affected by fever, headache, bone problems, inhalation problems, and other forms of problems. For example, 7% of the workers are facing ENT problems and 6% of the workers are facing some sort of eye problems. Overall, 17% of workers are affected by inhalation problems and 22% of the workers suffered headache and fever occasionally. A moderate amount population of the industry was affected by lung problems, 11% of the workers reported having kidney problems, a less amount of population (8%) had stomach problems, and 15% of the population expressed bone problems. Finally, a small population (4%) of the industry disclosed having heart deceases. Figure 10.8 explains clearly that no medical report has been collected from the workers to authenticate the above-mentioned medical problems. It is purely based on their responses given to us through the questionnaire data.

10.6.4   CAUSES FOR ACCIDENTS

Our utmost interest was to know how and on what occasions the explosion occurs, so that we could provide a solution to the fireworks industry. When we had discussions with them, every one of them provided various reasons for the explosive accidents. For example, from foreman's point of view, static electricity and human errors made while handling the substances are the major cause for the accidents. According to the survey and as Figure 10.9 represents, 5% employees mentioned the reason of temperature, 6% people mentioned human errors as the reason, 9% population mentioned the reason of impact friction, 3% people mentioned the reason of transportation of finished products, 9% people pointed

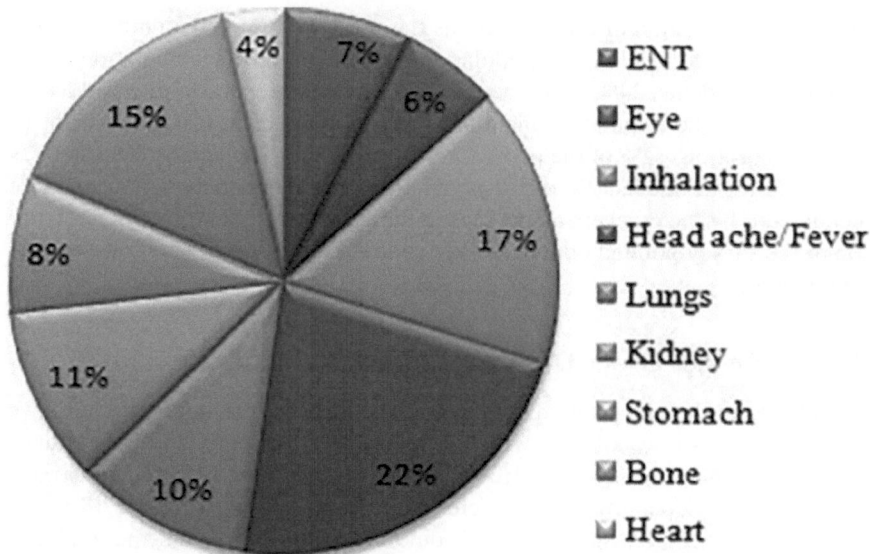

FIGURE 10.8   Health issues.

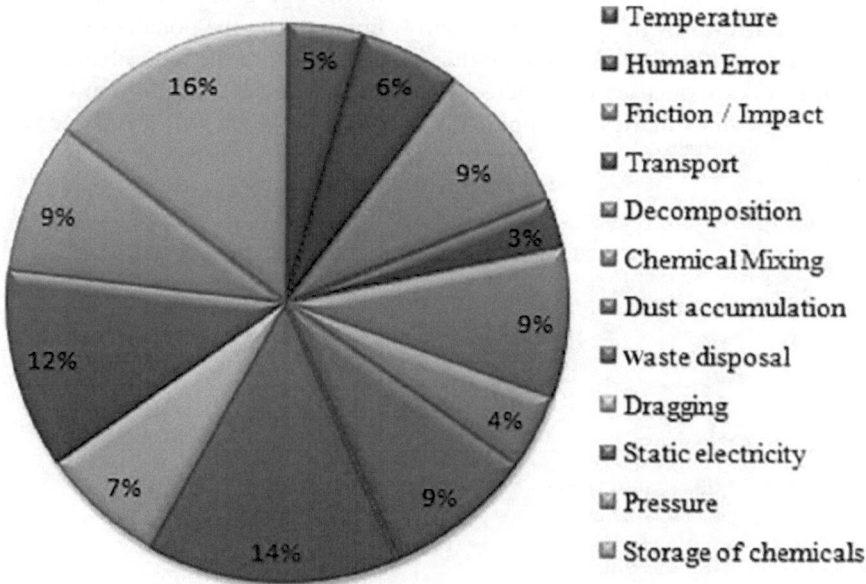

FIGURE 10.9   Causes for accidents.

out the decomposition of the crackers as the reason, 4% population blamed the chemical mixing duration, 9% population pointed out dust accumulation as the reason for accidents, 14% of the people blamed improper waste disposal as the reason, 7% of the population mentioned products or chemicals movement or dragging duration, 12% of the population strongly suggested static electricity as the reason for accidents, 9% of the population gave importance to the pressure encountered by the crackers as the reason for accidents, and 16% of the population of the survey pointed out the storage room of the crackers as a epicenter for the major accidents.

10.6.5   THE AGE GROUP DISPERSION IN FIREWORKS INDUSTRIES

To understand the age group dispersion of the fireworks industry workers, we have taken the survey of age, which reflects the experience and understanding of the workers toward the fireworks industry. Figure 10.10 demonstrates the age group dispersion and it is classified into five categories as 18 to 19, 20–29, 30–39, 40–49, and 50–59 for our understanding. From the analysis, it is observed that 2% population belong to the age group of under 20, 17% population belongs to the age group of 20–29 years, 30% population belongs to the age group of 30–39 years, 32% population belongs to the age group of 40–49 years, and 21% population belongs to the age group of 50–99 years. Thus, the fireworks industries are majorly supported by the population of the age group of 40–60 years, which indicates that in future the labor demand of the fireworks industries may increase, which can be balanced by the invention of new machineries to support the production, which will minimize the human loss.

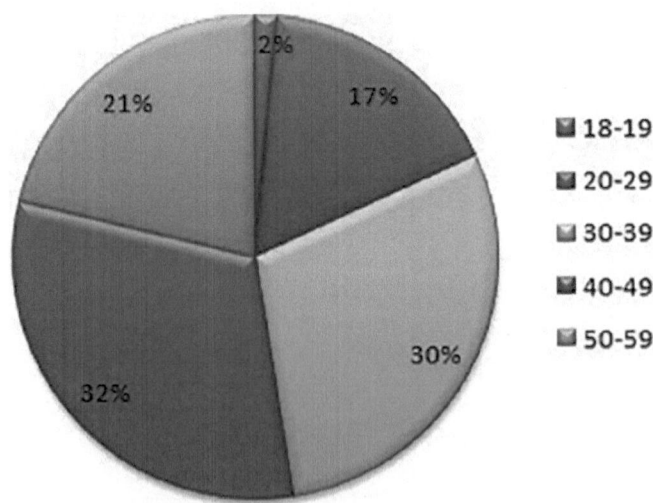

FIGURE 10.10   Age group dispersion of employees in fireworks industries.

10.7   CONCLUSIONS

Fireworks are a symbol of joy and happiness, and every age group of human population enjoys fireworks on various occasions. However, the production of fireworks involves many threat to humans working in these industries. Nevertheless, the accidents are not very common in these industries and often occur due to human errors and many unknown factors, which are yet to be identified. We have made this survey to evaluate the basic needs of the industry and also to recognize the potential risk factors associated with these industries. To minimize the risk factors in these industries, the industries should follow the safety measures as much as possible and substitute suitable machineries wherever the risks are high. Our recommendations for the safe running of these industries are given below:

- A good manufacturing protection guide should be prepared and implemented.
- Periodic safety training for fireworks producers and employees is required.
- Reorganization of danger spots for frequent safety inspections is required.
- Nanotechnology should be implemented for making environment-friendly crackers that may produce less pollution and a cleaner environment.
- Health issues can be improved through periodic medical camps to the workers.
- Storage places of the finished products should be monitored periodically.
- Accident-free manufacturing units should be encouraged by incentives.
- Automation should be encouraged for the safe production.
- Wastes disposed from the fireworks industries should be utilized for electricity or energy production.

REFERENCES

1. Rajathilagam, N., & Azhagurajan, A. (2012). "Accident analysis in fireworks industries for the past decade in Sivakasi". *International Journal of Research in Social Sciences*, 2(2), 170–183.
2. Rajanna, K. A. (2015). "Nature of work, working conditions and problems of women construction workers: A case study". *International Journal of Business Quantitative Economics and Applied Management Research*, 1(9), 70–85.
3. Ravi, A., & Gandhinathan, R. (2013). "Analysis of safety climate in fireworks industries in Tamilnadu". *International Journal of Science and Engineering Research*, 4, 760–764.
4. Ajith, S., Sivapragasam, C., & Arumugaprabu, V. (2019). "A review on hazards and their consequences in firework industries". *SN Applied Sciences*, 1(1), 120.
5. Sekar, T., Ramaswamy, S. N., & Nampoothiri, N. (2010). "Planning of industrial estate for fireworks industries in Sivakasi". *International Journal of Engineering, Science and Technology*, 2(6), 2207–2217.
6. Poulton, T. J., & Kosanke, K. L. (1995). "Fireworks and their hazards". *Fire Engineering*, 148(6), 49–66.
7. Katoria, D., Mehta, D., Sehgal, D., & Kumar, S. (2013). "A review of risks to workers associated with fireworks Industry". *International Journal of Environmental Engineering and Management*, 4(3), 259–264.

8. MuthuRajam, S.P., & Sathiabama, G., (2015). "A case study in environmental constraints, causes and remedies of industrial town Sivakasi". *International Journal of Humanities and Social Science Invention*, 4(1), 34–39.

9. Chen, X. L., Wang, Y. J., Wang, C. R., & Li, S. S. (2002). "Gunpowder explosion burns in fireworks factory: causes of death and management". *Burns*, 28(7), 655–658.

10. Surianarayanan, M., Sivapirakasam, S. P., & Swaminathan, G. (2008). "Accident data analysis and hazard assessment in fireworks manufacture". *Science and Technology Energetic Material*, 69(5), 161–168.

11. Pirone, A., Vallerotonda, M. R., & Bragatto, P. (2016). "Lessons learned from recent accidents in fireworks establishments". *Chemical Engineering Transactions*, 53, 259–264.

12. Husin, S. N. H., Mohamad, A. B., Abdullah, S. R. S., & Anuar, N. (2012). "Chemical health risk assessment at the chemical and biochemical engineering laboratory". *Procedia-Social and Behavioral Sciences*, 60, 300–307.

13. Freivalds, A., & Johnson, A. B. (1990). "Time-series analysis of industrial accident data". *Journal of Occupational Accidents*, 13(3), 179–193.

14. Al-Tamimi, E. R. (2014). "A peculiar case of a retained inert piece of fireworks as an intraocular foreign body in the anterior chamber". *Saudi Journal of Ophthalmology*, 28(3), 225–227.

15. Chalvatzaki, E., Chatoutsidou, S. E., Lehtomäki, H., Almeida, S. M., Eleftheriadis, K., Hänninen, O., & Lazaridis, M. (2019). "Characterization of human health risks from particulate air pollution in selected European cities". *Atmosphere*, 10(2), 96.

16. Azhagurajan, A., & Selvakumar, N. (2014). "Impact of nano particles on safety and environment for fireworks chemicals". *Process Safety and Environmental Protection*, 92(6), 732–738.

17. Taveau, J. (2014). "Application of dust explosion protection systems". *Procedia Engineering*, 84, 297–305.

18. Gouder, C., & Montefort, S. (2014). "Potential impact of fireworks on respiratory health". *Lung India: official organ of Indian Chest Society*, 31(4), 375.

19. Vijayaragavan T. (2014). "A study on perception of the employees relating to the job satisfaction at standard fireworks in Sivakasi". *International Journal of Management Research*, 363–374.

20. Selvakumar, M., & Jegatheesan, K. (2017). "Quality of work life of workers in fireworks industry–a study with reference to Sivakasi, Tamilnadu". *Humanities and Management*, 656–671.

21. Saravana Kumar, R.C., & Karunanidhi G. (2016). "A study on problems pertaining of women labourers in fireworks industry with special reference to Sivakasi". *Global Journal of Research Analysis,* 166–168.

22. Moshashaei, P., & Alizadeh, S. S. (2017). "Fire risk assessment: a systematic review of the methodology and functional areas". *Iranian Journal of Health, Safety and Environment*, 4(1), 654–669.

23. Sales, J., Mushtaq, F., Christou, M. D., & Nomen, R. (2007). "Study of major accidents involving chemical reactive substances: analysis and lessons learned". *Process Safety and Environmental Protection*, 85(2), 117–124.

24. Porchelvi, R. S., & Devi, P. J. (2015). Regression model for the people working in fire work industry-Virudhunagar district. *International Journal of Scientific and Research Publication*, 5, 1–6.

25. Rajathilagam, N., (2016). Analysis of safety in fireworks industries by Chi square analysis in Virudhunagar district of Tamilnadu. *International Journal of Management and Social Science Research*, 5, 32–37.

26. Singh, N.K., & Banodha, V. (2019). Job safety analysis in fire crackers industry. *International Journal for Science and Advance Research in Technology*, 5, 67–69.
27. Pandey, A., Tripathi, M. K., Chadha, S. S., Tripathi, S., & Yadav, B. P. (2019). Implementation of JSA Methodology in Fireworks Manufacturing. *i-Manager's Journal on Future Engineering and Technology*, 15(1), 11.

BOOK

1. Bauer, A. J. (2013). *Analysis of Pyrotechnic Components*. Amaravathi Research Academy.
2. Ghosh, K. N., & Khatsuria, H. (1987). *The Principles of Fireworks*. Economic Enterprises, Sivakasi.

CONFERENCE

1. Ananthi, S., (2017). A study on job satisfaction of firework industries employees with special reference to Sivaksi Taluk. *International Conference on Recent Trends in Engineering Science, Humanities and Management (RTESHM-17)*, Tamil Nadu, 771–778.

WEBSITE

1. Russell, Michael S., (2009). *The Chemistry of Fireworks*, 2nd Ed., Royal Society of Chemistry. https://www.bbc.com/news/world-asia-india-19596089.

11 High-Utility Itemset Mining

Fundamentals, Properties, Techniques and Research Scope

V. Jeevika Tharini and B.L. Shivakumar
Sri Ramakrishna College of Arts and Science

CONTENTS

11.1 Introduction 195
    11.1.1 Utility Mining... 196
    11.1.2 High-Utility Itemset Mining.. 196
11.2 Frequent Itemset Mining and High-Utility Itemset Mining................... 197
    11.2.1 Frequent Itemset Mining 197
11.3 High-Utility Itemset Mining.. 200
11.4 Comprehensive Analysis of HUIM Techniques 204
    11.4.1 Two-Phase Algorithm... 205
    11.4.2 Faster High-Utility Itemset Mining (FHM) 205
    11.4.3 Efficient High-Utility Itemset Mining (EFIM) 205
    11.4.4 High-Utility Itemset Miner (HUI-Miner)................................ 205
    11.4.5 High-Utility Pruning Strategy (HUP-Miner) 206
    11.4.6 Utility Pattern Growth (UP-Growth)....................................... 206
    11.4.7 Utility List Buffer (ULB-Miner) 206
    11.4.8 Hybrid Technique by the Integration of UP-Growth and
            FHM (UFH-Miner)... 207
    11.4.9 Direct Discovery of High-Utility Itemset (D2HUP) 207
    11.4.10 Optimization Approaches for HUIM 207
11.5 Conclusions... 208
References.. 209

11.1 INTRODUCTION

Traditional mining algorithms such as frequent itemset mining (FIM) or association rule mining (ARM) mine the frequencies of every individual item in the transaction database considering the confidence value or minimum threshold value [1,2]. FIM and ARM are insufficient for finding highly profitable itemsets with less frequency in the transaction. For example, in a supermarket sugar may be sold hundred or more than a hundred kilograms per day, while fewer kilograms of cashew nuts may be sold for a certain period of time. The sugar has a higher occurrence, but sold with a less profit value, while the cashew nuts have lesser frequency with a higher profit value for traders. Both FIM and ARM find the frequent item sugar with lower profit from the transaction data, and to overcome the limitations of conventional algorithms, HUIM was developed and later several algorithms were proposed in utility mining by several researchers [3,4].

11.1.1 UTILITY MINING

One of the emerging topics in data mining is utility mining. It is the process of enlightening HUIs from a database. In this, every item has its utility value and it depends on the nature of the dataset. The item with a utility value higher than the minimum utility threshold is termed an HUI. FIM considers only the presence of items in a dataset. It does not consider the profit or quantity of items. Hence, FIM cannot be used to retrieve profitable items from a dataset. To overcome the above-mentioned drawback, the utility mining concept is introduced. Utility mining considers all the factors such as the quantity of items and unit profit of itemsets [5]. Utility mining can be applied to a quantitative database, i.e., a transaction database with the quantity of items sold. It also takes the unit profit of items as input [6].

11.1.2 HIGH-UTILITY ITEMSET MINING

HUIM tasks are more problematic than those of FIM [1]. HUIM is employed for identifying the necessary patterns from the transaction of a customer [6]. It is comprised of identified itemsets that holds a high-utility value, i.e., profit, which are termed as high-utility items (HUIs) [7]. In the perspective of consumer transaction analysis, HUIM also has applications in other domains, namely clickstream analysis and biomedicine [8,9]. HUIM is developed to find the items with a high-utility value from a quantitative database. In utility mining, the count of items that occurred in a transaction is the internal utility (IU) and the profit of individual items is an external utility (EU). Initially developed algorithms have met with the combinational problem, and to overcome the issue, the two-phase transaction-weighted utility (TWU) model and the transaction-weighted downward closure (TWDC) property were designed for mining HUIs [7]. HUIs are effectively mined using the pruning strategy and indexed projection mechanism [8]. The former HUIM algorithms are established to handle a vast search space with unique items [9,10].

The most provoking task in recent research has been efficiently mining HUIs. The user preference determines the utility rate of the product. Several algorithms were projected to spot the most profitable items. Still, the profitable item discovery process meets with various limitations such as higher time and memory utilization, generation of various and repeated candidate items and, in certain cases, negligence of some of the needed items from the item generation. Optimization problems play an important role in many application fields in overcoming the mentioned limitations. This chapter surveys various HUIM algorithms and their significant aspects.

The remaining of the chapter is arranged as follows: The FIM and HUIM are illustrated in Section 11.2, HUIM algorithms are detailed in Section 11.3, and the article is concluded in Section 11.4.

11.2 FREQUENT ITEMSET MINING AND HIGH-UTILITY ITEMSET MINING

This section describes the issue in FIM and then presents how it simplified into HUIM. Then the main properties of HUIM are explained and contrasted with the FIM.

11.2.1 FREQUENT ITEMSET MINING

The FIM contains patterns retrieved from the database that comprises transaction [11,12]. Generally, the transaction database is described as follows. Let $I = \{m_1, m_2, m_3, \ldots, m_n\}$ be a set of n individual items in a quantitative data store with a transaction set $T = \{t_1, t_2, t_3 \ldots t_n\}$, where every transaction in the data store lies within $T_r \in DB$. In this, T_r is a subgroup of I and has a distinctive identifier r called TID. A running example is stated in Table 11.1 with seven transactions and five items D to H.

The main intent of FIM is to identify the itemsets with high support (incidence of frequent items). Primarily, the itemset S is a finite set of items $S \subseteq I$. The cardinality of the notion is specified as $|S|$, in other words the count of the items in S itemset. An itemset S is of length p, or a p-itemset consists of p items ($|S| = p$). For instance, $\{D, E, F, G, H\}$ is 5-itemset, $\{D, E, F, H\}$ is 4-itemset, $\{D, F, G\}$

TABLE 11.1

Transaction Database

T_{id}	Transaction Item
A_1	$\{D, E, G\}$
A_2	$\{D, H\}$
A_3	$\{D, F, G\}$
A_4	$\{E, F, G, H\}$
A_5	$\{F, H\}$
A_6	$\{D, E, F, H\}$
A_7	$\{D, E, F, G, H\}$

is 3-itemset, and {*D, H*} is 2-itemset. The significant properties of FIM are described as follows.

Definition 11.1

The support value of an itemset S in the transaction DB is represented as sup(S) and termed as sup(S) = $\{|T| S \subseteq T \wedge T \in DB\}$; that is, the count of the transaction is S.

For example, the itemset's support {*D, E*} in the transaction is 3. The itemset exists in three transactions (A_1, A_6 and A_7). This definition for support is determined as relative support. The support value is also expressed in percentage, which is the total count of the transactions. The support value representation with percentage is defined as absolute support, and {*D, E*} has 60% absolute support since it occurs in three out of five transactions.

Definition 11.2

The minimum support value is determined by the user, that is the threshold value (min_sup>0). An itemset S is considered as a frequent itemset if its support value sup(S) is not less than the threshold min_sup, that is $\sup(S) \geq \min\_\sup$. Otherwise, S is considered as an infrequent itemset.

Definition 11.3

The main objective of FIM is to spot all the frequent itemset with min_sup in the transaction *DB*.

In the transaction DB shown in Table 11.1 and with min_sup value 3, there are 14 infrequent and 16 frequent itemsets. The frequent itemsets are given in Table 11.2.

TABLE 11.2
The Frequent Itemset with a Minimum Support of 3

Itemset	Support	Itemset	Support
{*D*}	5	{*D, H*}	3
{*E*}	4	{*E, F*}	3
{*F*}	5	{*E, G*}	3
{*G*}	4	{*E, H*}	3
{*H*}	5	{*F, G*}	3
{*D, E*}	3	{*F, H*}	4
{*D, F*}	3	{*D, F, G*}	3
{*D, G*}	3	{*E, F, H*}	3

The issues of FIM have been examined for about two decades, and several algorithms are projected to identify the frequent itemsets effectively, which include FP-growth, Apriori, LCM, H-Mine and Eclat. Moreover, FIM has numerous applications; the robust statement about the FIM is that it is able to identify the frequent patterns that are beneficial and interesting to the user. To address the significant limitations of traditional FIM, the HUIM is developed and the items are interpreted with numerical values, where the itemsets are elected based on the utility function. The user will determine the utility function, or it will be estimated based on the nature of the dataset.

Property 11.1

The support value used in this approach is monotone. Let there be two itemsets S and R such that S belongs to R. It follows the support value $\sup(S) \geq \sup(R)$.

For example, in the transaction DB shown in Table 11.1, $\{D, E\}$ has the support value 3 and the supersets of this support is $\{D, E, F\}$:2, $\{D, G, E\}$:2, $\{D, E, H\}$:2, $\{D, G, E, F\}$:1, $\{D, H, F, F\}$:2, $\{D, G, H, E\}$:1 and $\{D, E, F, G, H\}$:1. The monotonicity property of the support measure assists in identifying the frequent patterns as it assures all the supersets of an infrequent item are also termed as infrequent. Thus, an FIM will discard infrequent itemsets superset from the search space. If the FIM spots the itemset $\{G, H\}$ as infrequent, then the supersets of all $\{G, H\}$ will be eliminated from the exploration, which will lessen the search space. The search space and the itemset generation for the transaction DB in Table 11.1 are depicted in Figure 11.1. The minimum support count is assigned as 3, and the itemset that satisfies the condition is highlighted in gray color. The anti-monotonicity property of the support can be perceived in Figure 11.1, where the frequent and infrequent itemsets

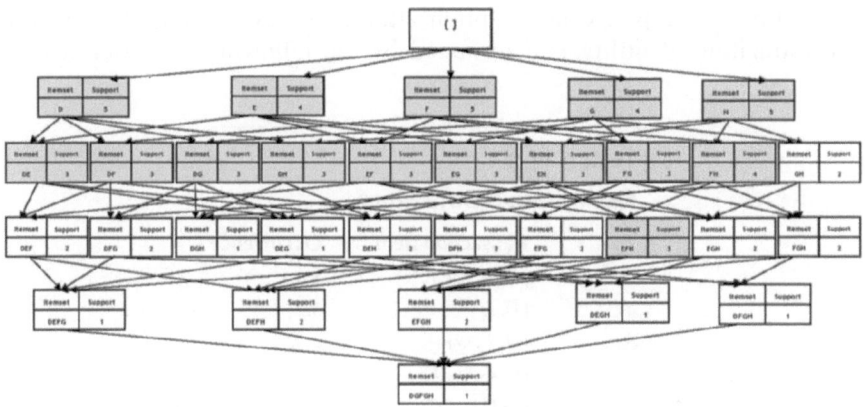

FIGURE 11.1 The search space of FIM for the transaction DB in Table 11.1 with a min_sup of 3.

are separated. This property is also stated as TWDC property or monotone or anti-monotonicity or Apriori property.

The frequent itemsets that satisfy the user-defined minimum support value 3 are $\{D\}$, $\{F\}$, $\{E\}$, $\{G\}$, $\{H\}$, $\{D, E\}$, $\{D, F\}$, $\{D, G\}$, $\{D, H\}$, $\{F, E\}$, $\{E, G\}$, $\{E, H\}$, $\{G, F\}$, $\{F, H\}$, $\{H, G\}$ and $\{E, F, H\}$, and these itemsets are enlightened in Figure 11.1. Among the 31 itemsets generated from the transaction DB in Table 11.1, 16 itemsets are frequent and 15 are infrequent itemsets.

11.3 HIGH-UTILITY ITEMSET MINING

The HUIM task identifies the useful patterns from the quantitative data store, where the additional details such as weight (relative significance) and quantity of every item are given to the user [13,14]. The quantitative data are given in Table 11.2, and the relative information is given in Section 11.1. The amount of the purchased items is characterized as an IU that is given in Table 11.1, and the profit of the item is characterized as an EU value that is specified in Table 11.3. According to the user preference, min_util value α is assigned.

To elucidate this definition, the transaction of the customer is given in Table 11.1 and it is utilized for the running example. The list of items $I = \{D, E, F, G, H\}$ denotes diverse products, namely soap, egg, bread, apple and biscuit. This example encompasses seven transactions $(A_1, A_2, A_3, A_4, A_5, A_6$ and $A_7)$, and transaction A_1 indicates the occurrence of purchased items D, E and G (IU D: 1, E: 4 and G: 1). The quantitative data store is specified in Table 11.3, and the unit profit of every individual item is specified in Table 11.4. The estimation of profit for every item and transaction is elucidated in the subsequent part.

The main intent of HUIM is to recognize the itemsets that occur in the quantitative database and have high profit or utility. The significance of an itemset in a transaction is estimated by the utility function. In this example, the utility is interpreted as the profit attained by every itemset. Generally, the estimation of utility is determined by the following definition and the

TABLE 11.3
Quantitative Data Store

T_{id}	Transaction Item	Occurrence
A_1	$\{D, E, G\}$	1,4,1
A_2	$\{D, H\}$	5,2
A_3	$\{D, F, G\}$	3,4,6
A_4	$\{E, F, G, H\}$	1,2,1,5
A_5	$\{F, H\}$	5,1
A_6	$\{D, E, F, H\}$	1,2,5,1
A_7	$\{D, E, F, G, H\}$	1,2,5,3,1

TABLE 11.4
Profit Values

Item	D	E	F	G	H
Profit	4	3	1	5	1

calculation for the running example using Tables 11.3 and 11.4 is given in the following section.

Definition 11.4

The utility of an item m_j in the transaction T_r is represented as $u(m_j, T_r)$ and indicated as

$$u(m_j, T_r) = \mathbf{r}(m_j, T_r) \times \mathbf{pr}(m_j) \tag{11.1}$$

The item D in the transaction A_1's value is computed as

$$u(D, A_1) = r(D, A_1) \times pr(D)$$
$$u(1, A_1) = r(1 \times 4) = 4$$

Definition 11.5

The utility of an item m_j in the DB is represented as $u(m_j)$ and indicated as

$$u(m_j) = u(m_j, A_p) + \cdots + u(m_n, A_n) \tag{11.2}$$

The item D in the DB value is computed as

$$\begin{aligned} u(D) &= u(D, A_1) + u(D, A_2) + u(D, A_3) + u(D, A_6) + u(D, A_7) \\ &= 4 + 20 + 12 + 4 + 4 = 44 \end{aligned}$$

Definition 11.6

The utility of an itemset L in the transaction T_r is represented as $u(L, T_r)$ and indicated as

$$u(L, T_r) = \sum_{m_j L v L \subseteq T_r} u(m_j, T_r) \tag{11.3}$$

The itemset D in the transaction A_1's value is computed as

$$u(DE, A_1) = u(D, A_1) + u(E, A_1) = 4 + 12 = 16$$

Definition 11.7

The utility of an itemset L in the data store DB is represented as $u(L)$ and indicated as

$$u(L) = \sum_{L \subseteq T_r \subseteq DB} u(L, T_r) \qquad (11.4)$$

The utility value of an itemset DE in the data store DB is computed as

$$U(DE) = u(DE, A_1) + u(DE, A_6) + u(DE, A_7) = 16 + 10 + 10 = 36$$

Definition 11.8

The transaction utility (*tu*) of a transaction T_r is represented as *tu* and indicated as

$$tu(T_r) = \sum_{L \subseteq T_r} u(L, T_r) \qquad (11.5)$$

The *tu* in the transaction T_r is computed as

$$tu(A1) = tu(D, A_1) + tu(E, A_1) + tu(G, A_1) = 4 + 12 + 5 = 21$$

Definition 11.9

The total transaction utility (tltu) of a transaction T_r is represented as tltu and indicated as

$$Tu = \sum_{T_r \in DB} Tu(T_r) \qquad (11.6)$$

The *tu* in the data store DB is computed as

$$Tu = 44 + 27 + 21 + 55 + 10 = 157$$

To retrieve the HUIs and their relative utility from the temporal transaction database, HUIM is employed. The estimated *tu* and tltu are specified in Table 11.5.

The issue of HUIM is interpreted in the identification of high-profit-yielding itemsets that satisfy min_util value. It is calculated from the TWU. It is calculated to evaluate the HTWUIs. Every item carries the distinct utility rate, and it is added together in every transaction to provide *tu*. The upper bound value of an item is estimated with the assistance of *tu*. The identified HTWUI values are shown in Table 11.5.

$$HTWUI(DB) = DB(A_1) + DB(A_2) + DB(A_3) + DB(A_6) + DB(A_7)$$

$$= 21 + 22 + 46 + 16 + 31$$

$$HTWUI(DB) = 136$$

TABLE 11.5

Transaction Utility (tu) and Total Transaction Utility (tltu)

Transaction	I-D	I-E	I-F	I-G	I-H	tu
A_1	4	12	0	5	0	21
A_2	20	0	0	0	2	22
A_3	12	0	4	30	0	46
A_4	0	3	2	5	5	15
A_5	0	0	5	0	1	6
A_6	4	6	5	0	1	16
A_7	4	6	5	0	1	31
tltu	44	27	21	55	10	157

The minimum threshold value is calculated from the tltu and the user-defined value is 0.6.

$$\text{Minimum threshold value } (\alpha) = 157 \times 0.2$$

$$\alpha = 31.4$$

The HUIM is more interesting than FIM. Any approach can be applied for the identification of HUIs and can be utilized to identify the frequent items in the transaction database; to accomplish that, the subsequent approaches are incorporated.

- The quantitative data store is generated from the transaction database. For every item i belongs to I, the EU is given in Table 11.4.
- The min_util value is assigned based on the nature of the dataset or assigned by the user, which is applied to the quantitative data store to retrieve the HUIs.

The transaction database in Table 11.1 is converted to the quantitative data store that is given in Table 11.3, and the HUIM algorithm is utilized for extracting the HUIs from the quantitative data. The items that satisfy the minimum threshold is termed as high-profit-yielding items and are given in Table 11.6 and Figure 11.2. The high-profit-yielding items are highlighted in Figure 11.2 in gray color. Among the 31 itemsets in the transaction, 17 items are high-profit-yielding items and these items satisfy the threshold value assigned by the user.

Property 11.2

The measure of utility value is neither monotone nor anti-monotone. For the two itemsets S and R such that $S \subset R$, the correlation among the utilities S and R is either util(S)<util(R), util(S)>util(R) or util(S)=util(R).

TABLE 11.6
HUIM with min_util=31

Item	Utility	Item	Utility
D	44	DEG	46
G	55	DEH	81
DE	36	EFG	36
DF	34	EFH	34
DG	70	EGH	35
DH	31	DEFH	52
EG	46	EFGH	42
FG	61	DEFGH	31
DFG	70		

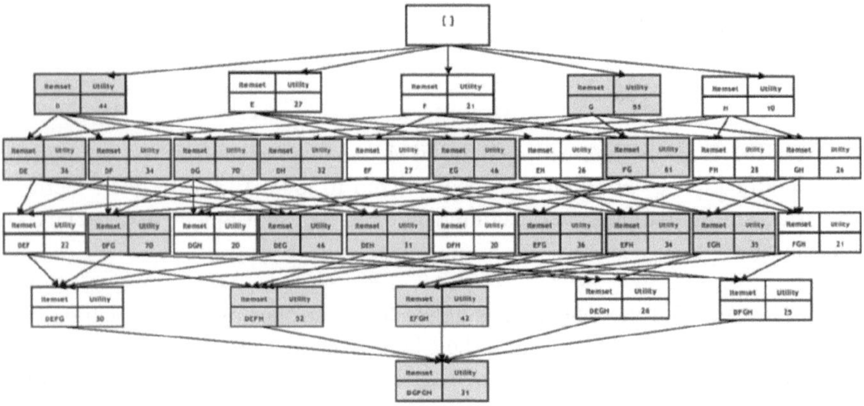

FIGURE 11.2 The HUIM search space with min_util=31.

This property states that the HUIs in the database are scattered across the quantitative data store and they can be observed in the search space in Figure 11.2. This is the key issue in the HUIM algorithm that is more complicated than the issues in the FIM, where the FIM has the effective monotone property that the support of the itemset is equal to or greater than the frequency of any supersets in the database.

11.4 COMPREHENSIVE ANALYSIS OF HUIM TECHNIQUES

HUIM algorithm is developed to identify the high-profit-yielding items, and it uses similar inputs that also produce similar output. The variation in the diversified forms of the HUIM algorithm lies in the exploration strategy and data structure applied in the HUI searching. The design choices such as search strategy, representation, search space exploration and computation of utility will reflect the performance of the algorithm in terms of memory usage, execution time and scalability. Typically, HUIM is inspired by the classical FIM and it initiates novel ideas to handle the facts, which is a measure of utility that is neither anti-monotone nor monotone. Numerous HUIM algorithms have been proposed, and their significance is detailed in this section.

11.4.1 Two-Phase Algorithm

It proficiently prunes the candidate items and specifically acquires the complete set of HUIs. The main issue of utility mining is in limiting the size of the set of candidate items and in simplifying the estimating process of the value of utility. This issue is effectively handled by the two-phase approach. In this algorithm, any superset of TWU itemset is considered as minimum-profit-yielding itemset. The minimum TWU itemset is eliminated by the TWDC property, which is stated above [15].

11.4.2 Faster High-Utility Itemset Mining (FHM)

FHM adopts the vertical data representation, and this approach reduces the number of join operations. FHM prunes the itemset and its supersets that fail to satisfy the minimum threshold value. The FHM uses the co-occurrence-based pruning mechanism, that is estimated utility co-occurrence pruning. EUCP eliminates the low-utility itemset and their transitive extensions without formulating utility-list [16]. FHM+ reduces the length of the upper bound (LUR) and prunes the search space by revising the values of TWU. The utility-list is generated by using the revised utility value. From the revised utility-list, the HUIs are identified effectively [17].

11.4.3 Efficient High-Utility Itemset Mining (EFIM)

The EFIM uses two new strategies in upper bounds, namely local utility and utility in revised sub-tree, which effectively prune the search space. The upper bound utility computation is initiated by a novel approach called array-based counting. The redundant database scan is minimized with the effective merging as well as data projection approach. EFIM effectively prunes the search space, and the pruned utility value is stored in the list structure. Further, HUI is estimated from the utility values stored in the utility utility-list. It is a one-phase approach, which eliminates the redundant candidate generation of the two-phase approach. In EFIM, the utility estimation operation is attained linearly on the search space with time [18].

11.4.4 High-Utility Itemset Miner (HUI-Miner)

The HUI-Miner and HUP-Miner utilize a list structure for storing the utility information of itemsets along with heuristic data of pruning search space.

The utility-list permits the direct discovery of itemsets from the list of utility items. This process is accomplished without any scan of the database. The speed of the process is increased by the utility-list*, and it uses the horizontal technique for the construction of list structure. This approach recursively generates the utility-list and utility-list*. For sparse databases, the construction of utility-list* is highly effective for identifying HUIs. This approach is entirely different from the HUIM, and candidate itemset generation is not done in HUI-Miner [19].

11.4.5   HIGH-UTILITY PRUNING STRATEGY (HUP-MINER)

The HUP-Miner uses the pruning strategy with a list structure, and it uses U-Prune, PU-Prune and LA-Prune strategies for pruning irrelevant items from the database. In U-Prune strategy, the summation of the entire utility and also the remaining utility of the generated itemset is compared with the threshold value assigned by the user and the itemsets with minimum threshold are eliminated along with the superset values. In PU-Prune and LA-Prune, two itemsets are considered for pruning. If the condition is satisfied, then the superset of the itemsets belongs to HUIs in the PU-Prune. The summation value of utility and the remaining utility is compared with the threshold value. The items that satisfy the condition are considered as HUIs. The pruned values are further utilized in the process of exploration, and partition list structure is constructed by the intersection of the set of items by merging the list with ID. This partition list is used for the computation of HUIs [20].

11.4.6   UTILITY PATTERN GROWTH (UP-GROWTH)

The UP-Growth approach uses a set pruning strategy and UP-tree where the generation of candidate itemset is accomplished in two scans of the database. The number of generations of candidate itemsets are effectively reduced with this approach. The significant information is stored in the UP-tree and the tree is created with two scans of the database and, after the construction, the potential HUIs are identified [21]. In the UP-Growth+ approach, the potential HUIs are identified from the UP-tree and local UP-tree. The incidence of unpromising items is eliminated during the process of construction of the local UP-tree [22].

11.4.7   UTILITY LIST BUFFER (ULB-MINER)

The utility-list is improvised to minimize the usage of memory, and the quick join operation is triggered, which is stated as utility-list buffer for HUIM. ULB-Miner utilizes the utility-list buffer structure to effectively store the item and extracts utility-lists, which reuses memory during the process of mining HUIs. This approach also initiates a linear time technique for the generation of utility-list segments in a utility-list buffer. ULB-Miner is effective in the identification of HUIs when compared to the utility-list-based approaches [23].

11.4.8 HYBRID TECHNIQUE BY THE INTEGRATION OF UP-GROWTH AND FHM (UFH-MINER)

The UFH-Miner is a recursive technique that uses two scans for the identification of HUIs. In the initial scan, the TWU value of a distinct itemset that occurs in the database is estimated. In the subsequent scan, the itemsets with less TWU are eliminated from the database, that is the itemsets that don't satisfy the minimum threshold value. The items in every transaction are arranged in subsiding order with the assistance of TWU and the itemsets are interleaved to create a complete UP-tree and the UP-Growth+ is initiated. The TWU of the itemset is estimated, and the TWU is checked with the threshold value. The items that satisfy the condition are used for the generation of prefix, and the FHM technique is invoked. The local tree is constructed and recursively called to create HUIs [24].

11.4.9 DIRECT DISCOVERY OF HIGH-UTILITY ITEMSET (D2HUP)

D2HUP is a single-phase approach without candidate itemset generation, and the itemset is generated by the prefix extension, which prunes the search space considering the upper bound and effectively maintains the utility information. D2HUP enables the data structure to estimate the tight bound value for effective pruning and directly recognize HUIs in a scalable and powerful way. This algorithm also uses two effective strategies: Recursive filtering of an irrelevant itemset is initiated for sparse data, and for dense data, a look-ahead strategy is used [25].

11.4.10 OPTIMIZATION APPROACHES FOR HUIM

Evolutionary computation is an effective way and has the capability to discover efficient optimal solutions considering the ethics of natural evolution [26]. Particle swarm optimization (PSO) is a bio-inspired approach for identifying the optimal solutions. Further, an OR/NOR tree is initiated to minimize the numerous database scans and eliminate invalid combinations by prompt pruning for identifying HUIs. This approach can effectively minimize complicated computations [27]. The crossover and mutation operations used in genetic algorithms (GAs) are not applicable in swarm intelligence, and the GA is incorporated with the OR/NOR tree for attaining efficiency [28].

Bio-inspired computation has grabbed wide attention and makes the development of new HUI mining algorithms possible. The HUI identification is not guaranteed, even if the efficiency is attained, and the quality of HUI identification is poor in terms of the count of the identified HUIs. This issue is rectified by the bio-inspired framework-based approach, and it uses genetic particle swarm as well as bat approaches. Bio-inspired framework technique fine-tunes the typical road map of bio-inspired approach by proportionally electing the identified HUIs as the objective values of the subsequent population, rather than preserving the optimal values in the subsequent population. Hence, the

TABLE 11.7

Comprehensive Analysis of HUIM

Algorithm	Database Representation	Search Type
Two-phase	Horizontal	Breadth-first
FHM	Utility-list and vertical form	Depth-first
FHM+	Utility-list and vertical form	Depth-first
EFIM	Horizontal with merging	Depth-first
HUI-Miner	Utility-list and vertical form	Depth-first
HUP-Miner	Prefix tree and horizontal form	Depth-first
UP-Growth and UP-Growth+	Prefix tree and horizontal form	Depth-first
IHUP	Prefix tree and horizontal form	Depth-first
mHUI-Miner	Utility-list and vertical form	Depth-first
HMiner	Utility-list and vertical form	Depth-first
ULB-Miner	Vertical (buffered utility-list)	Depth-first
UFH	Hybrid (tree and list)	Depth-first
D2HUP	Vertical form with hyperstructure	Depth-first
BPSO-tree	OR/NOR tree	–
GA-tree	OR/NOR tree	–
HUIF-PSO	Bitmap	–
HUIF-GA	Bitmap	–
HUIF-BA	Bitmap	–

diversity within the generated population can be enriched [29]. Artificial bee colony (ABC) is incorporated with the HUIM; a bitmap is employed to convert the original data that depict a nectar source and varieties of bees [30]. Moreover, the size of identifying the itemset is utilized to produce a fresh nectar source that has a huge opportunity of generating HUIs than producing a random fresh nectar source. HUIM-ABC mines HUIs within a minimum cycle of iteration that is composed of three varieties of honey bees [30]. However, HUIM has several mechanisms and their comprehensive summary is presented in Table 11.7.

11.5   CONCLUSIONS

HUIM is commonly seen as a tough problem, due to the utility estimation process employed in HUIM. This article has surveyed diverse HUIM techniques and their strategies of handling the search space. It is an active research arena that has diversified applications, and this study article has depicted the issues in HUI mining. This chapter has discussed the limitations and other significant aspects of HUIM approaches. Most of the researchers utilized list-based approaches for the identification of HUIs. However, the main drawbacks of algorithms based on the generation of utility-list are maintenance and generation, which are time-consuming and can use a vast quantity of memory. Several utility-lists are constructed during the HUI identification process, which is the main reason for high time and memory consumption. The utility-list operations such as join or intersection make the generation of a utility-list quite costly. The tree-based approach also has certain limitations in generating several candidate items, and most of the items are rejected for having a lower min_util. In certain contexts, the HUIs may be discarded from the above-discussed process and effective optimization is also necessary.

REFERENCES

1. Agrawal, R., & Srikant, R. (1994, September). Fast algorithms for mining association rules. In *Proceedings of 20th International Conference Very Large Databases, VLDB* (Vol. 1215, pp. 487–499). Morgan Kaufmann Publishers Inc., San Francisco, CA.
2. Chen, M. S., Han, J., & Yu, P. S. (1996). Data mining: an overview from a database perspective. *IEEE Transactions on Knowledge and data Engineering*, 8(6), 866–883.
3. Ahmed, C. F., Tanbeer, S. K., Jeong, B. S., & Lee, Y. K. (2009). Efficient tree structures for high utility pattern mining in incremental databases. *IEEE Transactions on Knowledge and Data Engineering*, 21(12), 1708–1721.
4. Yen, S. J., & Lee, Y. S. (2007, September). Mining high utility quantitative association rules. In *International Conference on Data Warehousing and Knowledge Discovery* (pp. 283–292). Springer, Berlin, Heidelberg.
5. Chan, R., Yang, Q., & Shen, Y. (2003). Mining high utility itemsets. In *The Third IEEE International Conference on Data Mining*, (pp. 19–26). IEEE, Melbourne, FL.
6. Liu, Y., Liao, W. K., & Choudhary, A. (2005, May). A two-phase algorithm for fast discovery of high utility itemsets. In *Pacific-Asia Conference on Knowledge Discovery and Data Mining*, (pp. 689–695). Springer, Berlin, Heidelberg.
7. Han, J., Pei, J., & Yin, Y. (2000). Mining frequent patterns without candidate generation. In *Proceedings of the ACM-SIGMOD International Conference on Management of Data*, (pp. 1–12). Association for Computing Machinery, New York, NY.
8. Lan, G. C., Hong, T. P., & Tseng, V. S. (2014). An efficient projection-based indexing approach for mining high utility itemsets. *Knowledge and Information Systems*, 38(1), 85–107.
9. Cattral, R., Oppacher, F., & Graham, K. L. (2009, May). Techniques for evolutionary rule discovery in data mining. In *2009 IEEE Congress on Evolutionary Computation* (pp. 1737–1744). IEEE, Trondheim, Norway.
10. Eberhart, R., & Kennedy, J. (1995, October). A new optimizer using particle swarm theory. In *MHS'95. Proceedings of the Sixth International Symposium on Micro Machine and Human Science* (pp. 39–43). IEEE Nagoya, Japan.
11. Grahne, G., & Zhu, J. (2005). Fast algorithms for frequent itemset mining using fp-trees. *IEEE Transactions on Knowledge and Data Engineering*, 17(10), 1347–1362.
12. Pei, J., Han, J., & Lakshmanan, L. V. (2001, April). Mining frequent itemsets with convertible constraints. In *Proceedings 17th International Conference on Data Engineering* (pp. 433–442). IEEE, Heidelberg, Germany.
13. Yun, U., Ryang, H., & Ryu, K. H. (2014). High utility itemset mining with techniques for reducing overestimated utilities and pruning candidates. *Expert Systems with Applications*, 41(8), 3861–3878.
14. Wu, J. M. T., Lin, J. C. W., & Tamrakar, A. (2019). High-utility itemset mining with effective pruning strategies. *ACM Transactions on Knowledge Discovery from Data (TKDD)*, 13(6), 1–22.

15. Liu, Y., Liao, W. K., & Choudhary, A. (2005, May). A two-phase algorithm for fast discovery of high utility itemsets. In *Pacific-Asia Conference on Knowledge Discovery and Data Mining* (pp. 689–695). Springer, Berlin, Heidelberg.

16. Fournier-Viger, P., Wu, C. W., Zida, S., & Tseng, V. S. (2014, June). FHM: Faster high-utility itemset mining using estimated utility co-occurrence pruning. In *International Symposium on Methodologies for Intelligent Systems* (pp. 83–92). Springer, Cham.

17. Fournier-Viger, P., Lin, J. C. W., Duong, Q. H., & Dam, T. L. (2016, August). FHM $$+ $$: faster high-utility itemset mining using length upper-bound reduction. In *International Conference on Industrial, Engineering and Other Applications of Applied Intelligent Systems* (pp. 115–127). Springer, Cham.

18. Zida, S., Fournier-Viger, P., Lin, J. C.-W., Wu, C.-W., Tseng, V. S. (2015). EFIM: a highly efficient algorithm for high-utility itemset mining. In *Proceedings of the 14th Mexican Intern. Conference on Artificial Intelligence (MICAI 2015)*, Springer LNAI. Morelos, Mexico.

19. Qu, J. F., Liu, M., & Fournier-Viger, P. (2019). Efficient algorithms for high utility itemset mining without candidate generation. *High-Utility Pattern Mining* (pp. 131–160). Springer, Cham.

20. Krishnamoorthy, S. (2015). Pruning strategies for mining high utility itemsets. *Expert Systems with Applications*, 42(5), 2371–2381.

21. Tseng, V. S., Wu, C. W., Shie, B. E., & Yu, P. S. (2010, July). UP-Growth: an efficient algorithm for high utility itemset mining. In *Proceedings of the 16th ACM SIGKDD International Conference on Knowledge Discovery and Data Mining* (pp. 253–262). Association for Computing Machinery, Washington, DC.

22. Tseng, V. S., Shie, B. E., Wu, C. W., & Yu, P. S. (2013). Fellow, efficient algorithms for mining high utility itemsets from transactional databases. *IEEE Transactions on Knowledge And Data Engineering*, 25(8), 1772–1786.

23. Duong, Q. H., Fournier-Viger, P., Ramampiaro, H., Nørvåg, K., & Dam, T. L. (2018). Efficient high utility itemset mining using buffered utility-lists. *Applied Intelligence*, 48(7), 1859–1877.

24. Dawar, S., Goyal, V., & Bera, D. (2017). A hybrid framework for mining high-utility itemsets in a sparse transaction database. *Applied Intelligence*, 47(3), 809–827.

25. Liu, J., Wang, K., & Fung, B. C. (2012, December). Direct discovery of high utility itemsets without candidate generation. In *2012 IEEE 12th International Conference on Data Mining* (pp. 984–989). IEEE, Brussels, Belgium.

26. Lin, J. C. W., Yang, L., Fournier-Viger, P., Wu, J. M. T., Hong, T. P., Wang, L. S. L., & Zhan, J. (2016). Mining high-utility itemsets based on particle swarm optimization. *Engineering Applications of Artificial Intelligence*, 55, 320–330.

27. Lin, J. C. W., Yang, L., Fournier-Viger, P., Hong, T. P., & Voznak, M. (2017). A binary PSO approach to mine high-utility itemsets. *Soft Computing*, 21(17), 5103–5121.

28. Kannimuthu, S., & Premalatha, K. (2014). Discovery of high utility itemsets using genetic algorithm with ranked mutation. *Applied Artificial Intelligence*, 28(4), 337–359.

29. Song, W., & Huang, C. (2018). Mining high utility itemsets using bio-inspired algorithms: A diverse optimal value framework. *IEEE Access*, 6, 19568–19582.

30. Song, W., & Huang, C. (2018, June). Discovering high utility itemsets based on the artificial bee colony algorithm. In *Pacific-Asia Conference on Knowledge Discovery and Data Mining* (pp. 3–14). Springer, Cham.

12 A Corpus Based Quantitative Analysis of Gurmukhi Script

Gurjot Singh Mahi and Amandeep Verma
Punjabi University

CONTENTS

12.1 Introduction 211
12.2 Data Collection and Pre-processing 213
12.3 Basic Concepts and Research Methods 214
    12.3.1 Sentences, Words, and Characters..214
    12.3.2 Method of Analysis..216
        12.3.2.1 Mean, Mode, and Median...216
        12.3.2.2 Standard Deviation 217
        12.3.2.3 Skewness...217
        12.3.2.4 Correlation 218
        12.3.2.5 Type Token Ratio 219
        12.3.2.6 Frequency..219
12.4 Results and Discussion 220
    12.4.1 Word.. 220
    12.4.2 Sentence... 223
12.5 Conclusions... 231
References... 233

12.1 INTRODUCTION

Computer-supported analysis of linguistic data reveals many undisclosed facts about a specific natural language. Natural language has fascinated computer research for a long. Patterns embedded in the natural language can be used in computer science research. A considerable amount of research has been performed in the area of Natural Language Processing (NLP),[1] i.e., to develop computer software's for the world's languages. Heterogeneous textual data analysis plays a vital role in gaining insights into a particular language, natural or computer (Fize, Roche, and Teisseire 2018; Kosmidis, Kalampokis, and Argyrakis 2006). Conventional statistical techniques deliver the understanding of the language statistics such as frequency analysis, word probabilities, and language modeling, which are further utilized for designing software systems such as word prediction system and sentence auto-completion system. On the contrary, an unconventional statistical approach such as natural language time series analysis shares the information about the hidden structural patterns in the natural language or written text documents. Time series mapping of natural language has drawn the interest of many scientists and researchers. The analysis of natural language using time series analysis gives relevant insights into a particular language. Time series can be defined as a collection of observations x_t, where time $t = 1, 2, 3 \ldots$ or $t > 0$. Time series has many applications in diverse fields of computer research, such as data compression (Chirikhin and Ryabko 2019), weather forecasting (Kumar and Jha 2013), stock market prediction (Kim 2003), and signal processing (Chan and Fu 1999).

The first preliminary analysis on Indian languages was done by Bharati et al. (2002). The researchers made statistical interpretation of ten Indian languages, which was limited to the frequency distribution examination. Mehta and Majumder (2016) performed a quantitative study on three Indo-Aryan languages, i.e., Hindi, Gujarati, and Bengali, to understand whether the stated languages hold power law or not. Other works by Lakshmi Priya and Manimannan (2014), Kumar et al. (2007), and Daud et al. (2017) used conventional statistical techniques to examine the statistical richness of natural languages of the Indian subcontinent. Jayaram and Vidya (2008) applied Zipf's law on two Indo-Aryan (Hindi and Marathi) and two Dravidian (Kannada and Telugu) set of languages. The analysis of the pattern of occurrence of words in the Hindi language was performed by Pande and Dhami (2013). The same authors also published a work on the occurrence of characters in the Hindi language (Pande and Dhami, 2015).

India is a multi-lingual country. The eighth schedule in the Constitution of India gives recognition to 22 languages, including Punjabi,[2] an Indo-Aryan language. Punjabi has more than 100 million users across the globe and uses Gurmukhi and Shahmukhi scripts for writing. The present study presents a novel work of analyzing the Punjabi language that is written using the Gurmukhi script. Owing to the case of the Gurmukhi script, the study cites the work of Singh and Singh Lehal (2010) and Goyal (2011). These works evaluate the character frequency, word length analysis, n-gram analysis, and frequency distribution of syllables occurring in the Gurmukhi script in general. However, these studies have analyzed the Gurmukhi script at a very fundamental level. The following points differentiate the present research from the other studies of similar nature:

1. For the first time, an extensive corpus (>6 million words and >440 thousand sentences) of Gurmukhi script text in seven distinct genres was extracted and statistically operated. An attempt has been made to reveal the physical nature of words in the Gurmukhi script in general and in individual genres in particular, mainly concentrating on the word length, character frequency, vowel usage, word frequency, word length frequency, and Type Token Ratio (TTR).

2. Additionally, an effort has been made to evaluate the Gurmukhi script sentence structure utilizing various statistical measures such as word usage, character usage, word usage after the removal of stop-words, and character usage after removing stop-words.
3. Sentence length time series (SLTS) study is also conducted to establish two SLTS distributions for an individual genre drawn from the same distribution in the Gurmukhi text and distinct genres. SLTS study was extended to understand the correlation between word usages between various separate genres. This established that the word count could not be used as a metric to distinguish the genres in the Gurmukhi script, which aligns with the previous study conducted for the Zhuang language (Wei et al., 2019).

Overall, in its true nature, the present analysis makes an effort to quantify the statistical richness of the Gurmukhi script to the best of the authors' knowledge. For the present study, a text corpus was extracted. The detailed collection process and pre-processing methodology are presented in Section 12.2. Section 12.3 gives details about the basic concepts and research methods of the study. The results are presented and discussed in Section 10.4. Lastly, Section 10.5 concludes the study.

12.2   DATA COLLECTION AND PRE-PROCESSING

English dictionary defines a "corpus" as *a collection of written or spoken material in machine-readable form, assembled for the purpose of studying linguistic structures, frequencies, etc.* Using a large amount of text corpus to find the underlying facts about the natural language is one of the most viable practices in contemporary linguistics. A substantial amount of machine-readable text data was collected to draw the statistical inference for the current study. The study used Punjabi Tribune,[3] a prominent newspaper website, for the corpus. A specialized website crawler program was designed to collect text in seven genres—*business, entertainment, international, kids, regional, special page,* and *sports.* The website crawler was designed using urllib[4] module and BeautifulSoup[5] library in Python[6] programming language. Figure 12.1 gives the details about the data collection process used in the crawler.

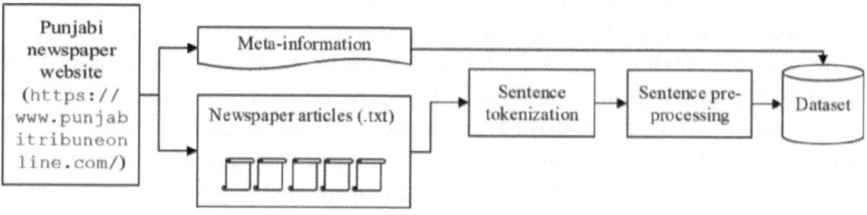

FIGURE 12.1   Corpus extraction methodology.

TABLE 12.1
Meta-Information for Various Genres

Genre	Size in MB	Total News Items	From	Till
Business	21.7	12,701	February 7, 2014	March 18, 2019
Entertainment	7.70	1,421	August 21, 2010	March 9, 2019
International	61.1	32,834	February 26, 2010	March 15, 2019
Kids	8.26	2,012	August 21, 2010	March 16, 2019
Regional	29.7	19,323	March 27, 2017	March 14, 2019
Special page	6.41	1,163	December 28, 2010	September 19, 2018
Sports	119	38,791	February 26, 2010	November 6, 2018

The crawler extracts the news articles in the text file format (.txt), utilizing specialized functions and maintaining the meta-information file containing title, date, month, and year for respective news genre. Collectively, 1,08,245 news articles were extracted. All the textual data obtained were in Unicode format, making it machine-readable. The data were collected for a total of 9 years, ranging from 2010 to 2019. The contemporary statistics of the extracted news items in various genres are shown in Table 12.1.

In the next phase, each news article was tokenized and converted into sentence form using the punctuation mark—"।" [ɖan ˜ɖiː]. Then, the extracted news articles in each genre were further forwarded to the second phase of pre-processing. Additionally, all the unwanted symbols were eliminated from the sentence in the pre-processing. In the final step of corpus creation, pre-processed sentences were saved using an Extensible Markup Language[7] (XML or .xml) schema utilizing the meta-information file maintained at the initial data collection stage. The formatted corpus in XML format is illustrated in Figure 12.2.

Some general statistics of the final corpus are given in Table 12.2. Finally, the corpus is exploited for making the statistical interpretation for the present study.

12.3   BASIC CONCEPTS AND RESEARCH METHODS

12.3.1   Sentences, Words, and Characters

Every written script constitutes three levels of linguistic structure—sentence, word, and character. In this study, symbolic representation to these various levels of abstraction in textual data is used. The study uses the symbol "Ş" to describe a sentence, "ω" to represent a word, and "Ć" to represent a character.

Gurmukhi text data can also be abstracted or decomposed in sentence, word, and character. While taking an abstract look at the text files generated during the data crawling process, it can be generalized that the text in news articles is written in paragraphs, which can be decomposed into sentences. The sentences are further decomposed into words, and words, into characters. These three general levels of abstraction are mentioned in Figure 12.3.

```xml
<?xml version="1.0"?>
<news_data>
<data>
    <sent_id>TRI_SP_0</sent_id>
    <news_paper>PUNJABI TRIBUNE</news_paper>
    <genre>SPORTS</genre>
    <sub_genre>WRESTLING</sub_genre>
    <title>ਖੇਡ ਦੰਗਲ:  ਕਮਲਜੀਤ ਝੁਮਡੇਗੀ ਨੇ ਝੰਡੀ ਦੀ ਝੁਲਟੀ ਜਿੱਤੀ</title>
    <sentence>ਇਸ ਦੰਗਲ ਝੰਡੀ ਦੀ ਝੁਲਟੀ ਦਾ ਮੁਕਾਬਲਾ ਕਮਲਜੀਤ ਝੁਮਡੇਗੀ ਤੇ ਵਾਰਨ ਰੁੱਜਰ ਵਿਚਕਾਰ ਹੋਇਆ ।</sentence>
    <date>6</date>
    <month>NOVEMBER</month>
    <year>2018</year>
</data>
<data>
    <sent_id>TRI_SP_1</sent_id>
    <news_paper>PUNJABI TRIBUNE</news_paper>
    <genre>SPORTS</genre>
    <sub_genre>WRESTLING</sub_genre>
    <title>ਖੇਡ ਦੰਗਲ:  ਕਮਲਜੀਤ ਝੁਮਡੇਗੀ ਨੇ ਝੰਡੀ ਦੀ ਝੁਲਟੀ ਜਿੱਤੀ</title>
    <sentence>ਸਾਬਕ ਕੈਬਨੇਟ ਮੰਤਰੀ ਤਾ ,  ਦਲਜੀਤ ਸਿੰਘ ਦੀਆ ਨੇ ਪਹਿਲਵਾਨਾਂ ਦੀ ਹੌਸਲੇਗੀ ਕਰਵਾਈ ।</sentence>
    <date>6</date>
    <month>NOVEMBER</month>
    <year>2018</year>
</data>
<data>
    <sent_id>TRI_SP_2</sent_id>
    <news_paper>PUNJABI TRIBUNE</news_paper>
    <genre>SPORTS</genre>
    <sub_genre>WRESTLING</sub_genre>
    <title>ਖੇਡ ਦੰਗਲ:  ਕਮਲਜੀਤ ਝੁਮਡੇਗੀ ਨੇ ਝੰਡੀ ਦੀ ਝੁਲਟੀ ਜਿੱਤੀ</title>
    <sentence>ਇਸ ਮੁਕਾਬਲੇ ਵਿੱਚ ਕਮਲਜੀਤ ਝੁਮਡੇਗੀ ਨੇ ਵਾਰਨ ਰੁੱਜਰ ਨੂੰ ਇੱਕ ਕਰਕੇ ਝੰਡੀ ਆਪਣੇ ਨਾ ਕੀਤੀ ।</sentence>
    <date>6</date>
    <month>NOVEMBER</month>
    <year>2018</year>
</data>
```

FIGURE 12.2   XML representation of the final corpus.

TABLE 12.2
Calculated Corpus Statistics (Number of Words and Sentences) in Various Genres

Genre	Total Words	Total Sentences
Business	471,540	34,828
Entertainment	285,111	21,317
International	1,463,278	99,973
Kids	323,997	26,062
Regional	548,964	37,758
Special page	223,828	17,009
Sports	2,980,330	203,142

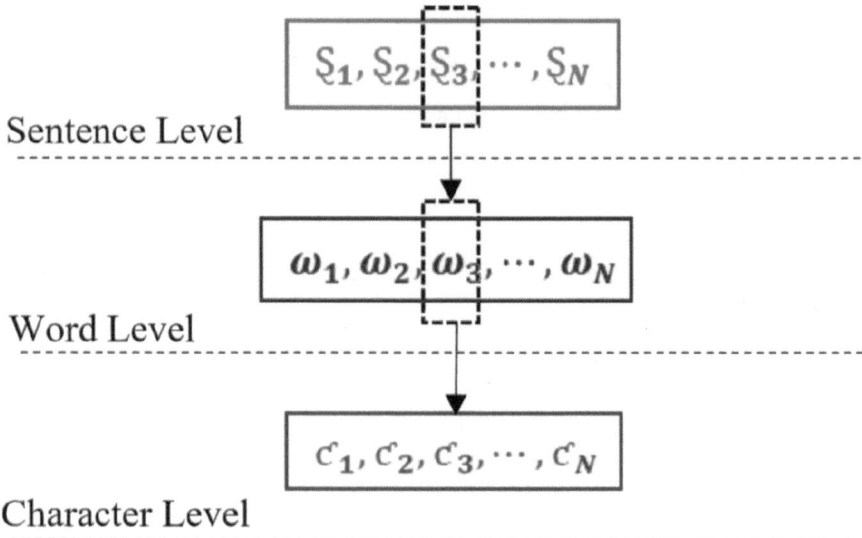

FIGURE 12.3   Syntactic levels of script.

The decomposition could be exemplified by using a sentence from the contemporary Gurmukhi script for demonstration:

ਮੈਂ ਪੰਜਾਬ ਦਾ ਵਸਨੀਕ ਹਾਂ ("I am a resident of Punjab")

This example can be defined in the form of the sentence set $Ş=$ਮੈਂ ਪੰਜਾਬ ਦਾ ਵਸਨੀਕ ਹਾਂ, which comprises a number of words. The sentence in $Ş$ is decomposed into words, which is represented in the form of words set $\omega=\{$ਮੈਂ, ਪੰਜਾਬ, ਦਾ, ਵਸਨੀਕ, ਹਾਂ$\}$, where $\omega_1=$ਮੈਂ and $\omega_2=$ਪੰਜਾਬ. Similarly, the word ω_2 is decomposed into a character set $c=\{$ਪ, $^\circ$, ਜ, ਾ, ਬ$\}$, where $c_1=$ ਪ, $c_2=$ $^\circ$, and so on.

12.3.2   Method of Analysis

An in-depth detailed quantitative analysis of the gathered linguistic corpus was conducted. The corpus was analyzed using the mentioned metrics by developing a Python function code, utilizing pandas[8] and SciPy[9] scientific libraries. Data visualization was performed using Microsoft Excel software, Matplotlib,[10] and the Seaborn[11] library. The analysis was performed on two syntactic levels of the script, i.e., words (ω) and sentences ($Ş$), to know the usage-based properties in various genres of written text individually and collectively. For the said purpose, the following statistical metrics were selected.

12.3.2.1   Mean, Mode, and Median

Mean is the measure of the center of distribution of central tendency (Barde and Barde 2012). Mean values in this paper are quantified for the two mentioned syntactic levels—words and sentences. Finding the word and sentence mean lengths is the fundamental approach used for studying the underlying syntactic structure of written texts by measuring the usage of characters and words, respectively. The calculated mean values of the word and sentence lengths are represented using symbols – μ_ω and $\mu_{\S}$, and are calculated using Formulas 12.1 and 12.2:

$$\mu_\omega = \frac{\sum_{i=1}^{N_\omega} len(\omega_i)}{N_\omega} \tag{12.1}$$

$$\mu_{\S} = \frac{\sum_{i=1}^{N_{\S}} len(\S_i)}{N_{\S}} \tag{12.2}$$

where word and sentence length functions $len(\omega_i)$ and $len(\S_i)$ are calculated using $len(\omega_i) = \sum_{j=1}^{N_c^{\omega i}} 1$ and $len(\S_i) = \sum_{j=1}^{N_\omega^{\S i}} 1$, respectively. $N_c^{\omega i}$ and $N_\omega^{\S i}$ represent the number of characters in the word ω_i and the number of words in the sentence $\S_i$, whereas N_ω and $N_{\S}$ represent the number of words and sentences in the corpus.

Median (med_i) value separates the dataset into two halves, i.e., lower half and higher half, whereas mode ($mode_i$) symbolizes the value with the highest frequency in the dataset, where $i = \{\omega, \S\}$ represents the calculated value for the words and sentences for the respective median and mode.

12.3.2.2   Standard Deviation

Standard deviation is the measure of variability or dispersion (Altman & Bland 2005). The larger the value of standard deviation, the higher will be the variability in the dataset (Isotalo 2014). Formulas 12.3 and 12.4 state the method used for investigating standard deviation. Symbols σ_ω and $\sigma_{\S}$ are used to represent the measured standard deviation valuation of words and sentences.

$$\sigma_\omega = \sqrt{\frac{\sum_{i=1}^{N_\omega} \left(len(\omega_i) - \mu_\omega\right)^2}{N_\omega}} \tag{12.3}$$

$$\sigma_{\S} = \sqrt{\frac{\sum_{i=1}^{N_{\S}} \left(len(\S_i) - \mu_{\S}\right)^2}{N_{\S}}} \tag{12.4}$$

The calculated values of σ_ω and $\sigma_{\S}$ are always positive, i.e., $\sigma_\omega \geq 0$ and $\sigma_{\S} \geq 0$.

12.3.2.3   Skewness

After the publication by Karl Pearson in 1985 (Pearson 1895), a lot of research has been conducted on the properties and characteristics of statistics. Kenney (1939) states the skewness as the *"lack of symmetry in a distribution."* The present study uses *Pearson's second skewness coefficient* indicated in Doanne and

Seward (2011), for statistical inference. sk_ω and $sk_\S$ mentioned in Formulas 12.5 and 12.6 are used for finding the skewness for words and sentences in the corpus.

$$sk_\omega = \frac{3(\mu_\omega - med_\omega)}{\sigma_\omega} \qquad (12.5)$$

$$sk_\S = \frac{3(\mu_\S - med_\S)}{\sigma_\S} \qquad (12.6)$$

The value of sk_ω and $sk_\S$ lies in the interval $[-3,3]$, i.e., $-3 \le sk_\omega \le 3$ and $-3 \le sk_\S \le 3$. Kalimeri et al. (2015) stated the distribution to be perfectly symmetric when the skewness value is realized to zero, or else the distribution is either left- or right-skewed (also called negatively or positively skewed).

12.3.2.4 Correlation

Linear correlation is stated as a rule when one random variable depends on another random variable in a bivariate environment. Statistical procedures have two broad classifications—parametric and non-parametric (Hoskin 2012). The study uses both parametric and non-parametric statistical procedures in statistical measurements, predominantly in the case of finding the linear correlation between two variables. Three diverse methods are used in the study: Pearson correlation coefficient (Bolboaca and Jäntschi 2006; Good 2009; Sedgwick 2012), Spearman rank correlation coefficient (Sedgwick, 2012), and Kendall rank correlation (Kendall 1938, 1975), for finding correlation coefficient covering both parametric and non-parametric tests.

Pearson made a great effort in the development and generalization of the concept of correlation introduced by Francis Galton in 1888 (Blyth 1994). Pearson gave the concept of Pearson correlation coefficient, which is also known as the product-moment correlation coefficient and is illustrated using Formula 12.7. The normal distribution of dataset is one of the conditions to conduct a parametric test.

$$r = \frac{\sum_{i=1}^{N}(x_i - \langle X \rangle)(y_i - \langle Y \rangle)}{\sqrt{\sum_{i=1}^{N}(x_i - \langle X \rangle)^2}\sqrt{\sum_{i=1}^{N}(y_i - \langle Y \rangle)^2}} \qquad (12.7)$$

Similar to the Pearson correlation coefficient, Ghasemi and Zahediasl (2012) argue that the distribution of a dataset or sample can be ignored if the sample size is large enough. The authors continue to state that the statisticians can follow the parametric procedure even if the data are not normally distributed.

The symbol r is used to represent the calculated Person correlation coefficient. $\langle X \rangle$ and $\langle Y \rangle$ are mean values of variables x and y, respectively, whereas N is the number of values in the dataset. The value of r always ranges between ±1, i.e., $[-1,1] = \{r \mid -1 \le r \le 1\}$. Spearman's rank correlation coefficient (ρ) is one of the non-parametric statistical parameters, stated in Formula 12.8. The symbol ρ is used to describe the calculated value for the Spearman rank correlation.

$$\rho = 1 - \frac{6\sum_{i=1}^{N} D_i^2}{N(N^2 - 1)} \tag{12.8}$$

D_i is the difference between two rankings, which can be written as $D_i = rg(X_i) - rg(Y_i)$, where rg is the rank. Following the Pearson correlation coefficient, Spearman rank correlation values tend to range between ±1, i.e., $[-1,1] = \{\rho \mid -1 \leq \rho \leq 1\}$. Similarly, N is the number of values in the dataset.

To be more specific about the findings, this research has used the Kendall rank correlation method (Formula 12.9), another non-parametric measure for finding the relationship between two sets of data. The main reason for using Kendall rank correlation is that Kendall coefficient approaches normality rapidly even for small values of N, as compared to Spearman's rank correlation coefficient (Kendall 1938; Colwell and Gillett 1982).

$$\tau = \frac{N_c - N_d}{\frac{1}{2}N(N-1)} \tag{12.9}$$

τ is used to represent the calculated value of the Kendall rank correlation coefficient. N_c and N_d define the number of concordant and discordant pairs in a bivariate environment. As followed by the other two mentioned correlation coefficients, τ value also ranges between ±1, i.e., $[-1,1] = \{\tau \mid -1 \leq \tau \leq 1\}$. The number of values in the dataset is defined using N.

12.3.2.5 Type Token Ratio

Lexical diversity or lexical richness is defined as the measure of distinct words used in a text or corpus (Johansson 2009). TTR is extensively used as the index of lexical diversity (Richards 1987). Biber et al. (2007) defined the TTR as the "ratio between the number of different lexical items in a text (the 'types') and the total number of words in that text."

$$TTR = \frac{T_1}{T_2} \times 100 \tag{12.10}$$

Formula 12.10 gives the TTR, a mathematical foundation for its usage in this research. T_1 is the total number of unique words (tokens), and T_2 depicts the total number of words in the corpus.

12.3.2.6 Frequency

Another metric used for finding the most frequent items in the corpus is by finding the frequency distribution. An instantaneous description of the corpus stated by means of finding the frequency distribution at syntactic levels of the corpus is performed in Mahi and Verma (2020). Let the corpus contain the set *VAR* of unique variables:

$$VAR = \{v_1, v_2, \ldots, v_n\} \tag{12.11}$$

The frequency of *i*th variable in the set *VAR* can be defined as:

$$f_1 + f_2 + \cdots + f_n = N \tag{12.12}$$

Subsequently, f_i corresponds to the frequency of the variable v_i, $i = 1, 2, \ldots, n$, in the set *VAR* and *N* is the total number of variables. Frequency graphs have been plotted according to the top ranks in various syntactic areas of the corpus.

12.4   RESULTS AND DISCUSSION

The subsequent sections assist in elaborating on the results and discussing the experiments performed to know the physical properties of the Gurmukhi script in detail.

12.4.1   WORD

The statistical quantification is performed in this section to reveal the underlying properties of words used in the Gurmukhi script. Tables 12.3 and 12.4 stage the calculated metrics records of an experiment on the combined and individual

TABLE 12.3
Values for the Detailed Analysis Performed on the Combined Corpus Using Various Statistical Moments

	μ_ω	mode_ω	med_ω	σ_ω	sk_ω
Combined	4.03	2	4	1.84	0.05

TABLE 12.4
Values for the Detailed Analysis Performed on the Individual Genres of Text Using Various Statistical Moments

Genre	μ_ω	mode_ω	med_ω	σ_ω	sk_ω
Business	4.03	2	4	1.83	0.05
Entertainment	3.89	2	4	1.74	−0.19
International	4.00	2	4	1.86	0.00
Kids	3.88	2	4	1.80	−0.20
Regional	4.06	2	4	1.86	0.10
Special page	4.02	2	4	1.95	0.03
Sports	4.06	2	4	1.82	0.10

genres of the extracted corpus by analyzing the characters in the word. The result shows the highest word mean length (μ_ω) values are recorded in the sports and regional genres (4.06 characters for both), which is also near to the mean value in the combined corpus (4.03). Subsequently, the lowest was recorded in the kids (3.88) and entertainment (3.89) sections, which displays the practice of using shorter words in the mentioned sections. Interestingly, $mode_\omega$ (2) and med_ω (4) of words are judged to be identical across all the genres of written text. The findings also reveal that $mode_\omega$ value of 2 has been established due to the fact that the writer of Gurmukhi text use words containing 2 characters with high frequency. For example, the words such as ਹੈ (ਹ + ੈ), ਦੇ (ਦ + ੇ), ਦੀ (ਦ + ੀ), ਨੇ (ਨ + ੇ), and ਦਾ (ਦ + ਾ) are commonly used in all forms of Gurmukhi text.

The statistical results also show the variation in the higher-order moment, i.e., standard deviation of words—σ_ω. The findings further show that the value of σ_ω in special page (1.95) section was higher as compared to the other sections, which indicates that the mentioned genre utilizes more variety in word length usage as compared to other sections. This is due to the fact that the special page section text is written by specialized persons with higher knowledge for extraordinary purpose using distinct word types. The lowest σ_ω was encountered in the entertainment section, whereas for other genres, the σ_ω kept revolving around the value of 1.80. The interpretation of skewness (sk_ω) results display the distribution to be nearly symmetrical and normally distributed, as all sk_ω values in Tables 12.3 and 12.4 rely on the open interval of (−1,1).

Lexical diversity is examined by measuring the type token ratio or TTR (given in Table 12.5). The combined corpus TTR is identified as 2.84%, which is comparatively low compared to the other genres. The lowest TTR is found in the sports (3.01%) text. Likewise, special page (10.95%) genre is acknowledged with the highest TTR, followed by the kids (9.69%) genre. It can be clearly noticed from the results that there is more variety of words utilized in the special page and kids genres and least in sports.

Interesting findings were revealed during the frequency distribution analysis performed to know the practice of usage of Gurmukhi characters and vowels in the

TABLE 12.5
TTR Values for Various Genres

Genre	TTR (%)
Business	7.12
Entertainment	7.84
International	4.61
Kids	9.69
Regional	5.18
Special page	10.95
Sports	3.01

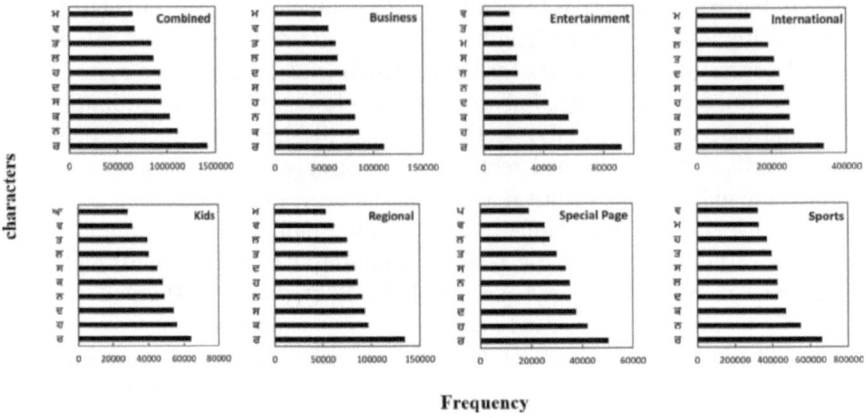

FIGURE 12.4   Calculated frequency distribution of top ten Gurmukhi characters in the combined corpus and distinct genres.

corpus. Figure 12.4 displays the most frequently used characters in the Gurmukhi text. Among all, character "ਰ" has emerged to be the most frequent character across all genres of text and also in the combined text corpus. Additionally, character "ਕ" keeps emerging in the higher ranks of mentioned genres, i.e., second rank in business and regional; third in entertainment, international, and sports; fourth in special page; and fifth in kids section. Fascinatingly, characters "ਹ" and "ਨ" are also used repeatedly with great frequency by the writers. Considering the combined corpus, "ਨ", "ਕ" and "ਸ" are the most used characters after "ਰ".

The present study has not only focused on the characters, but also on the usage of vowels (nasal sounds), also known as *laga matra* in the Gurmukhi script. Table 12.6 exhibits the usage of vowels according to their ranks, where rank 1 is the most frequently used vowel and rank 10 is the least used. During the inspection, no significant difference was found between the usage of vowels and their usage frequency. The vowel "ੋ" and "ੀ" emerged to be the top two vowels. Some ranks of vowels are unevenly distributed, which is only due to the corpus size effect (for example, vowels such as "ੇ" and "ਿ" appeared at ranks 3 and 4 interchangeably in the international and kids genres).

After determining the character and vowel usage frequency, further analysis was done toward knowing the word frequency distribution in the mentioned genres. The corresponding word frequency distribution is shown in Figure 12.5. During the examination, "ਹੈ" emerged as the word with the highest peak in majority of genres, which is used to exemplify the ongoing task or work. In sport and the combined text corpus, "ਨੇ" emerged with the highest peak. The word "ਨੇ" is used for stating the work done in the past by the doer, as most of the sports text is written in the form of work being performed in the past. However, the words "ਦੇ" and "ਦੀ" comprehend the second-highest position in the mentioned analysis, which is mainly used for mentioning the association between two events. The words "ਵਿੱਚ" and "ਨੂੰ" follow the previously mentioned words with the third position in the analysis.

TABLE 12.6

Ranks of Top 10 Vowel Symbols Usage

Rank	Combined	Business	Entertainment	International	Kids	Regional	Special Page	Sports
1	◌ਾ	◌ਾ	◌ਾ	◌ਾ	◌ਾ	◌ਾ	◌ਾ	◌ਾ
2	◌ੀ	◌ੀ	◌ੀ	◌ੀ	◌ੀ	◌ੀ	◌ੀ	◌ੀ
3	ਿ◌	ਿ◌	ਿ◌	ਿ◌	◌ੇ	ਿ◌	ਿ◌	◌ੇ
4	◌ੇ	◌ੇ	◌ੇ	◌ੇ	ਿ◌	◌ੇ	◌ੇ	ਿ◌
5	◌ੁ	◌ੁ	◌ੁ	◌ੁ	◌ੁ	◌ੁ	◌ੁ	◌ੁ
6	◌ੂ	◌ੂ	◌ੂ	◌ੈ	◌ੂ	◌ੂ	◌ੂ	◌ੂ
7	◌ੰ	◌ੈ	◌ੰ	◌ੂ	◌ੰ	◌ੰ	◌ੰ	◌ੰ
8	◌ੈ	◌ੰ	◌ੈ	◌ੰ	◌ੈ	◌ੈ	◌ੈ	◌ੈ
9	◌ੋ	◌ੋ	◌ੋ	◌ੋ	◌ੋ	◌ੋ	◌ੋ	◌ੋ
10	◌ੋ	◌ੋ	◌ੋ	◌ੌ	◌ੋ	◌ੌ	◌ੌ	◌ੋ

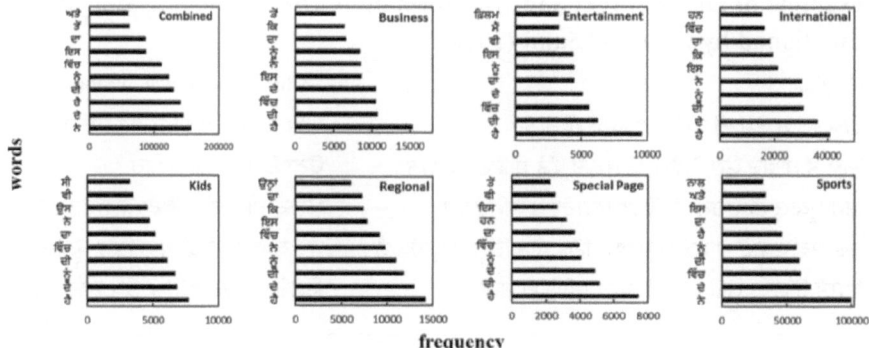

words

frequency

FIGURE 12.5 Calculated frequency distribution of top ten words.

In addition to the mentioned results, further analysis was done to show the top ten lengths of the words in the distinct genres. It can be concluded that words with two characters (ਹੈ, ਦੇ, ਦੀ, ਨੇ) are the most extensively used words in the various genres and the combined corpus, followed by words of length of 4 (ਵਿੱਚ, ਸਿੰਘ, ਫਿਲਮ, ਆਪਣੇ), 3 (ਨੂੰ, ਅਸਰ, ਰਹੇ, ਤੱਕ), 5 (ਗਿਣਤੀ, ਕੰਪਨੀ, ਫਿਲਮੀ, ਇਮਾਰਤ), and 6 (ਸੁਵਿਧਾ, ਸਥਾਪਿਤ, ਸਰਕਾਰੀ, ਹੁਲਾਰਾ) characters, which is presented in Figure 12.6.

12.4.2 SENTENCE

This section will emphasize another syntactic level of this study—sentence. The aim of performing the study on the sentence level is to find the underlying physical aspects of the sentence structure in the Gurmukhi script. Kosmidis et al. (2006) elaborated on the usage of time series in understanding the physical phenomenon

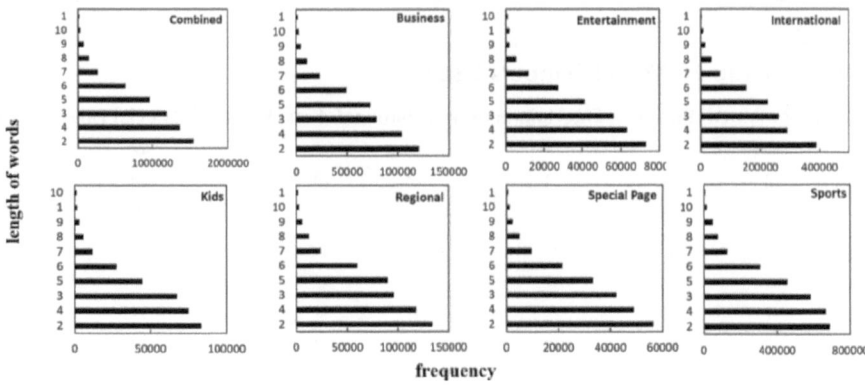

FIGURE 12.6   Calculated frequency distribution of top ten word lengths.

TABLE 12.7

Example of Original Text and Transformation After Removing Stop-Words Mentioned by Kaur and Saini (2016)

Original	After the Removal of Stop-Words
ਵੀਅਤਨਾਮ ਦੇਸ਼ ਦੀ ਕਰੰਸੀ ਸਿੱਕਿਆਂ ਵਿੱਚ ਨਹੀਂ ਹੈ	ਵੀਅਤਨਾਮ ਦੇਸ਼ ਦੀ ਕਰੰਸੀ ਸਿੱਕਿਆਂ
ਉਨ੍ਹਾਂ ਤੋਂ ਬਾਅਦ ਉਹ ਏਸ਼ੀਆ ਦੇ ਸਭ ਤੋਂ ਵੱਡੇ ਸ਼ਾਇਰ ਹੋਏ ਹਨ	ਉਨ੍ਹਾਂ ਏਸ਼ੀਆ ਵੱਡੇ ਸ਼ਾਇਰ
ਇਸ ਪ੍ਰਕਾਰ ਦੀ ਰਚਨਾ ਨੂੰ ਵਿਡਾਮਨਸਟੇਟਰ ਕਹਿੰਦੇ ਹਨ	ਪ੍ਰਕਾਰ ਦੀ ਰਚਨਾ ਵਿਡਾਮਨਸਟੇਟਰ ਕਹਿੰਦੇ
ਹਰ ਇੱਕ ਕਕਾਰ ਆਪਣੇ ਆਪ ਵਿੱਚ ਇੱਕ ਖਾਸ ਚਿੰਨ੍ਹ ਤੇ ਪ੍ਰਤੀਕ ਹੈ	ਹਰ ਕਕਾਰ ਆਪਣੇ ਖਾਸ ਚਿੰਨ੍ਹ ਪ੍ਰਤੀਕ
ਇਸਤੋਂ ਉੱਪਰੰਤ ਕਈ ਵਰ੍ਹਿਆਂ ਤੱਕ ਸ਼ਾਂਤੀ ਬਣੀ ਰਹੀ	ਇਸਤੋਂ ਉੱਪਰੰਤ ਵਰ੍ਹਿਆਂ ਤੱਕ ਸ਼ਾਂਤੀ ਬਣੀ

of language documents by giving meaning to the natural language in time series analysis. Vieira et al. (2018) mentioned that mapping the language text into the time series can make revelations about the hidden structural patterns in the natural language. These patterns help in understanding the physical phenomenon hidden in the Gurmukhi text. Collectively considering the mentioned points in the literature, the study implements a mixed approach adopted by Kosmidis et al. (2006), Vieira et al. (2018), and Hřebíček (1997) on the usage of SLTS.

To reveal the differences in the sentences of distinct genres, the study extracts four sentence time series: (i) the number of words (NW), (ii) the number of characters (NC), (iii) the number of non-stop-words (NSW), and (iv) the number of characters in non-stop-words (NSC). For the purpose of points (iii) and (iv), this study carried out the removal of stop-words (also known as function words), for which the list of stop-words mentioned in Kaur and Saini (2016) was used with little modifications; for example, "ਵਿੱਚ" was added into the stop-words list. The difference between the original sentences and variants is revealed in Table 12.7.

Further, the study examines the stated four sentence time series using the statistical formulas mentioned in Section 12.3.2. Tables 12.8 and 12.9 give the details about the results obtained for sentences in the combined corpus in the given time series. Tables 12.10–12.13 state the evaluated experiment statistical values for the mentioned four SLTS.

In Table 12.8, the combined corpus holds on an average of 14.31 words per sentence, which is supported by the median and mode value of 14 in both cases in NW SLTS. The standard deviation is discovered to be at 4.37, and the skewness, 0.21, which makes the dataset symmetrical. However, the mean sentence length values found in NSW SLTS is 9.56, similarly depicted by mode and median of nine words. The removal of stop-words from sentences decreases the variability to 3.45, compared to 4.37 in NW, but increases the skewness to 0.49 from 0.21 revealed in NW. This behavior uncovers that the removal of stop-words decreases the variability and increases the skewness. Likewise, Table 12.9 indicates that, on

TABLE 12.8
Statistical Values of NW and NSW SLTS Analyses for the Combined Corpus

Time Series	μ_ς	$mode_\varsigma$	med_ς	σ_ς	sk_ς
NW	14.31	14	14	4.37	0.21
NSW	9.56	9	9	3.45	0.49

TABLE 12.9
Statistical Values of NC and NSC SLTS Analyses for Combined Corpus

Time Series	μ_ω	$mode_\omega$	med_ω	σ_ω	sk_ω
NC	57.59	64	58	19.13	−0.06
NSC	45.22	44	44	17.27	0.21

TABLE 12.10
Statistical Values of NW SLTS Analysis for Distinct Genres

Genre	μ_ς	$mode_\varsigma$	med_ς	σ_ς	sk_ς
Business	13.54	13	14	4.64	−0.30
Entertainment	13.38	11	13	4.53	0.25
International	14.64	18	15	4.30	−0.25
Kids	12.43	9	12	4.71	0.27
Regional	14.54	18	15	4.55	−0.30
Special page	13.16	12	13	4.59	0.10
Sports	14.67	16	15	4.12	−0.24

TABLE 12.11
Statistical Values of *NSW* SLTS Analysis for Distinct Genres

Genre	$\mu_\S$	$mode_\S$	$med_\S$	$\sigma_\S$	$sk_\S$
Business	8.96	9	9	3.5	−0.03
Entertainment	8.24	8	8	3.4	0.21
International	9.59	10	10	3.4	−0.36
Kids	7.81	6	7	3.3	0.74
Regional	9.71	10	10	3.4	−0.26
Special page	8.52	7	8	3.4	0.46
Sports	10.06	9	10	3.4	0.05

TABLE 12.12
Statistical Values of *NC* SLTS Analysis for the Combined Corpus

Genre	μ_ω	$mode_\omega$	med_ω	σ_ω	sk_ω
Business	54.53	57	55	19.8	−0.07
Entertainment	52.06	49	51	18.9	0.17
International	58.57	65	59	18.9	−0.07
Kids	48.25	39	46	19.4	0.35
Regional	59.09	67	60	19.0	−0.14
Special page	52.87	41	52	19.6	0.13
Sports	59.53	62	60	18.5	−0.08

TABLE 12.13
Statistical Values of *NSC* SLTS Analysis for Distinct Genres

Genre	μ_ω	$mode_\omega$	med_ω	σ_ω	sk_ω
Business	42.61	38	42	17.3	0.11
Entertainment	38.53	32	37	16.5	0.28
International	45.66	42	45	17.0	0.12
Kids	36.20	33	34	16.6	0.40
Regional	46.55	49	46	16.6	0.10
Special page	40.74	37	39	17.0	0.31
Sports	47.44	44	47	17.1	0.08

average, 57.59 characters are used for writing a sentence in the case of combined corpus (the median value of 58 in Table 12.9 supports the claimed average character usage). However, the mode value of 64 indicates the usage of the mentioned number of characters with a relatively high frequency, which contradicts the average character usage in a sentence in the combined corpus.

The mean length of sentence words observed in Tables 12.10 and 12.11 was interpreted. It can be clearly understood from the mean word length values that writers of sports, regional, and international genres make more use of words as compared to other remaining genres. This elucidation was also verified by the median values for the mentioned genres, which is inferred as 15. Similarly, the mode values in Table 12.10 showcase us that the writers of both regional and international genres make use of 18 words in most sentences and the writers of the kids section make use of nine words for maximum time in sentences. While examining the difference between the NW and NSW, it was proven that the entertainment and international section writers make more usage of stop-words, due to the fact that the difference is realized to be 5.14 and 5.04 average words for both genres, which is more than the mean difference of the combined corpus (4.75 average words) implied in Table 12.8. It was also noted during the exploration of results in Tables 12.12 and 12.13 that despite using almost equal words in international and entertainment genres as shown in Tables 12.10 and 12.11, the writers of entertainment use a greater number of characters with a mean difference between NC and NSC of 13.53 characters in a sentence as compared to the international genre. The mode values exhibited by regional (67), international (65), and sports (62) in Table 12.12 give the detail of the fact that the writers of the above-mentioned genres utilized more characters as compared to other genres.

Furthermore, the examined standard deviation (σ_{ς}) values given in Tables 12.10 and 12.11 are almost similar in nature as it ranges between 4.12 and 4.71 for NW and between 3.3 and 3.5 for NSW, and no significant difference was observed in the respective values represented in tables individually. But, when the dissimilarity between NW and NSW is measured in Tables 12.10 and 12.11 by subtracting σ_{ς} values from each other, the sports and international genres showcased the lowest difference of 0.72 and 0.9, respectively, which clearly mentions the significant difference perceived in the sports and international genres variability compared to other five remaining genres, which have high variability between their corresponding NW and NSW values. Likewise, during the inspection of σ_{ω} in Tables 12.12 and 12.13, a similar phenomenon for characters was demonstrated (which ranges from 18.5 to 19.8 in NC and from 16.5 to 17.3 in NSC) as seen for words in this case, which also includes the low σ_{ω} variability issue identified in the sports (1.4) and international (1.9) genres. Finally, the skewness values display the symmetry witnessed in Tables 12.10–12.13, as most values rely on the acceptable open interval (−1,1) of symmetrical values.

Figure 12.7 depicts the sentence length frequency distribution for the combined corpus and individual genres. The results exhibit that sentence length in various genres is very dynamic in nature; no straightforward similarity can be established between the said genres for higher-frequency sentence lengths. The only similarity observed is that the minimum word usage in the sentence is either 4 or 22 across all genres.

The method given by Vieira et al. (2018) was implemented for identifying the correlation between the mentioned four SLTS inferred for various genres of Gurmukhi text. The three correlation functions mentioned in Section 12.3.2.4

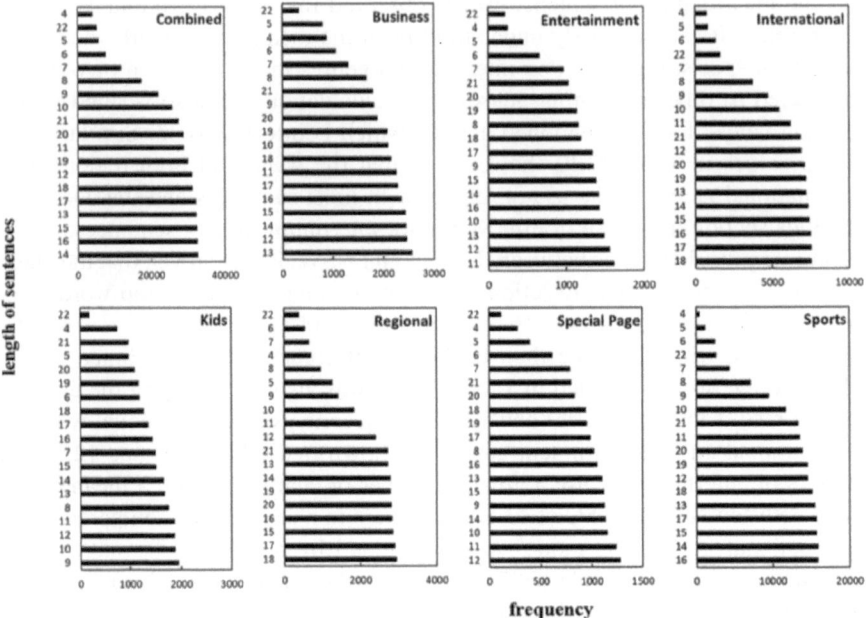

FIGURE 12.7 Sentence length frequency distribution for the combined corpus and discrete genres.

were used for the said purpose. Figure 12.8 illustrates the Pearson correlation coefficient results obtained for the combined corpus and the seven mentioned genres (illustrated using heat map charts). The highest correlation value of $r = 0.95$ has been observed for the $Business_{NSC} \times_{NSW}$; $Business_{NSC} \times_{NC}$; $Entertainment_{NC} \times_{NW}$; $Entertainment_{NSC} \times_{NSW}$; $International_{NSC} \times_{NC}$; and $Sports_{NSC} \times_{NC}$, whereas the weakest value of $r = 0.82$ was perceived particularly in the case of performing the correlation analysis on the NSC and NW SLTS of an individual genre, i.e., $International_{NSC} \times_{NW}$; $Kids_{NSC} \times_{NW}$; $Regional_{NSC} \times_{NW}$; and $Sports_{NSC} \times_{NW}$. In the case of the combined corpus, the highest correlation was obtained ($r = 0.95$) for NSC and NC time series and the weakest in the case of NSC and NW. Taking the combined corpus correlation r values into consideration, the value of 0.89 is attained on average. It can be evidently seen in the results that higher values have been achieved while calculating the Pearson correlation coefficient for NSC and NC sentence length time series, followed by NSC and NSW, and NC and NW. The lowest values have been achieved in the case of performing SLTS for NSC and NW. Still, looking at the bigger picture, the results illustrated in the distinct variants of heat maps clearly show the strong correlation observed between the combined corpus and four SLTS in individual genres of text while considering the Pearson's correlation scale (± 0.8 to ± 1.0 = high correlation, ± 0.6 to ± 0.79 = moderately high correlation, ± 0.4 to ± 0.59 = moderate correlation, ± 0.2 to ± 0.39 = low correlation, and ± 0.1 to ± 0.19 = negligible correlation).

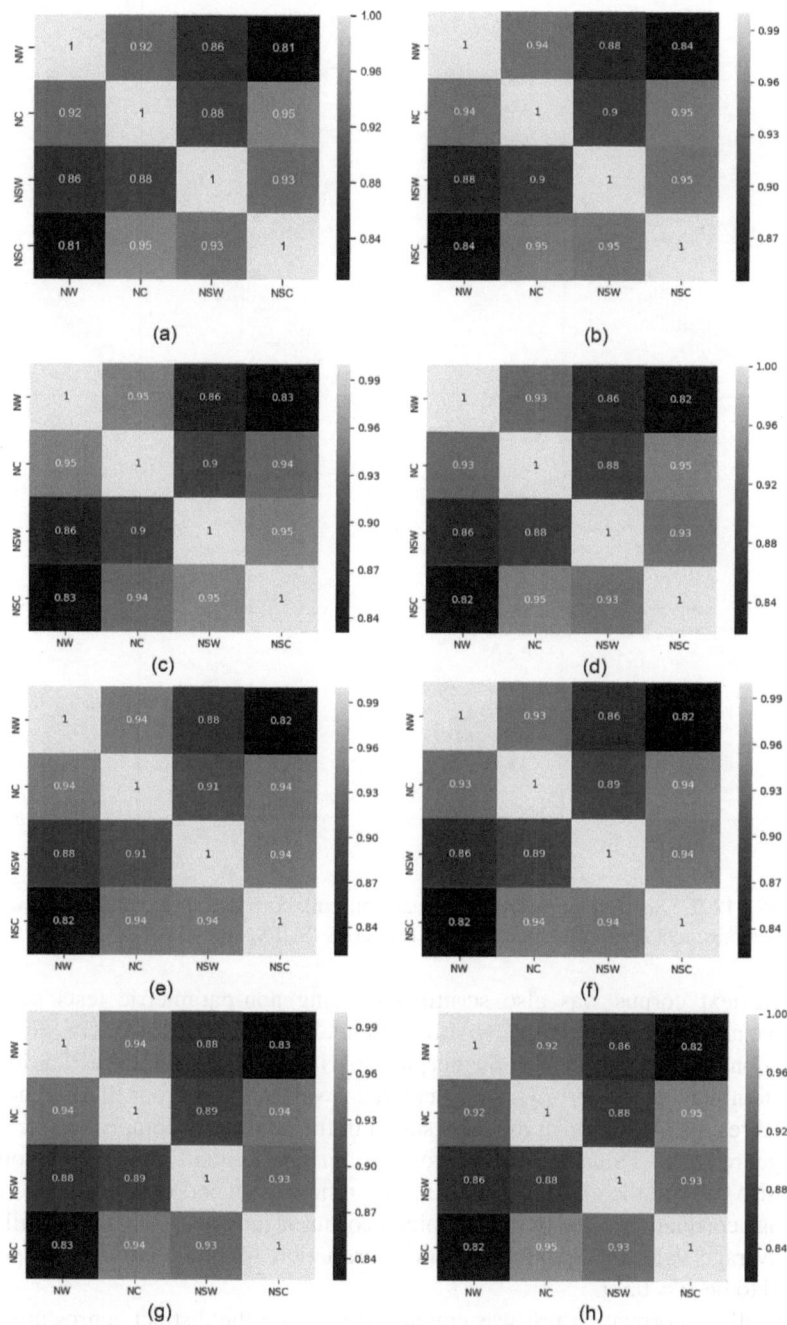

FIGURE 12.8 Pearson correlation coefficient value heat maps illustration of (a) the combined corpus, and other genres of text, i.e., (b) business, (c) entertainment, (d) international, (e) kids, (f) regional, (g) special page, and (h) sports.

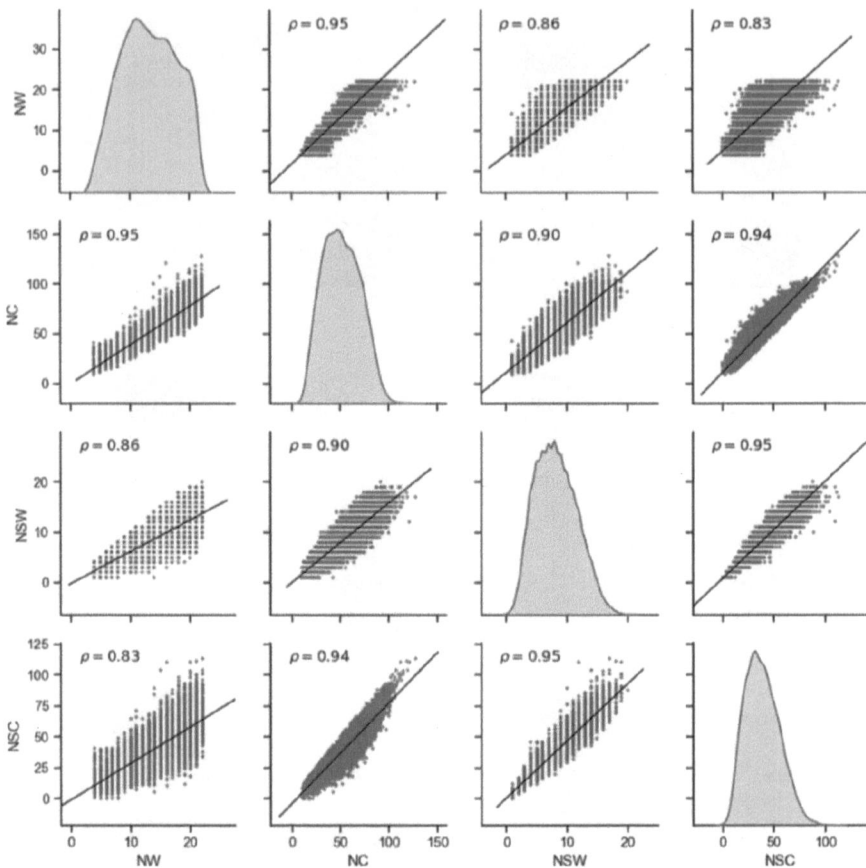

FIGURE 12.9 Scatterplot matrix for Spearman rank correlation (ρ) values obtained for the entertainment genre sentence length time series (SLTS) analysis.

The text corpus was also scrutinized using non-parametric tests such as Spearman's rank correlation (ρ) and Kendall rank correlation (τ) given in Equations 12.8 and 12.9. The Scatterplot matrix displayed in Figure 12.9 for the entertainment genre portrays the calculated ρ values for the four SLTS. The displayed results also confirm the facts stated in the analysis accomplished for various genres of SLTS using Pearson correlation. The Kendall rank correlation (τ) test also backed the claim made using the other two mentioned tests, where an average correlation value of the combined corpus is judged as $\tau = 0.76$. Similarly, the average value of Spearman's rank correlation for the combined corpus is found to be $\rho = 0.90$.

Finally, a correlation test was conducted between the distinct genres utilizing *NW* SLTS. A total of 500 words were taken for this task by using the law of randomization. Tables 12.14 and 12.15 depict the experimental results attained for this purpose utilizing Pearson's and Spearman's correlation coefficients. The findings

TABLE 12.14

Analysis of Pearson Correlation Coefficient Results between Distinct Genres' NW SLTS

		1	2	3	4	5	6	7
1	Business	1						
2	Entertainment	0.007	1					
3	International	−0.026	−0.057	1				
4	Kids	−0.013	0.006	−0.040	1			
5	Regional	−0.033	−0.010	−0.015	−0.001	1		
6	Special page	−0.057	0.075	0.021	−0.121	−0.006	1	
7	Sports	0.014	−0.129	0.034	−0.069	−0.007	0.004	1

TABLE 12.15

Analysis of Spearman's Correlation Coefficient Results between Distinct Genres' NW SLTS

		1	2	3	4	5	6	7
1	Business	1						
2	Entertainment	0.027	1					
3	International	−0.011	−0.060	1				
4	Kids	0.014	0.013	0.010	1			
5	Regional	0.010	−0.070	−0.037	0.020	1		
6	Special page	0.033	0.018	−0.051	−0.140	−0.020	1	
7	Sports	0.040	0.033	−0.003	0.006	−0.015	−0.001	1

show that in no case, the correlation existed between the genres of text, and it clearly indicates that every genre has its own distinct usage of words in writing. These results can be seen under the umbrella of comments made by Hyland (2011), which states that "writers in different disciplines represent themselves, their work and their readers in different ways."

12.5 CONCLUSIONS

In this linguistic research, an effort has been made to perform a quantitative analysis of the "Gurmukhi" script at two syntactic levels—words and sentences. A corpus for seven distinct genres (*business, entertainment, international, kids, regional, special page,* and *sports*) was collected, and also an analysis was performed on the combined corpus (taking all genres text into consideration) as an additional lead for the assessment purpose. This study follows the standard statistical measures such as *mean, mode, median, standard deviation, skewness,* TTR, *frequency distribution*, and *correlation* for the evaluation of results. Various distinct facts, similarities, and dissimilarities were observed in the Gurmukhi text, for which the detailed arguments were made in the previous section of this chapter.

In summary, a word length analysis of the extracted corpus was performed. It was revealed that, on average, the writer uses 4.03 characters in Gurmukhi words. The kids and entertainment genres use the least number of characters while composing the words; in contrast to this, the sports and regional genres use words with more characters. The similarity that is detected while examining the word length is that the words containing two characters are utilized with high frequency. While reviewing the gathered standard deviation values, the special page genre is encountered with high variability due to the fact that the writers of the said genre use distinct words in writing as compared to those of the other genres. Furthermore, the said claim of usage of distinct words was also supported in the TTR analysis, where the special page genre has achieved the TTR score of 10.95%, which is the highest among all the genres, followed by the kids genre with 9.69% TTR.

Also, frequency distribution analysis was performed for examining the practice of usage of common Gurmukhi characters and vowels in the corpus. During the study, the character "ਰ" occurred to be the most frequent character. It was also noticed that "ਕ" is the most frequently used character in the various mentioned genres. While performing the frequency distribution study on the vowels, it was revealed that in most cases, the vowels are used with the identical frequency in all of the mentioned genres of the text. However, vowels "ੀ" and "ੋ" emerged to be the two uppermost vowels. In addition, a review was carried out to know the detail about the words and word length frequency distribution of the script. It was found that "ਹੈ" appeared as the word with the highest peak across all genres, whereas the word "ਨੇ" is the most frequent word in the combined corpus. Various similarities were discovered while performing the frequency distribution study on the word lengths; for example, words with two characters are the most extensively utilized words, followed four-, three-, five-, and six-character words.

After making the statistical revelations about the "words" in the Gurmukhi script, the focus was shifted toward another syntactic level, i.e., sentences. Several physical aspects of Gurmukhi sentences were measured by means of pondering over four techniques to plot them to time series. Four SLTS, i.e., the number of words, the number of characters, the number of words without stop-words, and the number of characters without stop-words, are utilized. It was recognized that writers utilize 14.31 words on average in sentences. It was also shown during the analysis that the regional and international genres make more use of words in sentences and the kids genre makes least.

Four SLTS were used for individual genres to answer the null hypothesis question that two distributions are independent. The value of $r = 0.89$ is attained on average for the combined corpus using Pearson correlation. Correlation results of diverse genres visibly rejected the null hypothesis and established that two SLTS distributions for an individual genre are drawn from the same distribution. It was also established that higher values had been achieved between NSC and NC, followed by NSC and NSW, and NC and NW SLTS distributions. The lowest values have been attained in the case of performing a correlation test between NSC and NW. It was also established that the number of words (NW) as a metric cannot be used to distinguish the genres text in the Gurmukhi. These particular results follow the Wei et al. (2019) study on word length distribution of the Zhuang language, in which the authors have stated that word length distribution is not a good index to differentiate between Zhuang language genres. Experiment values reveal that the removal of stop-words decreases the variability and increases the skewness in general Gurmukhi sentences.

Our study is the first of its kind performed on the Gurmukhi script and makes a significant contribution to the field of quantitative linguistics. Although similar kind of research in which quantitative aspects of Indian languages such as Hindi, Bengali, and Gujarati has been carried out by other researchers, the Gurmukhi script has remained untouched mostly due to ignorance, and more focus remains on the development of NLP software and applications for other languages. However all the facts and experiments cannot be stated and performed in one paper due to limited time and resources. The extracted corpus is distinctive in nature because of the fact that meta-information (date, month, year, and title) has been maintained for each genre and can be used in future experiments. Future work can be oriented toward finding the usage of syntactic types with respect to the time it occurred or utilizing the corpus for genre identification, and analysis can be performed to discover the utilization of distinct parts of speech (PoS) categories, verification of Zipf's law for the mentioned genres, etc. Also, the stated SLTS distributions could further be used to perform the detrended fluctuation analysis and further investigate the long-range correlations. The authors are willing to provide the corpus on request.

REFERENCES

Altman, D. G. and, J. M. Bland, 2005. "Standard deviations and standard errors" *BMJ* 331(7521):903. doi: 10.1136/bmj.331.7521.903

Barde, M.P. and, P.J. Barde, 2012. "What to use to express the variability of data: Standard deviation or standard error of mean?" *Perspectives in Clinical Research* 3(3):113. doi: 10.4103/2229-3485.100662.

Bharati, A., P. Rao, R. Sangal, and S.M. Bendre, 2002. "Basic statistical analaysis of corpus and cross comparision basic statistical analysis of corpus and cross comparison among corpora." pp. 18–21 in *ICON-2002: International Conference on Natural Language Processing*. Mumbai.

Biber, D., U. Connor, and T.A. Upton. 2007. *Discourse on the Move: Using Corpus Analysis to Describe Discourse Structure*. John Benjamins Pub. Co, Amsterdam.

Blyth, S. 1994. "Karl Pearson and the Correlation Curve." *International Statistical Review* 62(3):393–403. doi: 10.2307/1403769.

Bolboaca, S.D. and L. Jäntschi. 2006. "Pearson versçus Spearman, Kendall's Tau correlation analysis on structure-activity relationships of biologic active compounds." *Leonardo Journal of Sciences* 5(9):179–200.

Chan, K.-p., and A.W.-c. Fu. 1999. "Efficient time series matching by wavelets." in *Proceedings of 15th International Conference on Data Engineering*, pp. 126–33. IEEE, Manhattan, NY.

Chirikhin, K.S., and B. Ya Ryabko. 2019. "Application of data compression techniques to time series forecasting." pp. 2–6. Retrieved (http://arxiv.org/abs/1904.03825).

Colwell, D.J., and J.R. Gillett. 1982. "Spearman versus Kendall." *The Mathematical Gazette* 66(438):307–9.

Daud, A., W. Khan, and D. Che. 2017. "Urdu language processing: a survey." *Artificial Intelligence Review* 47(3):279–311. doi: 10.1007/s10462-016-9482-x.

Doanne, D.P., and L.E. Seward. 2011. "Measuring skewness: a forgotten statistic? David." *Journal of Statistics Education* 19(2):1–18.

Fize, J., M. Roche, and M. Teisseire. 2018. "Matching heterogeneous textual data using spatial features." in *2018 IEEE International Conference on Data Mining Workshops (ICDMW)*. Vols. 2018-Nov, pp. 1389–96. IEEE, Manhattan, NY.

Ghasemi, A., and S. Zahediasl. 2012. "Normality tests for statistical analysis: a guide for non-statisticians." *International Journal of Endocrinology and Metabolism* 10(2):486–89. doi: 10.5812/ijem.3505.

Good, P. 2009. "Robustness of Pearson correlation." *Interstat* 15(5):1–6.

Goyal, L. 2011. "Comparative analysis of printed Hindi and Punjabi text based on statistical parameters." in *Communications in Computer and Information Science*. Vol. 139, edited by C. Singh, G.S. Lehal, J. Sengupta, D.V. Sharma, and V. Goyal, pp. 209–13. Springer, Berlin, Heidelberg.

Hoskin, T. 2012. "Parametric and nonparametric: demystifying the terms." *Mayo Clinic* 1–5. doi: 10.1016/s0924-977x(16)31643-1.

Hřebíček, L. 1997. "Persistence and other aspects of sentence-length series." *Journal of Quantitative Linguistics* 4(1–3):103–9. doi: 10.1080/09296179708590083.

Hyland, K. 2011. "Disciplines and discourses: social interactions in the construction of knowledge." in *Writing in the Knowledge Society*, edited by D. Starke-Meyerring, A. Paré, N. Artemeva, M. Horne, and L. Yousoubova, pp. 193–214. Fort Collins, Colorado: The WAC Clearinghouse and Parlor Press.

Isotalo, J. 2014. *Basics of Statistics*. CreateSpace Independent Publishing Platform, Scotts Valley, CA.

Jayaram, B. D., and M. N. Vidya. 2008. "Zipf's Law for Indian Languages." *Journal of Quantitative Linguistics* 15(4):293–317. doi: 10.1080/09296170802326640.

Johansson, V. 2009. "Lexical diversity and lexical density in speech and writing: a developmental perspective." *Lund University, Department of Linguistics and Phonetics Working Papers 53:61–79*.

Kalimeri, M., V. Constantoudis, C. Papadimitriou, K. Karamanos, F.K. Diakonos, and H. Papageorgiou. 2015. "Word-length entropies and correlations of natural language written texts." *Journal of Quantitative Linguistics* 22(2):101–18. doi: 10.1080/09296174.2014.1001636.

Kaur, J., and J.R. Saini. 2016. "Punjabi stop words: a Gurmukhi, Shahmukhi and Roman scripted chronicle." in *Proceedings of the ACM Symposium on Women in Research 2016*, pp. 32–37. Association for Computing Machinery, New York, NY.

Kendall, M.G. 1938. "A new measure of rank correlation." *Biometrika* 30(1/2):81. doi: 10.2307/2332226.

Kendall, M.G. 1975. *Rank Correlation Methods*. Griffin, London.

Kenney, J.F. 1939. *Mathematics of Statistics Part I*. Chapman & Hall Ltd., London.

Kim, K.-j. 2003. "Financial time series forecasting using support vector machines." *Neurocomputing* 55:307–319 doi: 10.1007/s00357-015-9167-1.

Kosmidis, K., A. Kalampokis, and P. Argyrakis. 2006. "Language time series analysis." *Physica A: Statistical Mechanics and Its Applications* 370(2):808–16. doi: 10.1016/j.physa.2006.02.042.

Kumar, G.B., K.N. Murthy, and B.B. Chaudhuri. 2007. "Statistical analysis of Telugu text corpora." *International Journal of Dravidian Languages* 36(2): 71–99.

Kumar, N., and G.K. Jha. 2013. "Time series ANN approach for weather forecasting." *International Journal of Control Theory and Computer Modeling* 3(1):19–25. doi: 10.5121/ijctcm.2013.3102.

Lakshmi Priya, R., and G. Manimannan. 2014. "A study of ambiguous authorship in Tamil articles using multivariate statistical analysis." *International Journal of Computer Applications* 86(1):21–25. doi: 10.5120/14951-3112.

Mahi, G.S., and A. Verma. 2020. "PURAN: word prediction system for Punjabi language news." *Advances in Intelligent Systems and Computing* 1042: 383–400.

Mehta, P., and P. Majumder. 2016. "Large scale quantitative analysis of three Indo-Aryan languages." *Journal of Quantitative Linguistics* 23(1):109–32. doi: 10.1080/09296174.2015.1071151.

Pande, H., and H. S. Dhami. 2013. "Mathematical modelling of the pattern of occurrence of words in different corpora of the Hindi language." *Journal of Quantitative Linguistics* 20(1):1–12. doi: 10.1080/09296174.2012.754596.

Pande, H., and H.S. Dhami. 2015. "Analysis and mathematical modelling of the pattern of occurrence of various Devanāgari letter symbols according to the phonological inventory of Indic script in Hindi language." *Journal of Quantitative Linguistics* 22(1):22–43. doi: 10.1080/09296174.2014.974457.

Pearson, K. 1895. "Contributions to the mathematical theory of evolution, II: skew variation in homogeneous material." *Philosophical Transactions of the Royal Society of London. A* 186(1895):343–414.

Richards, B. 1987. "Type/token ratios: what do they really tell us?" *Journal of Child Language* 14(2):201–9. doi: 10.1017/S0305000900012885.

Sedgwick, P. 2012. "Pearson's correlation coefficient." *BMJ* 345:1–2 doi: 10.1136/bmj. e4483.

Singh, P., and G.S. Lehal. 2010. "Corpus based statistical analysis of Punjabi syllables for preparation of Punjabi speech database." *International Journal of Intelligent Computing Research* 1(3):138–42. doi: 10.20533/ijicr.2042.4655.2011.0015.

Vieira, D.S., S. Picoli, and R.S. Mendes. 2018. "Robustness of sentence length measures in written texts." *Physica A: Statistical Mechanics and Its Applications* 506:749–54. doi: 10.1016/j.physa.2018.04.104.

Wei, A., Q. Lu, and H. Liu. 2019. "Word length distribution in Zhuang language." *Journal of Quantitative Linguistics* 1–28. doi: 10.1080/09296174.2019.1678225.

NOTES

1 "Natural Language Processing," available at https://en.wikipedia.org/wiki/Natural_language_processing.

2 "Punjabi," available at https://en.wikipedia.org/wiki/Punjabi_language.

3 https://www.punjabitribuneonline.com/.

4 https://docs.python.org/2/library/urllib.html

5 https://www.crummy.com/software/BeautifulSoup/bs4/doc/

6 "Python"—https://www.python.org/

7 https://en.wikipedia.org/wiki/XML

8 "Pandas"—https://pandas.pydata.org/

9 "SciPy"—https://www.scipy.org

10 "Matplotlib"—https://matplotlib.org/

11 "Seaborn"—https://seaborn.pydata.org/

13 An Analysis of Protein Interaction and Its Methods, Metabolite Pathway and Drug Discovery

P. Lakshmi and D. Ramyachitra
Bharathiar University

CONTENTS

13.1 Introduction 237
    13.1.1 Related Works 239
13.2 Methodology 240
    13.2.1 The Rosetta Stone Method 240
    13.2.2 Yeast Two-Hybrid Method 240
    13.2.3 Sequence Alignment 241
    13.2.4 Docking and Drug Discovery 241
    13.2.5 Metabolite–Protein Interactions 243
    13.2.6 Protein Function Prediction 244
    13.2.7 Pathway of the Protein Interaction Network 245
    13.2.8 The Two-Hybrid System 245
    13.2.9 Perception of Protein Interaction Methods 246
13.3 Conclusions 247
References 247

13.1 INTRODUCTION

Proteins are present in all living systems. More than 650,000 human protein interactions are estimated to introduce various curated drug targets [1]. Protein interaction helps to identify the performance of cellular functions [2]. Cellular system's workhorse structured by proteins and the design of the protein complex underpins various cellular systems. Inappropriate time and location of the protein complex will lead to some kind of diseases such as autoimmune disorders and cancer to design the target region of the drugs, protein interaction used to understand the function of the protein [3,4]. The collection of interacting and non-interacting proteins helps to classify the feature structures of the protein interaction to determine the binding and non-binding behaviours of the protein [5]. The features of the proteins/genes classification are used to detect diseases from the human genome structure [6]. The functional and structural properties belong to the characteristics of protein interaction represented in the appearance of hubs, from this, nodes and edges are the proteins and interacted between proteins, respectively. Two types of hub proteins are classified: one is party hub and the other is date hub. Party hubs interact with their neighbour on the same time and space, but date hubs interact with different times and different spaces with their neighbour [7,8].

Figure 13.1 shows that the structure of the interacted protein complex is classified into two types. Protein function lifetime is permanent is known as stable and another one is transient, that is interacting partners with a short lifetime and the flow of the function belongs to the stable one [9]. PPIN contains some duplicate pre-processes such as gene duplication, mutation and duplication model to describe the DNA sequence evolution [10]. Inside the biological network pathways, protein interactions are experimentally determined and evaluated. To analyse the metabolic pathways and signalling pathways helps to know the special attention to reconstruct the cellular processes. This direction leads to understanding the transcription factor and moving information between them [11–13].

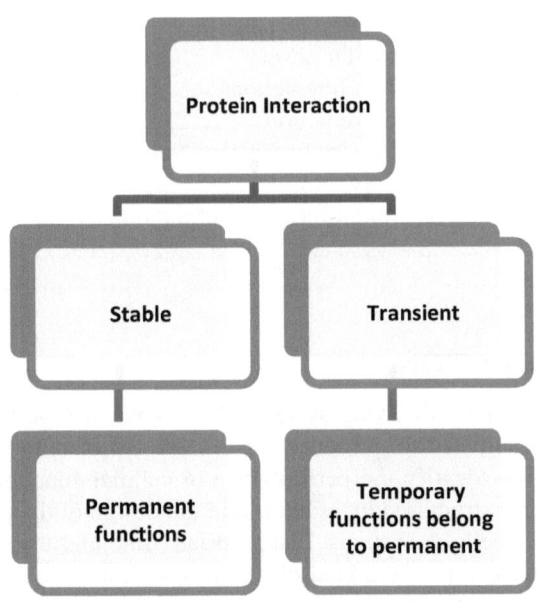

FIGURE 13.1   Types of interacting protein complexes.

13.1.1 Related Works

Proteins are the collection of 20 amino acids with the composed polypeptide amino acid sequences. They contain $-NH2$ (amine) or $-COOH$ (carboxyl) with side chain R group. The protein's primary structure is represented in Figure 13.2 [14].

From different features of the interacted proteins, particular protein affinity with its solid surface evaluation changes during the process of adsorption with the composition and the density of the protein changes. This process happens in multiple proteins and is called **Vroman effect** [15]. Commonly, proteins are in the form of a folder structure. Surfactant interactions of the hydrophobic and electrostatic power of the protein interactions direct to folding or misfolding the protein structures. This kind of misfolding may lead to diseases such as cancer, mutations on genes and neurodegenerative diseases. Drug discovery is one of the ways to overcome these kinds of problems and necessitates new inventions in protein interaction [16–18]. SIM (Spatial Interaction Map) tool uses to hot spot residue prediction from protein interaction. Protein complex functions lead to getting an idea for a therapeutic molecule design to the target proteins [19]. The classification of structural features of proteins helps in detecting the binding and unbinding behaviours of the proteins from the protein complexes to predict the interacting and non-interacting proteins among them [20]. The discovery of disease and non-disease genes depends on the structural characteristic of the PPIN [21]. Different kinds of metabolite factors are related to proteins. Finding biological pathways helps to know the transport factors, metabolites, codons, receptors, structures, alterations and functions of the PPIN [22]. The predicted protein interaction quality measured by benchmark datasets was analysed with the study about the techniques, tools, methodologies, algorithms, and architecture [23]. The amino acids in the protein interaction network are evaluated and categorized into two types, namely single and multi-hubs. The single and multi-interfaces, respectively, contain the hydrophobic and electrostatic forces and describe the different topologies in PPIN [24]. The performance of the homo- and hetero-oligo metric or motif of the PPIN helps to describe the drug invention of their target proteins [25].

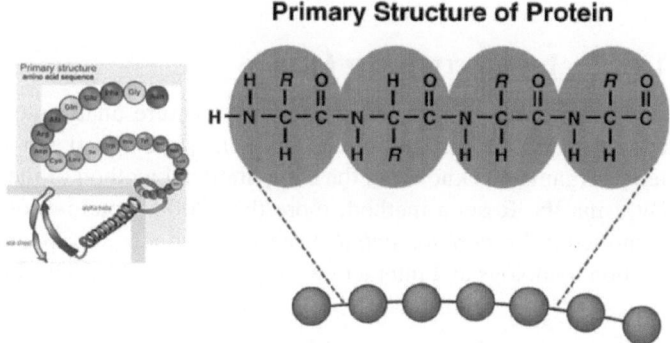

Primary Structure of Protein

FIGURE 13.2 Primary structure of the protein.

Protein function is the interface of the protein interaction networks and the evolutionary trace technique use to know the events and outline of the protein functions to indicate the analytical strategy of both prokaryotic and eukaryotic functions of the protein interactions. In the protein interaction network, the process of holding functions has an effect on other components of the cell [26]. In the life cycle of a cell, the interaction of the proteins plays a vital role in specificity changes of the protein, reproduction and growth of the organism, transduction of the cell signal, gene expression and its regulation, genetic material duplication, apoptosis and cell necrosis [27–31]. From the protein sequences, constituent amino acid residue locations can be identified with three findings. They are as follows:

1. Residue interaction, in the form of clusters in sequences with immediately related residues on both sides.
2. The presence of one additional interface residue.
3. Extra four residue interfaces [32,33]. Applications of the interacting proteins are docking, identification of disease pathogens, new therapeutic drug development with hotspots residues, ligand binding protein docking, homology detection, alignment of the sequences, prediction of the protein function, structure prediction of the protein, gene expression, gene annotation, etc. [34–42].

13.2 METHODOLOGY

Many methodologies have been discovered to predict protein interaction networks. The perspective of these methods depends on the circumstances of the genomic and biological environment of the proteins, and their protein interaction predictions are used to find whether they are based on the physical, genetic and functional relationships between the proteins and their participation in the biological pathways [43]. Information on protein and gene sequences is stored in databases. The databases are located in different places. Day to day, the information is exchanged, updated and synchronized [44]. The pictorial representation of the protein interaction network is given in Figure 13.3.

13.2.1 The Rosetta Stone Method

At the approach of protein interaction, structure analysis of gene fusion is the main task and the pair of proteins, homologs are bound and it is fused into a different organism is known as the computational method of the Rosetta Stone [45]. By using the Rosetta method, more than 7000 protein–protein interactions are identified in *Escherichia coli* and yeast. It produces approximately more than two million homologs and interactions of proteins from various organisms [46].

13.2.2 Yeast Two-Hybrid Method

We can simply say that the hybrid inheritance of two protein interactions in yeast is to be tested and analysed for protein fusion [47]. The aim of Y2H method

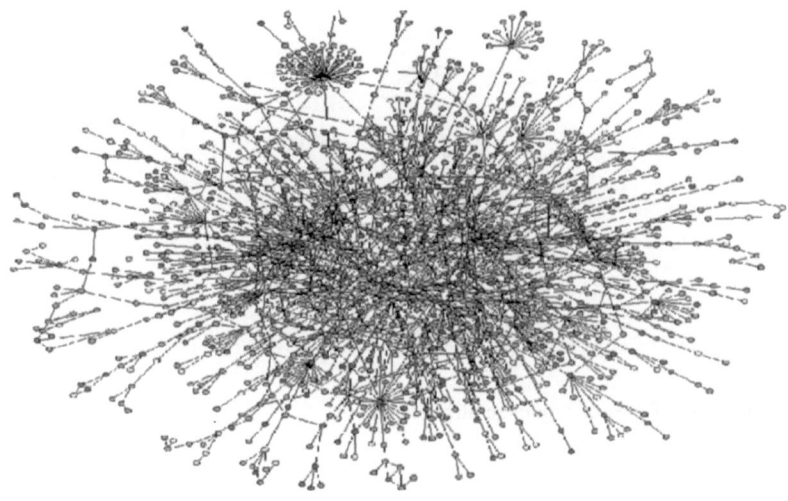

FIGURE 13.3   Combination of protein interaction networks.

is to reconsider the interacting protein transcription factor binding. From the hybrid inherited protein A, a transcription factor of the binding DNA field is compounded from the activation domain of the protein B. Suppose in the case of these proteins have interacted, then the reporter gene spot of the activation domain in a suitable location to activate transcription [48]. In computational biology, sequence analysis is the most primitive operation. This operation performs the similarity of the biological sequences, variation of the medical analysis and genome mapping processes [49].

13.2.3   SEQUENCE ALIGNMENT

Protein sequence alignment is used to know the importance of the homology detection, to predict the various features of a protein and to know the homologous structure. This alignment helps to predict the difference between the structure and the template of the sequence. In sequence alignment, BLAST and FASTA are the basic operations. The operations required to be performed level-wise are sequence identification, searching data in database, detection of homology, alignment of the sequences and updation of the structural information. Figure 13.4 shows the categorization of the sequence alignment. The recent versions of the instrument experiments with the help of NMR and X-ray crystallography are used to store the information of the isolation. These data are input to various algorithms to align the sequences effectively. Three types of alignments are available: single, pairwise and multiple sequence alignments.

13.2.4   DOCKING AND DRUG DISCOVERY

Docking use to conform ligand binding to the receptor; usually, the receptor is bigger than the molecules. This information includes the coordination of the ligand

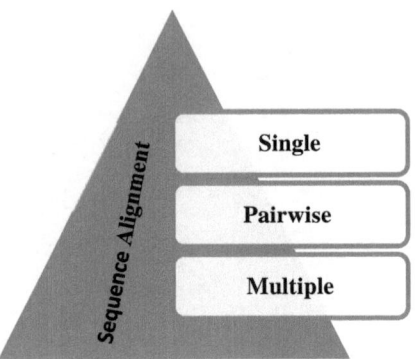

FIGURE 13.4 Types of sequence alignments.

atoms, to find the lowest energy binding site of the docking configuration. FLEX and AutoDock example programs will be shown in a later chapter of this chapter. The aim is to confirm the binding affinity and the bind. The overall minimum energy of complex formation can be found with the exact position and direction of the binding ligand that belongs to the interacting molecule activation within that. Various bioinformatics tools are helpful in disease management, diagnosis and drug discovery. Sequencing enables identifying the disease and drug discovery by scientists. Mutation and drug and all identified and experimented by utilizing different computational tools. Drug targets decide the suitable drug entry into the pipeline of drug development with the help of bioinformatics tools. The process of designing a medication with the target molecule is known as drug designing. The smallest molecule is ligand that switches on the biological target molecule output in the therapeutic effect [50]. However, the approach of single regulatory is a difficult task in marketing authorization application in all countries. Figure 13.5 represents the levels of drug discovery with regulations. Figure 13.6 shows the drug designing methods and their types.

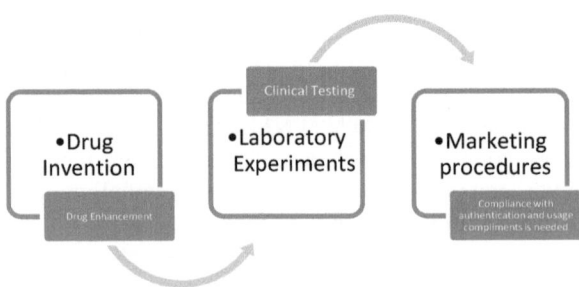

FIGURE 13.5 Level of drug discovery.

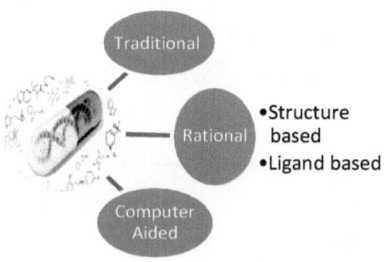

FIGURE 13.6 Design of drug discovery and its types.

13.2.5 METABOLITE–PROTEIN INTERACTIONS

The combination of chemicals known as "metabolome" appears in cells, organs, tissues and biological fluids in the form of molecules. Metabolites determine the information that belongs to the cellular process relationship with protein and gene expression, as shown in Figure 13.7. This information represents the molecular surface data, and it directs categorization and evaluation of the features of the metabolites which help to invent the new prediction of the biological and medicine.

The analysis of structural features and partition status methods help to improve the new techniques and experiments and obtain high-accuracy data from various

GENERAL PATHWAYS OF AMINO ACIDS METABOLISM

```
                        Proteins of food
                              |
                              v
                                              Metabolites of
          Amino acids  <---------------       glycolysis and
                                              Krebs cycle
            |
            v
      Anabolic ways                      Catabolic ways
            |                                  |
   +--------+--------+              +------+------+------+
   |                 |             Trans-   Deami-   Decar-
Synthesis of    Synthesis of       ami-     nation   boxila-
cell and        peptide            nation            tion
extracell       physiologi-                            |
proteins        cally active                         Amines
                substances
   |                 |                  |
   v                 v                  v
   Proteins and peptides          Urea, CO2, H2O
      of the organism
```

Urea, CO_2, H_2O

FIGURE 13.7 Proteins–amino acids metabolism pathway.

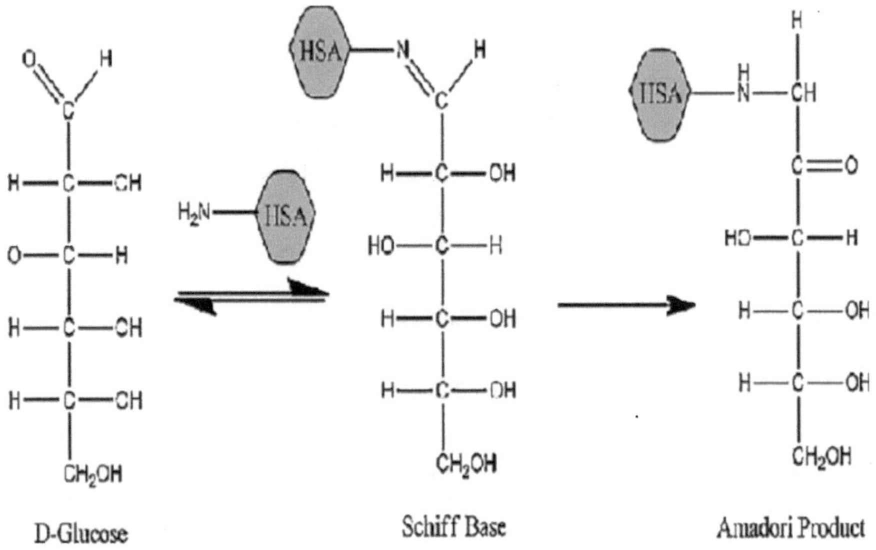

FIGURE 13.8 Example of HSA (human serum albumin) involved in the reactions of glycation of a protein.

combinations of complexes that belong to cells and tissues [51,52]. There are two types of methods: targeted and untargeted techniques. The targeted method categorizes the structural behaviours using NMR (nuclear magnetic resonance) and MS (mass spectrometry). This approach helps to know the particular class of proteins, metabolites and biochemical pathways. The untargeted method analyses the whole group of complexes with their chemical metabolites. It places in pathways of biochemical surroundings with increase the coverage of the metabolites [53].

For the purpose of binding metabolites with proteins contains various methods that can engage with the assessment of binding proteins has fewer mass drugs, metabolites with their involvement of hormones with relevant molecules shown in Figure 13.8. They are classified into three types, namely in silico, in vivo and in vitro [54–57]. A modified structure of the human serum albumin (HSA) may cause diabetes. Abnormally, disorder groups can be categorized with hyperglycaemias, from this result leads to deficiency and/or resistance of the insulin [58]. Nerve and heart problems may occur from the critical terms of diabetes; it belongs to the non-enzymatic glycation of proteins [59].

13.2.6 PROTEIN FUNCTION PREDICTION

Protein is used to build and repair tissues, and it is the basic structure of all cells in our body. It is a macromolecule that helps in the effective functioning of cells, and it performs a specific function in the body. The biological or biochemical role of a protein is assigned using protein function prediction [60]. Heterogeneous data are used in computational methods for protein function prediction. To create a protein interaction network, a number of top threshold measures are considered, such as protein sequences, building blocks of the protein structures and expressions of the genes [61]. In cell life cycle, with the combination of unknown proteins with its amino acids, photo- and chemical cross-linking is the broad application of protein interactions. It captures the highlights and maps the surface of the protein interactions, which can examine the analysis of photo-cross-linking and mass spectrometric combination. Some of the applications and improved areas of protein interactions with ncAAs are protein stapling, protein conformation with photo-control, cross-linking with two-dimensionality, transient stabilization and less affinity of protein interactions [62].

13.2.7 Pathway of the Protein Interaction Network

A protein–protein interaction network is a combination of protein interactions; also, this related information is stored in online databases. The protein–protein interactions are represented as a network where nodes and edges are denoted as protein and protein interactions, respectively. It contains the wellspring of related data in cell biology [63]. These databases are typically vast since information of the data is amassed at the time of experiments. Hence, revelation with database learning has turned into a trend of the bioinformatics. The important challenge in establishing the signalling pathways or networks is to convey the information from the recognized source to target [64]. The undirected PPI network is oriented and then identifies the directed pathways from the source to target. It is a complicated problem that several paths are linked with two proteins in interaction networks [65]. The reconstruction of biological pathways has received a lot of attention, such as the modifications of the biological networks [66–70], invention of the network signalling pathways [71–73] and the analysis of metabolic networks [74–76]. Protein interaction network data are undirected, and by using signalling networks the data can be made directed to find the relation within that. In the construction of the network, even though a set of interacting proteins with right direction are available, assembling protein pathways is a tedious task. Some of the signalling pathways such as Cell Database of the science signalling and KEGG [77,78]. Evaluation of protein interaction technologies has a challenging task to orient the given network [79,80]. Recent proteomics studies have inspected integrations between cellular proteins and molecules.

13.2.8 The Two-Hybrid System

The two-hybrid system is the widely used technique in screening and protein interaction prediction. In yeast, various transcription factors of eukaryotic domain interfere to properly activate the process of transcription and binding of DNA. To map the interaction of the genome, Y2H has been improved and modified with the required information. [81] Y2H is reasonably priced to use, and it is an in vivo method and necessitates small optimization for protein interaction prediction [82].

By using the Y2H method belongs to the nucleus for protein interaction prediction is complex, when compared to the cytoplasm-based protein interaction prediction [83,84].

13.2.9 Perception of Protein Interaction Methods

According to the environment of a broad area of protein, various efficient methods are to be improved to predict protein interactions. Some of the new ethics belong to the existing methods with developed techniques. It helps to know how to use the interpreted methods in experimental processes for the protein interaction prediction [85]. Figure 13.9 shows the encoded genetic method for cross-linking studies of ncAAs for protein interaction with ribosomal interpretation of live cells. It contains the suppressor tRNA with orthogonal AARS charges of ncAAs, that delivers to the ribosome, which incorporates to a responsible in-frame amber codon with nascent protein. In this way cross-linking ncAA characterize the protein features. Protein interaction helps to identify the mechanism of infection, drug development and the solution to the infection with treatment [86]. In protein–protein interaction relates to target regions and it helps to identify the functions and drug design of the proteins [87]. The SIM tool is used to find the site of the protein interaction with the unbounded protein structures [88]. By using interacting and non-interacting pairs of proteins, structural features are classified into binding or unbinding behaviour of proteins [89].

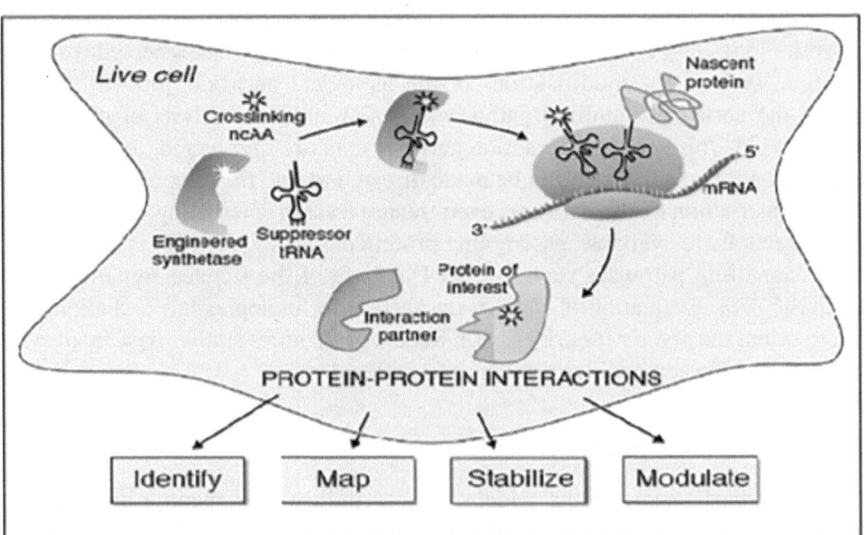

FIGURE 13.9 Cross-linking analysis of protein interactions with genetically encoded ncAAs.

13.3 CONCLUSIONS

Currently, the area of protein interaction has enormous growth in biological activities. This review focuses on various aspects of protein interaction prediction techniques. This review reveals the various aspects of protein interaction. The first section gives an overview of the methods used to characterize protein interactions. The second section discusses the techniques of protein interactions. A list of databases helps to get the input of the protein information, and screening techniques are represented to know the protein activities and structures in various organisms. Sequence alignment helps to know the importance of the structural aspects of proteins; FASTA and BLAST are the major operations of the protein sequence alignment. Bioinformatics tools provide additional data on protein interaction. This study investigates the different types of drug molecules and docking strategy. Nowadays, protein interaction prediction is developed in the area of protein interaction network pathway and protein function prediction. Some of the methods, tools and algorithms used in protein interaction prediction are reviewed. This review gives an idea of improvement of drug discovery with the help of protein interaction prediction.

REFERENCES

1. Stumpf, M.P.H., Thorne, T., de Silva, E., Stewart, R., An, H.J., Lappe, M. and Wiuf, C. 2008. Estimating the size of the human interactome. *Proc. Natl. Acad. Sci* 105(19): 6959–6964.
2. Gavin, A.C., Bosche, M., Krause, R., Grandi, P., Marzioch, M., Bauer, A. et al. 2002. Functional organization of the yeast proteome by systematic analysis of protein complexes. *Nature* 415: 141–147.
3. Thanos, C.D., DeLano, W.L. and Wells, J.A. 2006. Hot-spot mimicry of a cytokine receptor by a small molecule. *Proc. Natl. Acad. Sci. USA* 103 (42): 15422–15427.
4. Bullock, B.N., Jochim, A.L. and Arora, P.S. 2011. Assessing helical protein interfaces for inhibitor design. *J. Am. Chem. Soc.* 133 (36): 14220–14223.
5. Planas-Iglesias, J., Bonet, J., García-García, J., Marín-López, M.A., Feliu, E. and Oliva, B. 2013. Understanding protein–protein interactions using local structural features, *J. Mol. Biol.* 425: 1210–1224.
6. Salau, A.O. and Jain, S. 2021. Adaptive diagnostic machine learning technique for classification of cell decisions for AKT protein. *Inform. Med. Unlocked* 23 (1): 1–9. doi: 10.1016/j.imu.2021.100511.
7. Relating three-dimensional structures to protein networks provides evolutionary insights. *Science.* 314: 1938–1941.
8. Kim, P.M., Sboner, A. and Gerstein, M. 2008. The role of disorder in interaction networks: a structural analysis. *Mol. Syst. Biol.* 4: 179.
9. Nooren, I.M. and Thornton, J.M., 2003a. Diversity of protein-protein interactions. *EMBO J.* 22: 3486e3492.
10. Messer, P.W., Lässig, M. and Arndt, P.F. 2005. Universality of long-range correlations in expansion–randomization systems. *J. Stat. Mech: Theory* 10: P0004.
11. Bebek, G. and Yang, J. 2007. Pathfinder: mining signal transduction pathway segments from protein-protein interaction networks. *BMC Bioinf.* 8: 335.

12. Gitter, A., Klein-Seetharaman, J., Gupta, A. and Bar-Joseph, Z. 2011. Discovering pathways by orienting edges in protein interaction networks. *Nucleic Acids Res.* 39: e22.

13. Umezawa, T., Sugiyama, N., Takahashi, F., Anderson, J.C., Ishihama, Y., Peck, S.C. and Shinozaki, K. Genetics and phosphoproteomics reveal a protein phosphorylation network in the abscisic acid signaling pathway in Arabidopsis thaliana. *Sci. Signal.* 6(270): rs8.

14. Kessel, A. and Ben-Tal, N. 2010. *The Importance of Proteins in Living Organisms, in Introduction to Proteins: Structure, Function, and Motion*, CRC Press, Boca Raton, FL, pp. 1–59.

15. Vroman, L., Adams, A.L., Fischer, G.C. and Munoz, P.C. 1980. Interaction of high-molecular-weight kininogen, factor XII, and fibrinogen in plasma interfaces. *Blood* 55: 156–159.

16. Lee, H.J., McAuley, A., Schilke, K.F. and McGuire, J. 2011. Molecular origins of surfactant mediated stabilization of protein drugs, *Adv. Drug Deliv. Rev.* 63: 1160e1171.

17. Kishore, R.S., Kiese, S., Fischer, S., Pappenberger, A., Grauschopf, U. and Mahler, H.C. 2011. The degradation of polysorbates 20 and 80 and its potential impact on the stability of biotherapeutics, *Pharm. Res. (N. Y.)* 28: 1194e1210.

18. Khan, T.A., Mahler, H.C. and Kishore, R.S. 2015. Key interactions of surfactants in therapeutic protein formulations: a review. *Eur. J. Pharm. Biopharm.* 97: 60–67.

19. Agrawal, N.J., Helk, B. and Trout, B.L. A computational tool to predict the evolutionarily conserved protein-protein interaction hot-spot residues from the structure of the unbound protein. *FEBS Lett.* 588(2): 326–333.

20. Planas, J., Bonet, J., Garcia, J. Marin-Lopez, M.A., Feliu, E. and Oliva, B. 2013. Understanding protein-protein interactions using local structural features. *J. Mol. Biol.* 425: 1210–1224.

21. Wu, S.-Y., Shao, F.-j., Sun, R.-c., Sui, Y., Wang, Y. and Wang, J.-l. 2014. Analysis of human genes with protein-protein interaction network for detecting disease genes. *Physica A* 398: 217–228.

22. Matsuda, R., Bi, C., Anguizola, J., Sobansky, M., Rodriguez, E., Badilla, J.V., Zhang, X., Hage, B. and Hage, D.S. 2014. Studies of metabolite–protein interactions: A review, *J. Chromatograph. B*, 966: 48–58.

23. Lage, K. 2014. Protein-protein interactions and genetic diseases: The interactome. *Biochimica et Biophysica Acta* 1842: 1971–1980.

24. Peleg, O., Choi, J.-M. and Shakhnovich, E.I. 2014, October. Evolution of specificity in protein-protein *Interact. Biophys. J.* 107: 1686–1696.

25. Giordanetto, F., Schafer, A. and Ottmann, C. 2014, November 11. Stabilization of protein-protein interactions by small molecules, *Drug Discovery Today* 19: 1812–1821.

26. Lua, R.C., Marciano, D.C., Katsonis, P., Adikesavan, A.K., Wilkins, A.D. and Lichtarge, O. 2014. Prediction and redesign of protein-protein interactions, *Progress Biophys. Molecular Biol.* 116: 194e202.

27. De Las Rivas, J. and Fontanilla, C. 2012. Protein-protein interaction networks: unraveling the wiring of molecular machines within the cell. *Briefings Funct. Genomics* 11(6): 489–496.

28. Liu, G.-H., Shen, H.-B. and Yu, D.-J. 2016. Prediction of protein-protein interaction sites with machine-learning-based data-cleaning and post-filtering procedures. *J Membr Biol* 249:141–153.

29. Hayat, M. and Khan, A. 2011. Predicting membrane protein types by fusing composite protein sequence features into the pseudo amino acid composition. *J Theor Biol.* 271: 10–17.

30. Hayat, M. and Khan, A. 2013. WRF-TMH: predicting transmembrane helix by fusing composition index and physicochemical properties of amino acids. *AminoAcids* 44: 1317–1328.
31. Hayat, M. and Khan, A. 2013. Prediction of membrane protein types using pseudo-aminoacid composition and ensemble classification. *Int J ComputElectr Eng* 5: 456.
32. Ofran, Y. and Rost, B. 2003. Predicted protein-protein interaction sites from local sequence information. *FEBS Lett.* 544: 236–239.
33. Yan, C., Dobbs, D. and Honavar, V. 2004. A two-stage classifier for identification of protein-protein interface residues. *Bioinformatics* 20: i371–i378.
34. Torchala, M., Moal, I.H., Chaleil, R.A.G., Fernandez-Recio, J. and Bates, P.A. 2013. Swarm-Dock: a server for flexible protein-protein docking. *Bioinformatics* 29: 807–809.
35. Ghoorah, A.W., Devignes, M.-D., Smaïl-Tabbone, M. and Ritchie, D.W. 2011. Spatial clustering of protein binding sites for template-based protein docking. *Bioinformatics* 27: 2820–2827.
36. Tuncbag, N., Gursoy, A. and Keskin, O. 2013. Identification of computational hotspots in protein interfaces: combining solvent accessibility and inter-residue potentials improves the accuracy. *Bioinformatics* 25: 1513–1520.
37. Grove, L.E., Hall, D.R., Beglov, D., Vajda, S., Kozakov, D. and Flex, F.T. 2013. Accounting for binding site flexibility to improve fragment-based identification of druggable hot spots. *Bioinformatics* 29: 1218–1219.
38. Navlakha, S. and Kingsford, C. 2010. The power of protein interaction networks for associating genes with diseases. *Bioinformatics* 26: 1057–1063.
39. Mørk, S., Pletscher-Frankild, S., PallejaCaro, A., Gorodkin, J. and Jensen, L.J. 2014. Protein- driven inference of miRNA–disease associations. *Bioinformatics* 30: 392–397.
40. Zinzalla, G. and Thurston, D.E. 2009. Targeting protein–protein interactions for therapeutic intervention: a challenge for the future. *FutureMed.Chem.* 1: 65–93.
41. Johnson, D.K. and Karanicolas, J. 2019. Druggable protein interaction sites are more predisposed to surface pocket formation than the rest of the protein surface. *PLoS Comput.Biol.* e1002951.
42. Mignani, S., ElKazzouli, S., Bousmina, M.M. and Majoral, J.-P. 2014. Dendrimer space exploration: an assessment of dendrimers/dendritic scaffolding as inhibitors of protein-protein interactions, a potential new area of pharmaceutical development. *Chem.Rev.* 114: 1327–1342.
43. Skrabanek, L., Saini, H.K., Bader, G.D. and Enright, A.J. 2008. Computational prediction of protein-protein interactions. *Mol Biotechnol* 38: 1–17.
44. Altschul, S.F., Madden, T.L., Schaffer, A.A., Zhang, J., Zhang, Z., Miller, W., et al. 1997. Gapped BLAST and PSI-BLAST: a new generation of protein database search programs. *Nucleic Acids Res* 25(17): 3389–3402. [PMC free article] [PubMed].
45. Marcotte, E.M., Pellegrini, M., Ng, H.L., Rice, D.W., Yeates, T.O. and Eisenberg, D. 1999. Detecting protein function and protein-protein interactions from genome sequences. *Science* 285: 751–753.
46. Kamburov, A., Goldovsky, L., Freilich, S., Kapazoglou, A., Kunin, V., Enright, A.J., et al. 2007. Denoising inferred functional association networks obtained by gene fusion analysis. *BMC Genomics* 8: 460.
47. Ito, T., Ota, K., Kubota, H., Yamaguchi, Y., Chiba, T., Sakuraba, K., et al. 2002. Roles for the two-hybrid system in an exploration of the yeast protein interactome. *Mol Cell Proteomics* 1: 561–566.
48. Fields, S. and Song, O. 1989. A novel genetic system to detect protein-protein interactions. *Nature* 340: 245–246.

49. Lassmann, T. and Sonnhammer, E.L. 2005. Kalign–an accurate and fast multiple sequence alignment algorithm. *BMC Bioinf.* 6: 298.

50. Chen, R., Li, L. and Weng, Z. 2003. ZDOCK: an initial-stage protein-docking algorithm. *Proteins* 52: 80–87.

51. Kaddurah-Daouk, R., Kristal, B.S. and Weishiboum, R.M. 2008. Metabolomics: a global biochemical approach to drug response and disease. *Annu. Rev. Pharmacol. Toxicol.* 48: 653–683.

52. Kuehnbaum, N.L. and Mckibbin, P.B. 2013. New advances in separation science for metabolomics: resolving chemical diversity in a post-genomic era. *Chem. Rev.* 113: 2437–2468.

53. Lee, D.Y., Bowen, B.P. and Northen, T.R. 2010. Mass spectrometry-based metabolomics, analysis of metabolite-protein interactions, and imaging. *Biotechniques* 49: 557–565.

54. Yang, G.X., Li, X. and Synder, M. 2012. Investigating metabolite-protein interactions: an overview of available techniques. *Methods* 57: 459–466.

55. Sudlow, G., Birkett, D.J. and Wade, D.N. 1976. Further characterization of specific drug binding sites on human serum albumin. *Mol. Pharmacol.* 12: 1052–1061.

56. Clarke, W., Choudhuri, A.R. and Hage, D.S. 2001. Analysis of free drug fractions by ultra-fast immunoaffinity chromatography. *Anal. Chem.* 73: 2157–2164.

57. Clarke, W., Schiel, J.E., Moser, A. and Hage, D.S. 2005. Analysis of free hormone fractions by an ultrafast immunoextraction/ displacement immunoassay: Studies using free thyroxine as a model system. *Anal. Chem.* 77: 1859–1866.

58. Anguizola, J., Matsuda, R., Barnaby, O.S., Hoy, K.S., Wa, C., DeBolt, E., Koke, M. and Hage, D.S. 2013. Review: Glycation of human serum albumin. *Clin. Chim. Acta* 425: 64–76.

59. Hartog, J.W.L., Voors, A.A., Bakker, S.J.L., Smit, A.J. and Veldhuisen, D.J.V. 2007. Advanced glycation end-products (AGEs) and heart failure: pathophysiology and clinical implications. *Eur. J. HeartFail.* 9: 1146–1155.

60. Ruepp, A., and Mewes, H.W. 2006. Prediction and classification of protein functions. *Drug Discov. Today Technol.* 3(2): 145–151.

61. Cingovska, I., Bogojeska, A., Trivodaliev, K. and Kalajdziski, S. 2016, June. Protein function prediction by spectral clustering of protein interaction network. *Database Theory and Application, Bio-Science and Bio-Technology. BSBT 2011, DTA 2011. Communications in Computer and Information Science Series*, edited by T. Kim, vol 258, Springer, Berlin, Heidelberg.

62. Coin, I. 2018. Application of non-canonical cross-linking amino acids to study protein-protein interactions in live cells. *Curr. Opin. Chem. Biol.* 46: 156–163.

63. Chen, Y. and Xu, D. 2005. Understanding protein dispensability through machine-learning analysis of high-throughput data. *Bioinformatics.* 21: 575–581.

64. Segal, E., Shapira, M., Regev, A., Peer, D., Botstein, D., Koller, D. and Friedman, N. 2003. Module networks: identifying regulatory modules and their condition-specific regulators from gene expression data, *Nat. Genet.* 34: 166–176.

65. Grzegorczyk, M. and Husmeier, D. 2011. Improvements in the reconstruction of time-varying gene regulatory networks: dynamic programming and regularization by information sharing among genes. *Bioinformatics* 27: 693–699.

66. Liu, G.X., Feng, W., Wang, H., Liu, L. and Zhou, C.G. 2009. Reconstruction of gene regulatory networks based on two-stage Bayesian network structure learning algorithm. *J. Bionic Eng.* 6: 86–92.

67. Ravcheev, D.A., Best, A.A., Sernova, N.V., Kazanov, M.D., Novichkov, P.S., Rodionov, D.A. 2013. Genomic reconstruction of transcriptional regulatory networks in lactic acid bacteria. *BMC Genomics* 14: 14–94.

68. Margolin, A.A., Nemenman, I., Basso, K., Wiggins, C., Stolovitzky, G., Favera, R.D. and Califano, A. 2006. Aracne: an algorithm for the reconstruction of gene regulatory networks in a mammalian cellular context. *BMC Bioinf.* 7: S7.

69. Bebek, G. and Yang, J. 2007. PathFinder: mining signal transduction pathway segments from protein-protein interaction networks. *BMC Bioinformatics* 8: 335.

70. Gitter. A., Klein-Seetharaman, J., Gupta, A. and Bar-Joseph, Z. 2011. Discovering pathways by orienting edges in protein interaction networks. *Nucleic Acids Res.* 39: e22–e22.

71. Scott, J., Ideker, T., Karp, R.M. and Sharan, R. 2006. Efficient algorithms for detecting signaling pathways in protein interaction networks. *J. Comput. Biol.* 13: 133–144.

72. Kitagawa, J. and Iba, H. 2003. Identifying metabolic pathways and gene regulation networks with evolutionary algorithms. *Evol. Comput. Bioinforma.* 255–275.

73. Fischer, E. and Sauer, U. 2005 .Large-scale in vivo flux analysis shows rigidity and suboptimal performance of Bacillus subtilis metabolism. *Nat. Genet.* 37: 636–640.

74. Ruppin, E., Papin, J.A., Figueiredo, L.F. and Schuster, S. 2010. Metabolic reconstruction, constraint-based analysis, and game theory to probe genome-scale metabolic networks. *Curr. Opin. Biotechnol.* 21: 502–510.

75. Steffen, M., Petti, A., Aach, J., D'haeseleer, P. and Church, G. 2012. Automated modeling of signal transduction networks. *BMC Bioinform.* 3: 34.

76. Kanehisa, M. and Goto S. 2000. KEGG: Kyoto encyclopedia of genes and genomes. *Nucleic Acids Res.* 28: 27–30.

77. Medvedovsky, A., Bafna, V., Zwick, U. and Sharan, R. 2008. An algorithm for orienting graphs based on cause-effect pairs and its applications to orienting protein networks. In *Proceedings of the 8th International Workshop on Algorithms in Bioinformatics*. Karlsruhe, Germany, pp. 222–232.

78. Xiong, W., Xie, L., Zhou, S. and Guan J. 2014. Active learning for protein function prediction in protein-protein interaction networks. *Neurocomputing* 145: 44–52.

79. Kohli, R., Krishnamurti, R. and Mirchandani, P. 1994. The minimum satisfiability problem. *SIAM J. Discret. Math.* 7: 275–283.

80. Shlomi, T., Segal, D., Ruppin, E. and Sharan, R. 2006. QPath: a method for querying pathways in a protein-protein interaction network. *BMC Bioinformatics* 7: 199.

81. Walhout, A.J., Sordella, R., Lu, X., Hartley, J.L. et al., 2000. Protein interaction mapping in C. Elegans using proteins involved in vulval development. *Science* 287: 116–122.

82. Estojak, J., Brent, R., Golemis, E.A. 1995. Correlation of two-hybrid affinity data with in vitro measurements. *Mol. Cell. Biol.* 15: 5820–5829.

83. Aronheim, A., Zandi, E., Hennemann, H., Elledge, S.J. and Karin M. 1997. Isolation of an AP-1 repressor by a novel method for detecting protein-protein interactions. *Mol. Cell. Biol.* 17, 3094–3102.

84. Broder, Y.C., Katz, S. and Aronheim, A. 1998. The RAS recruitment system, a novel approach to the study of protein-protein interactions. *Curr. Biol.* 8: 1121–1124.

85. Tord B., Linse, S. and James, P. 2007. Methods for the detection and analysis of protein-protein interactions. *Proteomics* 7: 2833–2842.

86. Shatnawi, M. 2015. Review of recent protein-protein interaction techniques. *Emerging Trends in Computational Biology, Bioinformatics, and Systems Biology,* edited by Quoc Nam tran and Hamid R. Arabnia, Morgan Kaufmann, Burlington, MA, pp. 99–122.

87. Thanos, C.D., DeLano, W.L. and Wells, J.A. 2006. Hot-spot mimicry of a cytokine receptor by a small molecule. *Proc. Natl. Acad. Sci. USA* 103 (42): 15422–15427.

88. Agrawal, N.J., Helk, B. and Trout, B.L. 2014. A computational tool to predict the evolutionarily conserved protein-protein interaction hot-spot residues from the structure of the unbound protein. *FEBS Letters* 588: 326–333.

89. Planas-Iglesias, J., Bonet, J., García-García, J., Marín-López, M.A., Feliu, E. and Oliva, B. 2013. Understanding protein-protein interactions using local structural features. *J. Mol. Biol.* jmb article, 0022–2836/$ - see front matter © 2013 Elsevier Ltd. All rights reserved.

14 Biosensors for Disease Diagnosis

Ramneet Kaur
Regional Institute of Management
and Technology University

*Dibita Mandal, Juveria Ansari, Prachi R.
Londhe, Vedika Potdar, and Vishakkha Dash*
Mumbai University

CONTENTS

14.1 Introduction 253
    14.1.1 Disposable Immunosensors 255
    14.1.2 Point-of-Care Diagnostics 256
14.2 Biosensors in the Diagnosis of Alzheimer's Disease 257
14.3 Biosensors in Diagnosis of Cancer 259
14.4 Biosensor in Detection of Hepatitis 260
14.5 Biosensors in Diagnosis of HIV 261
14.6 Biosensors in Diagnosis of SARS-CoV-2 263
14.7 Conclusion 264
Acknowledgments 264
References 265

14.1 INTRODUCTION

Biological sensors (biosensors) are small analytical devices that incorporate a sensing element within close or integrated proximity that is biological in nature along with a signal transducer which is a physicochemical detector able to detect and quantify chemical substances. They are highly selective and sensitive. The basic fundamental framework of a biosensor includes three segments: a sensor which is a naturally responsive part, a detector to receive the signal and change it into the result for an easy display format, and an amplifier that amplifies the signals. The basic underlying principle of the biosensors is that whenever the active biological entity under consideration passes through a permeable membrane in the outer part, it gets detected by the recognition system which consists of immobilized enzymes or cells or antibodies. This is followed by the interaction of the biological material and the immobilized detectors under consideration that produces a chemical or gas

DOI: 10.1201/9781003224068-14

TABLE 14.1
Different Types of Biosensors

Calorimetric biosensors	Based on the interaction between the enzymes linked to the thermal detectors that detect heat changes occurring corresponding to the concentration of analyte.
Piezoelectric biosensors	Based on the affinity interaction of analyte and the sensing element coated on the piezoelectric material, causing a change in the oscillation of the material bound. It corresponds to the mass of the target analyte.
Electrochemical biosensors	On the basis of oxidation–reduction reactions that occur when an analyte reacts with a bioreceptor, leading to the alteration of the electrochemical properties of the solution.
Optical biosensors	Emitted or absorbed light owing to the biochemical reaction taking place between the receptor and the analyte is measured. These sensors directly use the samples with minimum wastage, and high-throughput analysis renders high sensitivity and capability to mask interference.

or heat or electric current which is transduced into a conventional electrical signal and then passed to the amplifier which amplifies it to give the resulting output signal. Biosensors are broadly classified into three categories on the basis of the method of transduction: (i) mass-based biosensors, (ii) electrochemical biosensors and (iii) optical biosensors. The materials used for biosensors are mostly carbon nanotubes. Single-walled carbon nanotubes (SWCNTs) have boosted the detection capabilities of electrochemical biosensors which provide increased sensitivity to the reactions. Other carbon nanomaterials including carbon graphene (CGR) and carbon fullerene (CFR) are also used for diagnostic purposes (Table 14.1).

Since the detection of biomarkers is expensive and time-consuming, and requires sophisticated technology, nanotechnology is employed to provide an economical, gradual and accurate substitute for point-of-care diagnostics. Nanostructured materials are a good alternative or a combination for diagnostics as they can be developed by incorporation of genomics, proteomics and molecular machine systems for efficient diagnosis and treatment. Nanoscience has its applications at a nanoscale level of 1–100 nm. It helps the nanostructures to integrate themselves in the atoms and molecules to provide enhanced developments in diagnostic devices. They have a small structure and high surface area that help them to display useful features. New therapies and diagnostic methods are under development due to the blend of nanotechnology with traditional diagnosis techniques. Some important biomedical applications include the increase in drug efficacy and drug delivery, the development of scaffolds and biosensors and the development of bioavailable implants. With the help of nanotechnology, developing biosensors with increased sensitivity and better performance is possible. Below are the types of nanostructures used in the construction of biosensors [1] (Figure 14.1 and Table 14.2).

This chapter discusses various applications of biosensors that have been employed for the detection of desired analytes for a number of diseases which

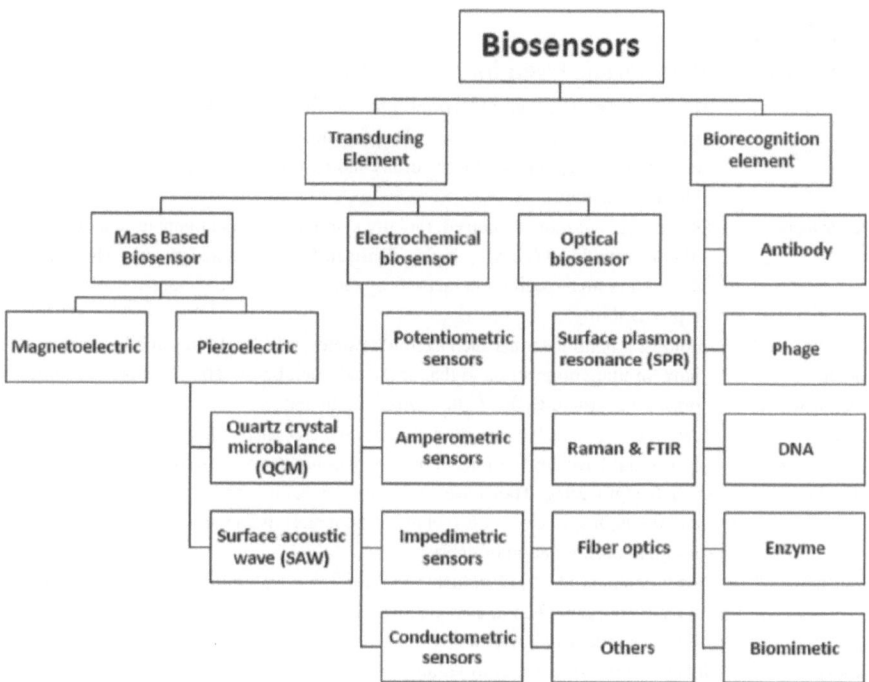

FIGURE 14.1 Classification of biosensor elements. (Biosensors: Principle, types and applications.)

pose a threat to public health worldwide. It also focuses on the need to develop and incorporate biosensors as they alleviate the difficulties in the overall improvement in disease diagnostics in an economical and time-saving manner and, to some extent, have diminished the need to use sophisticated traditional diagnostic tools.

14.1.1 DISPOSABLE IMMUNOSENSORS

Biosensors that are based on the transduction of signals produced in antigen–antibody interactions are known as immunosensors. The rudimentary requirement for the development of immunosensors is the precise design and building of a perfect framework between the biomaterial and the detector. The target antigen is detected by the labeled antibodies after the formation of antigen–antibody complexes. Immunosensors are very responsive and have operational clarity. They are divided into two categories: voltammetric immunosensors and amperometric immunosensors, and they incorporate two different types of immunoassays with them: sandwich-type immunoassays and competitive-type immunoassays, respectively. In sandwich-type immunosensors, first the primary antibodies attach to the transducer material, then there is addition of antigens, and finally, the enzyme-labeled antibodies are added to the solution. There is subsequent reaction

TABLE 14.2

Nanostructured Materials Used in Biosensors

Quantum dots	They are able to transport electrons upon interaction with UV light and range from 2 to 8 nm diameters in size. They are fluorescent, spherical nanocrystals derived from semiconductor materials and have important applications in bio-analytic units.
Dendrimers	Dendrimers are monodispersed, radially symmetrical, homogenous star-shaped 3D macromolecules having highly dense surface functional groups. They help in decreasing non-specific interactions by enhancing stability, sensitivity and reproducibility.
Biological and other nanomaterials	Biological nanomaterials such as lipid vesicles, thin lipid films and liposomes have similar composition to that of the cell membrane. Their composites form organized structures due to their amphiphilic nature.
Carbon nanotubes (CNTs)	They are cylindrical-shaped hollow tubes which consist of two or many graphite layers and have applications in biosensor construction, disease diagnosis and tissue engineering. They have very good mechanical and electrical properties, and the high surface area-to-volume ratio makes them highly stable with minimal accumulation of debris.
Chitosan	It is biocompatible, biodegradable and non-toxic with minimal implications on the target organ when in use, making it one of the best nanostructured materials for use in medical devices.
Graphene	It has a high surface area-to-volume ratio along with high electrical conductivity, which makes it an appropriate biosensor material for the detection of small molecules. It is used to develop electrochemical and biosensor-based diagnostic tools and also optical and potentiometric biosensors.

between the substrate and the antibodies, thus giving rise to the formation of antigen–antibody complexes. There are competitive interactions among labeled antigen and unlabeled antigens separately with the target molecule in the substrate in competitive-type immunosensors for the binding. The primary antibodies bind to the transducer material, followed by the addition of enzyme-labeled antigens, hence forming antigen–antibody complexes. The signal quantification is eventually carried out by precise addition of labeled and unlabeled antigens [2].

14.1.2　POINT-OF-CARE DIAGNOSTICS

Point-of-care testing (POCT) is a quick diagnostic testing that is carried out near the site of inpatient care or at the same place to obtain results in a short span of time. This is the widely used type of diagnostic tool in health management as it ensures rapid disease diagnosis. For this purpose, disposable electrochemical sensors and biosensors are generally used as POCT diagnostic tools. Disposable biosensors are used for the point-of-care assessment of acetylcholinesterase, bacteria, cancer biomarkers and DNA targets. Electrochemical biosensors comprising an array of 32 gold electrodes are used for point-of-care diagnostics in blood serum samples [2].

14.2 BIOSENSORS IN THE DIAGNOSIS OF ALZHEIMER'S DISEASE

Alzheimer's disease (AD) is a chronic neurodegenerative disease, distinguished by cognitive derailment or cerebral functional disability and personality changes, and is the main cause of dementia. The initial phase of AD is distinguished by unclear memory, memory- and non-memory-related damaging implications, deterioration of cognitive functions, difficulty in performing daily tasks, etc. AD is a rapidly evolving multifaceted disease and progresses quickly wherein the initial symptoms of dementia appear at a changeable clinical degree of an affected patient. The traditional detection systems are more likely to misdiagnose, thus delaying medication administration. This makes efficient diagnostics that favors precise detection and better treatment a high priority. This is where biomarkers play a very important role and biosensors are used to detect the biomarkers [3].

Bioreceptors are a component of the biosensor that takes part in the interaction between the biomarkers for disease detection and produce a signal by the biochemical reaction. The bioreceptors are very specific and exhibit high selectivity toward target detection. The most commonly used bioreceptors in the development of AD biosensors are aptamers and antibodies. Aptamers are also known as "chemical antibodies" as they resemble the functions of conventional antibodies and have an edge over the antibodies by having characteristics such as low molecular weight, good stability, low toxicity, reproducible in vitro synthesis, low immunogenicity and swift tissue penetration. Antibodies however, also have some beneficial properties such as reducing the amyloid plaques. The transducers are part of the biosensors that work in an optical or electrochemical fashion and are classified on the basis of the transducers used in it. They help in the interaction of the analyte with the detector element and subsequently convert it into measurable and quantified signals. The transducers used in the detection of AD are electrochemical and optical transducers, wherein the electrochemical bio-transducers transform chemical signals into a detectable signal and the optical bio-transducers use signal transduction, to collect information about the analyte [3].

The biosensors in AD detection are specially designed and consist of two parts: (i) a screen-printed screening probe and (ii) an electronic circuit. The electronic circuit is accountable for the acclimatization and accretion of the electric signal that would be used for the screening of the disease. For an efficient printing process, the biosensors are designed according to the organ, target disease and also the track they would be tracing for the conduction. Biosensors used in the preliminary detection of AD consist specifically of silver, carbon and carbon–silver chloride since they show compatibility with the signal conduction path. The electronic biosensor consists of reference electrode (RE), working electrode (WE) and counter electrode (CE). This is built keeping the anodic stripping voltammetry (ASV) measurements in consideration, rendering a three-electrode configuration based on ionic concentration. The acclimatization and accretion of electronic circuits were devised in a particular way to promote the fabrication of an estrade for a complete POCT. The electric signal generates a potential difference in the WE concerning the RE, which facilitates the movement of the electrons due to applied electromotive force in the track path, and thus, the current generated is detected in the counter electrode. First, the electrodes are conditioned to acclimatize to the detection conditions. The primary test was performed using different concentrations of NaCl to accommodate the change in conductivity of the solution and to quantify the changes in the peak between the WE and CE. The subsequent test consisted of the application of the electrochemical impedance spectroscopy principle for measuring changes in system impedance. The capture and detection antibody formed immune complexes, which characterized the quantification by a specific functionalization of the WE. The kit DuoSet® development system for Human CXCL8/IL-8 was used for the purpose. The secondary antibodies were detected in the presence of alkaline phosphatase (AP) to label them in the detection step. This propagated a cautious silver deposition on the kit, which was proportional to the amount of proteins detected. The buffer solution creates the ramp input on the WE and causes the oxidation of the deposited silver. This allows the current flow between WE and CE, which was subsequently measured, and it also allows the sensitive quantification of the proteins [4].

The circuit response results were recorded and evaluated according to the changes in the saline solution conductivity. Slopes that were formed in the concentration range of NaCl lower and higher than 1.0 mg/ml were observed for differences. A lower concentration of NaCl [0.2 mA/(mg/mL)] exhibited a higher sensitivity, and a higher concentration of NaCl [20 μA/(mg/mL)] exhibited a lower sensitivity. This may be owing to the specific feature of the materials chosen to build the biosensor, which makes it sensitive to detect small changes in the current and give precise reading for the same. The EIS measured the current emanating amidst the CE and WE, and the results generated a peak that showed a proportional decrease. This indicated the increase in the electrical resistance of the system due to the increasing concentration of antibodies coated on the WE surface. This led to an overall reduction in the exchange of electrons between the WE surface and the electrolytic solution. The biosensor showed a sensitivity of detecting a change in antibody coating concentration at 80 μA/(μg/mL). ASV measurement indicated elevation in the peak of current corresponding to the increase in interleukin concentration. Linear peak was observed at a sensitivity of 45 μA/(ng/mL) [4].

Early detection of AD plays a pivotal role in delaying the occurrence of the severity of the disease and also helps in combating the degenerative effects and neurological impairments. The recent research has focused on developing drugs with increased efficiency that can detect the emergence of AD. Diagnostics of AD involves performing cerebral tests such as magnetic resonance imaging (MRI), positron emission tomography (PET) and near-infrared spectroscopy (NIR), or the examination of cerebrospinal fluid (CSF) and blood plasma biomarkers by immunohistochemistry (IHC), enzyme-linked immunosorbent assay (ELISA), mass spectrometry (MS), western blot and flexible multi-analyte profiling (xMAP). These techniques not only are slow moving, expensive and intrusive, but also do not facilitate early-stage disease detection. Hence, the biosensors are used since the electrochemical biosensors can facilitate better diagnosis countering most of the drawbacks. Electrochemical biosensors detect these biomarkers and provide quantitative results in a short duration of time. Introduction of nanomaterials in the biosensors as labels or electrode modifiers has proven to increase the efficiency of biosensors for biomarkers detection. The frequently detected biomarkers in AD are peptides (Aβ peptide), followed by proteins. Nanocomposites can be used as electrode modifiers in a combination with electrochemical biosensors since they immediate a synergistic effect [5].

14.3   BIOSENSORS IN DIAGNOSIS OF CANCER

Cancer is considered to be one of the most serious and life-threatening diseases all around the globe. It is believed that the sooner the detection of cancer is, the higher are the chances of survival. Cancer can almost start in any part of the human body that is made up of trillions of cells. In spite of there being various developments in the recent advanced technologies, there are many late diagnoses that lower the survival rates of cancer patients. Cancer can sometimes be a multi-stage disease. Various traditional methods are still used to diagnose this disease, such as MRI, biopsy and ultrasound, which usually do not give the best results in the early stages of cancer. The abnormal divisions of cells which are uncontrollable, list of genetic defects and epigenetic mutations can interfere with the cellular signaling process which is related to its onset and progression and ends in one of the two forms of mutation, tumorigenic or malignant mutation [6].

In cancer, biosensors detect an analyte that is a cancer biomarker (CB) produced by tumor cells or by any other cells of the body in response to the tumor. There are various biosensors for cancer biomarkers. Biomarkers such as α-fetoprotein (AFP), cancer antigen 125 (CA-125), human chorionic gonadotropin (hCG), IL-8 and prostate-specific antigen (PSA) are targets for various types of cancer and have bio-recognition elements antibody competitive assay and antibody sandwich assay [7]. The majority of the biomarkers do not have sufficient specificity and sensitivity for converting into clinical use or for monitoring the treatment. Such an area is a good opportunity for biosensor technology to potentially improve upon. Each cell in the human body has its own identifiable characteristics such as activities of genes, proteins or other molecular features and molecular signature. New analytical techniques are being developed for clinical cancer diagnosis, which is efficient in various delicate, and simultaneous detection of the biomarkers exhibits useful POCT. In the case of detecting cancer monoclonal antibodies, antigens won't bind to the miRNAs corresponding to ssDNA [8]. The transducer is a device that converts a recognition signal into an electrical signal. It can be optical, colorimetric, electrochemical, or may even be based upon the mass changes. Transducers are useful as they are needed to provide high noise signals and radio wave signals; they give high levels of performance and also exhibit a superior resolution along with providing consistent results.

Reduced levels of present biomarkers are more often quantified by the biosensors in the physiological samples provided, which may assist with cancer diagnosis at a very early stage due to the limitation in their lower detection levels. Biosensors have the ability to test for various markers at one go, which assists greatly in diagnosis as well as helps in saving time and also financial resources. Currently, in the treatment of cancer, various different approaches are included, such as chemotherapy, radiotherapy, surgery and pharmacogenomics. Many biomarker detection systems integrated with smart wearables such as ISWEBDS that can help in early detection, diagnosis and also the treatment of cancer have been laid out, but are not yet in action. Sensors of this type are safe to use and easy to access as they utilize biological compounds or molecules as the bio-recognition entity. Various kinds of biosensors have been created and tested for betterment. Since cancer persists in various forms, individual forms have their own kind of biomarkers, thus making versatility as the most important aspect of a biosensor. Cancer is a phenomenon that happens at the nanoscale level; thus, it will only make sense when the tackling of this disease also occurs at the nanoscale level. This comes with lots of complexities in how the cancer is staged and the diversity of various kinds of cancer as it has already created a lot of challenges in the field of medicine. Biosensors are definitely a technology that has a good potential for providing factual and high accuracy in results while sustaining cost-effectiveness [9].

14.4   BIOSENSOR IN DETECTION OF HEPATITIS

The word is drawn out from the Greek word hepar, meaning "liver," and -itis, meaning "inflammation." Hence, it is the inflammation of the liver due to viral infections. There are many different types of viruses that can cause hepatitis, but the very common and deliberate types of hepatitis are with the hepatitis viruses that are labeled from A to E. Also, there are different viruses that cause hepatitis in humans. These include herpes simplex virus, cytomegalovirus, and Epstein–Barr virus, but most cases of hepatitis in humans are from hepatitis viruses. The symptoms produced are very diverse, and they depend upon whether the infection was acute or chronic, which depends upon many factors such as the type of the virus, the age of the patient, the level of the immune system of the patient, and the common health of the patient. The detection of DNA hybridization is important for its application to the diagnosis of pathogenic and genetic diseases. They are sensitive, of inherent miniaturization, compatible, and of low cost, and modern electrochemical DNA biosensors are extremely attractive. The hepatitis B virus is the causative agent of viral hepatitis, and infection with HBV comprises a public health problem of worldwide importance with cirrhosis and hepatocellular carcinoma. Acute and chronic clinical consequences, causing acute and chronic hepatitis, ELISA is the main method used in detecting HBV. The DNA hybridization is converted into an analytical electronic signal using electrochemical DNA biosensors to obtain sequence-specific information. They are majorly used for determining the rapid and precise recognition of infectious agents. Some of these biosensors usually need sophisticated pre-treatments and expensive advanced electrodes, and metal complexes are used for recognition.

The electrochemical indicator used is a manganese(II) complex, and the biosensor is the DNA probe-modified carbon paste electrode (DNA/CPE). This complex was assembled electrochemically and executed in the immobilized dsDNA layer ideally than in the single-stranded DNA (ssDNA) layer. For the detection of oligonucleotides corresponding to HBV, the manganese complex was cast off as an electrochemical hybridization indicator. Due to the difference between the reduction signals, hybridization was seen in the manganese(II) complex, which was attached to the DNA probe with a target sequence using a different pulse mode. Various factors affect the hybridization of oligonucleotides as well as the indicator's accumulation and immobilization. The non-complementary and mismatched sequence showed the better selectivity of the biosensor. Implementing this approach, the HBV target oligonucleotide's sequence can be evaluated over the range from 5.40 to 0.22 ng/L, with the limit discernment of 0.07 ng/L and a linear correlation coefficient of 0.9994 [10].

This biosensor procures an accurate, expeditious with a short manipulation time, sensitive, and cost-effective platform for a variety of practical applications as it is a general platform for the detection of hepatitis virus DNA in real samples and serves as a useful tool in the diagnosis of clinical genetic analysis, medical diagnostics, forensic identification, and environmental monitoring. Electrochemical transducers give fair advantages, for example, simplicity, high specificity, good sensitivity, and friendly materials such as manganese (II) complex, which is an electrochemical indicator used for converting nucleic acid hybridization into analytical signals. The complex generated from the bioreceptor and biomolecular binding at the sensor surface results in conversion into a detectable change, quantitative amperometric, impedimetric, or potentiometric signal. Electrochemical indicators are often used for detecting minute size proportions, affinity with microfabrication technology, and DNA hybridization, due to their high sensitivity and low cost [11].

14.5 BIOSENSORS IN DIAGNOSIS OF HIV

Lentivirus is a genus of retroviruses that cause chronic diseases that have a characteristic long incubation period. Human immunodeficiency virus or HIV is the most commonly known lentivirus, which causes acquired immunodeficiency syndrome (AIDS). HIV known as the immunodeficiency virus suppresses the immune system and affects its ability to combat ordinary diseases such as the common cold. HIV infects type CD4 T helper cells affecting the body's immune system. Around 38 million people were predicted to contract HIV in 2019. The death toll reached 69,000, and 1.7 million new cases were found in 2019. Looking at the impact of HIV on public health, scientists all over the world are developing new kits to detect HIV using antibodies. However, great limitations are faced during the detection of HIV at a very early stage of infection. Methods such as the use of metal nanoparticles (NPs) are employed, trying to increase the sensitivity and accuracy of the detection tests. Point-of-care (POC) diagnostic methods are being developed, since they have a very high sensitivity and can be applied in early-stage detection of HIV in children in mother-to-child transmission, antiret-roviral therapy, etc., by using nanoscale technologies. Some emerging methods which qualitatively and quantitatively detect HIV are listed below [12].

Electrical sensing-based methods or electrical sensing-based biosensors such as electrochemical biosensors use electrodes to convert the chemical signals into electrical signals. In this method, enzymes or proteins are immobilized on the transducer and analytes are measured. In amperometric/voltammetric biosensors, a sandwich immunoassay is used for HIV protein detection. The electrodes are covered in anti-p24 antibodies to trap HIV p24 proteins. Subsequently, horserad-ish peroxidase secondary antibodies are labeled. Potentiometric biosensors are devices comprising a biological sensing device attached to an electrochemical potential transducer. The signal generated by potentiometric biosensors is in the form of electrical potential. Thus, HIV-1 integrase activity can be measured using signal transduction method. The presence of *de novo* infections can be detected in tissues by measures of active integration, traversing probable causes of HIV-1 intransigence. Impedance biosensors can detect the change in the interfacial impedance after analyte binding and are developed by disabling a bio-recognition molecule on a biocompatible, conductive electrode. Similarly, a microdevice can be developed for the detection of HIV nano-lysates [13].

Optical assays-light absorbance and transmission, fluorescence, etc., are some target-binding changes which are broadly used for the detection of biological interactions. They provide a unique advantage of cheap and convenient mobile devices. They can be utilized to determine the presence of an analyte and its concentrations with high sensitivity by reacting an analyte of interest to produce an electrical signal. Fluorescent-based assays are one of the most commonly used methods for disease diagnosis due to their high susceptibility, accessibility, selec-tion of a wide range of fluorophores and various readout modes. Antibodies can be coupled with fluorescent markers to furnish selectivity in tagging proteins or molecules. Lens-free imaging platforms can be used to precisely count CD4+ cells in newly contracted HIV patients in un-equipped environments. Nanoplasmonic systems calculate the combined oscillations of electrons on nanoparticles by tracking ocular vibrations. Nanoplasmonic assessments for HIV can be used to compute HIV from the patient's blood at the POC [13].

The use of microfluidics methods such as ELISA uses a solid-phase type of method to detect the presence of particles of interest using the antibodies directed against the particle to be measured. ELISA-based tests recognize T cell CD4 pro-teins of the blood and could be used to measure the response to ART since they have a very high sensitivity. Another method is an automated imaging system called microfluidic ELISA, which captures particles from multiple banks with the help of magnetic beads [13].

Liat™ uses a system that withdraws and magnifies RNA of HIV from the blood. It is automated, user-friendly and of minimum operating requirements and has a detection limit of 57 copies/mL. Paper-based microfluidics can be used in clinical as well as in home settings to detect anti-HIV antibodies and also for HIV p24 antigen tests. They are simpler and more easily accessible than lateral flow immunoassays; use small samples; and are cheap and compatible with electro-chemical, microelectromechanical and chemiluminescence methods [13].

14.6 BIOSENSORS IN DIAGNOSIS OF SARS-COV-2

Severe acute respiratory syndrome coronavirus 2 (SARS-CoV-2) is a single-stranded, positive (+)-sense RNA, enveloped virus, belonging to the family Coronaviridae, beta-coronavirus (CoV) genera. Its genome includes four structural proteins - spike, envelope, membrane and nucleocapsid-16 non-structural proteins and five to eight accessory proteins; the S protein plays an important role in attaching, fusion, entry and transmission . N-terminal S1 subunits are for virus and receptor binding, and a C-terminal S2 subunits, for virus and cell membrane fusion. S1 includes an N-terminal domain and a receptor-binding domain. At infection, CoV binds to the host cell through an interaction between its S1-receptor-binding domain and the cell membrane receptor, causing structural changes in the S2 subunit resulting in its fusion and invasion into the target cell. Biosensors convert biological reactions into signals that are measurable. Over a transducer, the biological material which can be antibodies, cell receptors, biomimetic components, etc., is immobilised interacting with the analyte in the solution, giving a biochemical response. The transducer then converts this bio-chemical response into a signal which is measured by the digital detector module. Electrochemical, optical and piezoelectric transducers are the main transducing system. The most important application of biosensors is POCT, where a diagnostic test near the patient is done to give rapid results; appropriate, convenient care to patients; and more effective treatment for rapidly progressing infections. Saliva plays a vital role in non-invasive diagnostics, an economic and reliable POCT platform for instant detection, and helps increase the survival chances of those prone to COVID-19 disease. On the basis of analytes or reactions that the biosensors monitor, they can be classified as immunosensors, enzymatic biosensors, DNA biosensors and whole-cell biosensors.

Immunosensors are biological sensors in which there is an interaction between antibodies and antigens. Upon contact of the host with an antigen, lymphocyte B produces the antibody. Following the apoptosis of effector lymphocytes and memory B cells, the antigens are eliminated. Enzymatic biosensors use bio-recognition, the catalytic property of enzymes or molecules with high chemical specificity and efficient selectivity for the target substrate. The DNA biosensors or genosensor discriminate between organisms and detect various diseases and human pathogens through specific nucleic acid sequences. Whole cells act as recognition elements. Surface antigens present on the cell envelopes serve as targets for bio-recognition [14]. The platform for the detection of SARS-CoV-2 is based on three important aspects, which include the targets which are viral RNA, proteins or human immunoglobulins; the identification methods which are based on aptamers, antibodies, nucleic acid probes or receptors; and the amplification of signals and transduction systems based on electrical signals, surface plasmon resonance and fluorescent signals [15].

Fast, easy and broad diagnosis is essential for treating and managing COVID-19 to reduce and control its spread. Traditional techniques are not economic and require much labor and time; they also require recent special equipment and also expertise. Biosensors can be used for the detection of SARS-CoV-2 and other viral infections as they are sensitive, selective and economically analytical diagnostic systems [15].

14.7 CONCLUSION

The incomparable potential of biosensors in the diagnostics and care sector has driven scientists to develop new and innovative biological tools with enhanced properties. The research and development in the biosensor technologies have led to the designing of precise and innovative diagnostic tools. The keystone of the emerging popularity of biosensors in detecting an extensive variety of biomolecules and other biomarkers in medical diagnostics is attributable to their ingenuousness in operation, ability to perform complex analysis, higher sensitivity and capability to be integrated efficiently with different functions of the discrete chip. Traditionally, the detection of pathogenic agents has mainly been done by microscopic analysis, staining, identification of the microbes by cytopathic effects and analyzing their microscopic morphology. Biosensors are devices that combine a biological component with a physicochemical detector which is used for the detection of the desired analyte. A biosensor comprises a biological receptor (antibodies, enzymes, cells, etc.), a transducer component and an electronic system that consists of an amplifier, a processor and a display. Thus, biosensors owing to their high sensitivity, selectivity and simple operation give us an edge over the traditionally used analytical methods. Biosensors play a vital role in the detection of biomarkers in the early onset of the disease. So to facilitate faster recovery and improve the efficacy of the prescribed medication, biosensors should be largely incorporated in the diagnostic processes. Nano-biosensors are bioreceptor probes that selectively target the analyte molecules. Nanoscale materials provide novel quality since they can achieve rapid detection at low cost and can be applied in various arrangements. In recent years, significant discoveries have been concentrated on the detection of biomarkers of lung cancer and on general cancer detection using biosensors. It is still a massive challenge to fulfill all the aspects of efficient performance that would yield accurate results and the whole system be simple and affordable. The main goal of developing a biosensor-based diagnostic system is to make effective POCT services and equipment available to remote areas all over the globe, especially the developing countries.

ACKNOWLEDGMENTS

We would like to express our gratitude to all the individuals for supporting us throughout the process of writing this book chapter, especially to our instructor Dr. Ramneet Kaur for the keen interest she showed while writing this chapter. She explained to us the solutions to the problems that provided us with tremendous help as well as cleared all the doubts that arose during the process. Without her help, it was a matter of acute impossibility to propel in this endeavor. It is our privilege and pleasure to pay our sincere and heartfelt thanks to her. We are very grateful for receiving such an amazing opportunity and being able to add our inputs concerning the biosensors for disease detection.

REFERENCES

1. R. Nagraik, A. Sharma, D. Kumar, S. Mukherjee, F. Sen, and A.P. Kumar, (2021), Amalgamation of biosensors and nanotechnology in disease diagnosis: Mini-Review, *Sensors International*, 2, 100089, doi: 10.1016/j.sintl.2021.100089.
2. B.C. Janegitz, J. Cancino, and V. Zucolotto, (2014), Disposable biosensors for clinical diagnosis, *Journal of Nanoscience and Nanotechnology*, 14, 378–389, doi:10.1166/jnn.2014.9234
3. B. Shui, D. Tao, A. Florea, J. Cheng, Q. Zhao, Y. Gu, W. Li, N. Jaffrezic-Renault, Y. Mei, and Z. Guo, (2018), Biosensors for Alzheimer's disease biomarker detection: A review, *Biochimie*, 147, 13–24, doi: 10.1016/j.biochi.2017.12.015.
4. S. Tonello, M. Serpelloni, N. F. Looms, G. Abate, D. L. Uberti and E. Sardini, (2016), Screen-printed biosensors for the early detection of biomarkers related to Alzheimer disease: preliminary results. *30th Eurosensors Conference, Procedia Engineering*, 168 (147–150), doi: 10.1016/j.proeng.2016.11.182
5. C. Toyos-Rodríguez, F.J. García-Alonso, and A.d.l. Escosura-Muñiz, (2020), Electrochemical biosensors based on nanomaterials for early detection of Alzheimer's disease, *Sensors*, 20, 4748, doi:10.3390/s20174748.
6. M. Mascini and S. Tombelli, (2008), Biosensors for biomarkers in medical diagnostics, *Biomarkers*, 13, 7–8, doi: 10.1080/13547500802645905.
7. D. Kumawat, S. Ujjawane and H. Salunke, Biosensors in cancer, Article, https://www.researchgate.net/publication/349534398_biosensors_in_cancer.
8. B. Bohunicky and S.A Mousa, (2010), Biosensors: The new wave in cancer diagnosis, *Nanotechnology, Sciences and Applications*, 11, 4, doi: 10.2147/NSA.S13465.
9. N.A. Mungroo and S. Neethirajan, (2014), Biosensors for the detection of antibiotics in poultry industry, *Biosensors*, 4, 472–493, doi: 10.3390/bios4040472.
10. Department of Analytical Chemistry, Aristotle University of Thessaloniki, Panepistimioupoli Thessaloniki 54124, Greece Received 23 January 2014; Revised 4 April 2014; Accepted 9 April 2014.
11. S. N. Azizi, S. Ranjbar, J. B. Raoof and E. Hamidi-Asl, (2013), Preparation of Ag/NaA zeolite modified carbon paste electrode as a DNA biosensor, *Sensors and Actuators B: Chemical*, 181, 319–325. doi: 10.1016/j.snb.2013.02.026.
12. Y. Saylan, Ö. Erdem, S. Ünal and A. Denizli, (2019, May 21), An alternative medical diagnosis method: biosensors for virus detection, *Biosensors* 9, 65. doi: 10.3390/bios9020065.
13. M.A. Lifsona, M.O. Ozena, F. Incia, S.Q. Wanga, H. Inana, M. Badaya, T.J. Henrichf and U. Demircia, (2016 August 1), Advances in biosensing strategies for HIV detection, diagnosis, and therapeutic monitoring, *Advanced Drug Delivery Reviews*, 103, 90–104. doi: 10.1016/j.addr.2016.05.018.
14. B.V. Ribeiro, T.A.R. Cordeiro, G.R.O. e Freitas, L.F. Ferreira and D.L.F. Atlanta, (2020), Biosensors for the detection of respiratory viruses: A review *Open* 2, 100007.

15. S. Iravani, (2020), Nano- and biosensors for the detection of SARS-CoV-2: Challenges and opportunities, *Materials Advances* 1, 3092, doi: 10.1039/d0ma00702a.

16. S. Malhotra, A. Verma, N. Tyagi and V. Kumar, Biosensors: Principle, types and applications, *International Journal of Advance Research and Innovative Ideas in Education* ISSN(O)-2395-4396.

Index

4G 155–159, 161, 162, 167
5G 155, 157–162, 165–168
5G Numerology 161
6G 155, 158, 162, 164–168

Abnormality 24–26, 35
accounting and finance 75–77
Adaboost random forest ensemble 99
Adaptive Boosting 11
Alzheimer's disease 257
analytic artificial intelligence (AAI) 69
analytical diag-nostic systems 264
ANN 124
Aptamers 257
Area Under ROC 103
Artificial Intelligence 64–65, 169
Artificial Intelligence as a Disruptive
    Technology 74–75
Artificial Intelligence for Modern Businesses 67
Autocorrelation 38
Average characters; Mean sentence length 226

Backpropagation 34–35, 40
Bagging and Boosting 8, 101
Big data 169
Biological sensors 253
biomarkers detection 259
biosensor-based diagnostic system 264
BLAST 241
Blood pressure 93
Blood tests 24–26
Boosted regression 97
Business Benefits of Adopting AI 82–84

Cancer 259
Cardiovascular Disease 2
CDR 93
Cellular system 237
Characters; Corpus TTR 221
Chemical Reactivity 183
Chi-Square Test 185
Class-Balanced 58
Classification accuracy 128
Classification and Regression Tree (CART) 98
Clustering techniques 25
Cohort study 56
Colour space 25, 28, 30, 37
Communication 143
CoMP (coordinated multipoint) 169
Components of Artificial Intelligence 65–67

contact center 81
Correlation; pearson correlation; spearman
    rank correlation; kendall rank
    correlation 218
Cross Recurrence Plots (CRP) 114
Cross Recurrence Quantification Analysis
    (CRQA) 114
Cross-linking 245

Data Collection 149
Daubechies 30–31, 37
Decision Trees 14
Decomposition 182–190
Delta 35, 43
Demographic 25–29
Determinism 115
Device to Device (D2D) 157, 160, 170, 171
Diagnostic features 27–28, 34, 48–49
Diagnostic imaging 24–25
Diagonal Line Length 115
digital transformation 70–71
Direct Discovery of High-Utility Itemset
    (D2HUP) 207
Discriminant analysis 35
Discriminant classifier 26, 34, 42
Discriminant type 35, 43
Disruptive Technology 69–70
Docking 242
Drug discovery 242

e-commerce 79–81
Edge computing 168
EEG 111
Efficient High-Utility Itemset Mining
    (EFIM) 205
electrochemical biosensors 254
Electrochemical indicators 261
Enhanced Mobile Broad Band (eMBB) 157,
    158, 160, 162, 164, 166
Ensemble Methods 8
Environmental Conditions 182
examples of disruptive technology 71–73
Explosives Act 178
Extra Trees 10

Factories Act 178, 181–186
False Nearest Neighbors 113
Faster High-Utility Itemset Mining (FHM) 205
Feature Engineering 54
Female infertility 24

Filter-based 54
Fireworks 178
FLEX 242
F-measure 28, 36–37, 47
Follicle-stimulating hormone 24
Frequent Itemset Mining 196
Friction 182–188
functional artificial intelligence (FAI) 68

Gamma 35, 43
Generalized Linear Model 97
Geometric features 27
Gini index 98
Gradient 35–36, 40–42
Gradient boost random forest ensemble 99
Gradient Boosting 9
Grid Search CV 14
Gumurkhi characters frequency distribution;
      Gurmukhi vowel usage 222
Gumurkhi sentence analysis; SLTS 224
Gumurkhi word analysis; Word mean length 220
Gurmukhi Script; Punjabi; Corpus 212

Hazardous 182
hepatocellular carcinoma. 260
High-Utility Itemset Miner (HUI-Miner) 205
High-Utility Itemset Mining 196
High-Utility Pruning Strategy (HUP-Miner) 206
HIV nano-lysates 262
Homo & hetero-oligometric 239
Hormones 24
HSA (Human Serum Albumin) 244
Hub(intro) 238
Human Error 178–191
Hybrid Technique by the Integration
      of UP-Growth and FHM
      (UFH-Miner) 207

ICM 109–110
immunosensors 255
impact of big data in disruptive technology
      73–74
Infertility 24, 26
Information and communications technologies
      (ICT) 145
Intelligent system 27, 34, 42, 49
Intensity features 32, 38
interactive artificial intelligence (IAI) 68
Interconnection of massive devices 168
Interfollicular 26, 32, 38–39
IoP 93
Job Safety Analysis 186

Kernal based SVM 96
Kernel 35–36, 44
K-means 25, 27, 32, 38

KNN 98
knowledge-driven 60
K-SOM 124
Kutty Japan 178

LDPC 1, 61, 171
Learning rate 34–35, 40–42
Lentivirus 261
Lesions 25–26, 33
Ligand (docking) 242
limited memory 67
Linear 31, 35–36, 40–44
Logistic Regression 8
Long Term Evolution (LTE) 156–158, 160, 162
Luteinizing hormone 24

Machine learning 96
Magazine 181
marketing 77–79
Marketing Authorization Application
      (MAA) 242
mass-based biosensors 254
Massive Machine Type Communication
      (mMTC) 157, 158, 159, 162, 164, 166
Mean; Mode; Median 216
Medical Images 25, 32, 35
Metabolites 243
Metabolome 243
MLP 124
Model selection 95
Momentum 34–35, 40–42
monoclonal antibodies 259
MS (Mass Spectrometry) 244
Multilayer Perceptron 13
Music 108–110

Nano-biosensors 264
Nanotechnology 191
Natural language; linguistic; Textual data
      analysis 211
Network parameters 34–35, 42
Neural network 25, 27, 40
NMR (Nuclear Magnetic Resonance) 244

Occupational Hazards 183
OCSVM 56
optical biosensors 254
optimized feature set 60
Ovarian Classifier 28, 34, 42, 47, 49
Ovaries 24, 26–27, 34, 36, 42, 47–48
Over fitting 96
Ovulation process 24
Ovulatory disorders 24–25

Particulate Matter (PM) 183
Performance metrics 28, 36, 46

Personal Protective Equipment (PPE) 187
Phase Space Plots 112–114
Phychological 184
Pixel 31–33
Point-of-care testing 256
Precision 28, 36–37, 47
Preprocessing 25, 28, 37, 48, 93
Principal Component Analysis 7
Prognostic model 53
Prolactin 24, 28, 29, 34
Proteins (Related) 239
Pseudolinear 43

Ragas 109–110, 112
Random Forest 8, 99
Raw dataset 59
reactive machines 67
Recurrence Rate 115
Regression analysis 54, 185
Regularization 35, 43
Risk Assessment 182
Robotics 168
Rosetta stone method 240

SARS-CoV-2 263
Satellite imageries 123
self-awareness 67
Sensitivity 28, 36, 47
Sigmoid 35, 40–42
SIM (Spatial Interaction Map) 239
Smart Environment 166
Solver 35, 36, 44
Specificity 28, 36, 47
Speckle 25, 28, 37, 48
Spirometry 54
Standard Deviation; Skewness 217
Statistical 25, 32–33
Statistical measures; Sentence length time
    series; News Corpus; corpus
    extraction 213
Sum average 38
Sum variance 38
Support Vector 25, 35
SVM 124

Tactile Internet 168
telecommunications 81–82
Telemedicine 105, 146
text artificial intelligence (TAI) 69
Texture features 25–26, 28, 32–34, 38–40
theory of mind 67
Third Generation Partnership Project (3GPP)
    158, 160, 170
Thyroid-stimulating hormone 24
Time-sensitive networking (TSN) 169
Two-Phase Algorithm 205
Type Token Ratio; Frequency 219
Types of Artificial Intelligence 67

uHDD (ultrahigh data density) 164
uHSLLC (ultra high-speed and low-latency
    com-munication) 164
Ultra Reliable Low latency
    Communication (URLLC)
    157–159, 162, 164, 166
Ultrasound 24–26, 28, 31–33, 36
uMUB (ubiquitous mobile
    ultra-broadband) 164
Under fitting 96
Unicode; Extensible Markup Language;
    XML format 214
Utility List Buffer (ULB-Miner) 206
Utility Mining 195
Utility Pattern Growth (UP-Growth) 206

Virtual reality/augmented reality
    (VR/AR) 169
visual artificial intelligence (VAI) 69
Vroman effect (Rel) 239

Wavelet transform 28, 30, 31
Word and Sentence Representation 215
Word length 223

XGBoost 101

Yeast two Hybrid (Y2H) 240

Zipf's law 233